The ATLAS of ARCHAEOLOGY

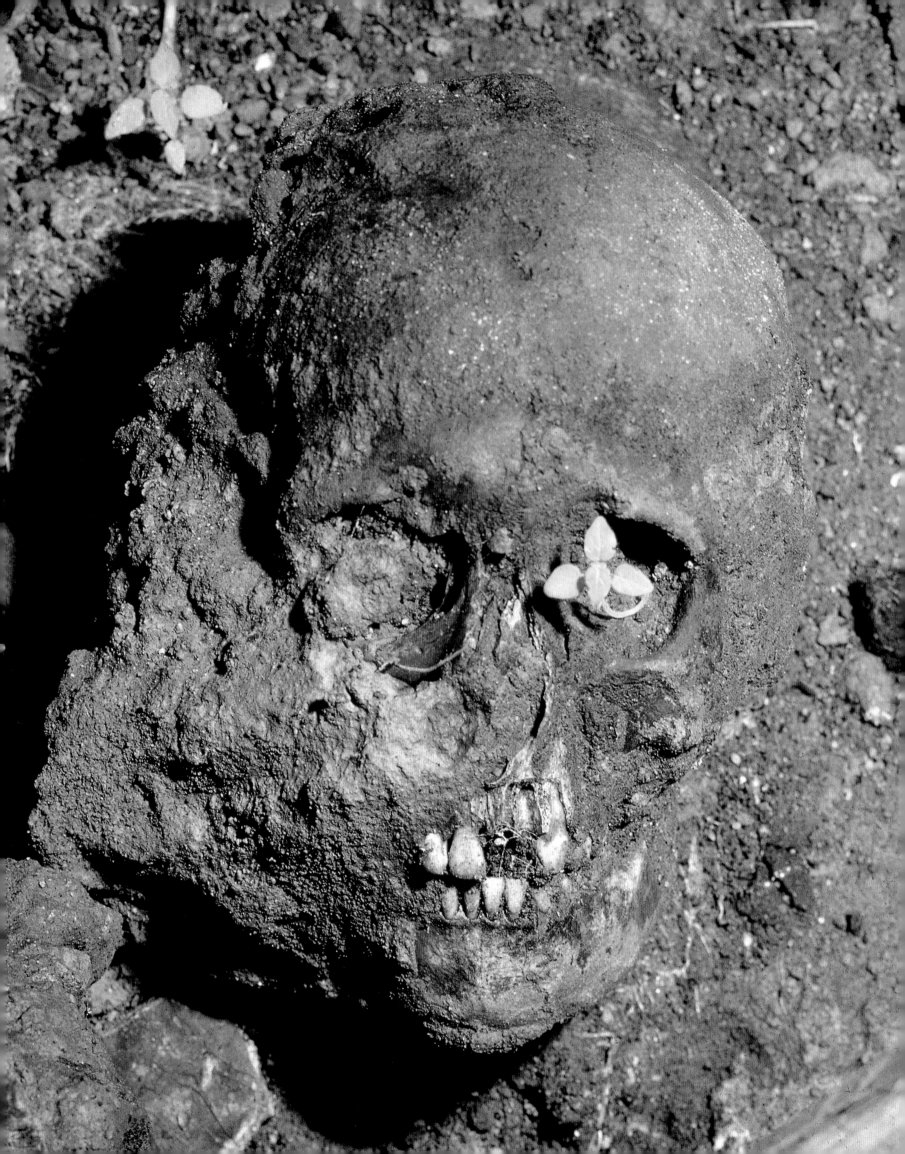

The ATLAS of ARCHAEOLOGY

MICK ASTON & TIM TAYLOR

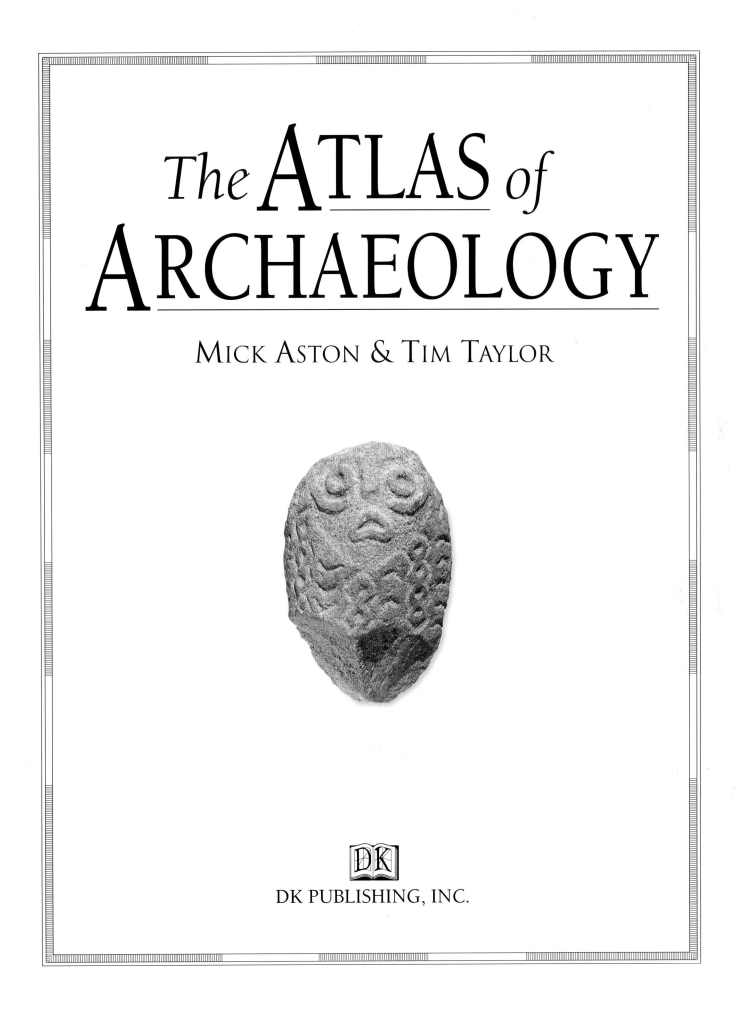

DK PUBLISHING, INC.

DK

A DK PUBLISHING BOOK

PROJECT EDITOR
JOANNA WARWICK

EDITOR
DAVID WILLIAMS

SENIOR EDITOR
CHRISTINE WINTERS

SENIOR MANAGING EDITOR
SEAN MOORE

US EDITOR
JILL HAMILTON

DESIGNER
JOANNE MITCHELL

ART EDITOR
PAUL GREENLEAF

SENIOR DESIGNER
HEATHER MCCARRY

DEPUTY ART DIRECTOR
TINA VAUGHAN

MANAGING ART EDITOR
TRACY TIMSON

SPECIAL PHOTOGRAPHY
STEVE GORTON

CARTOGRAPHERS
JOHN PLUMER, DALE BUCKTON

PRODUCTION
SARAH COLTMAN

PICTURE RESEARCH
JO CARLILL

CONTRIBUTING CONSULTANT
JENNI BUTTERWORTH

First American Edition, 1998
2 4 6 8 10 9 7 5 3 1
Published in the United States by DK Publishing, Inc.,
95 Madison Avenue, New York, New York 10016
Copyright © 1998 Dorling Kindersley Limited, London
Text copyright © 1998 Mick Aston & Tim Taylor
Maps copyright © 1998 Dorling Kindersley Limited

Visit us on the World Wide Web at http://www.dk.com

Library of Congress Cataloging-in-publication Data

Aston. Mick.
   The atlas of archaeology / by Mick Aston and Tim Taylor.
     p. cm.
   Includes index.
   ISBN 0-7894-3189-0
   1. Archaeology 2. Excavation (archaeology) 3. Antiquities.
4. Extinct cities 5. Civilization, Ancient. I. Taylor. Tim
II. Title.
CC165.T36 1998
930.1—DC21   97–32262
                           CIP

Reproduced in Italy by GRB. Printed in Italy by Mondadori.

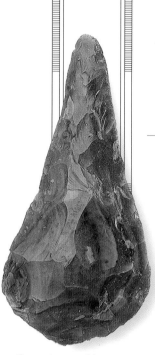

PALEOLITHIC HAND-AX

CONTENTS

MOHENJO-DARO

Roman Fish-head Spout

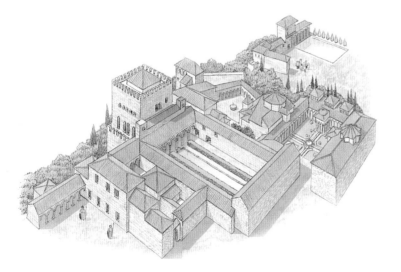

The Alhambra, Granada

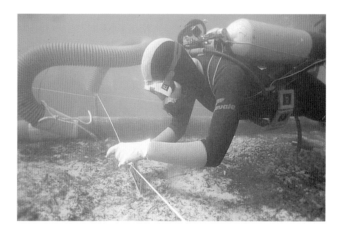

Marine Archaeology

INTRODUCTION

THE MYSTERIOUS SIGNS left behind by our ancestors offer us a tantalizing glimpse into the past and help fuel our instinctive desire to learn more about those who have preceded us. The process of interpreting these signs is the fascinating job of archaeologists around the world, and in the following chapters, we will lead you on a journey of discovery through a selection of key archaeological sites. Unlike in most traditional archaeology books we have attempted to present the process of archaeology – from excavation to analysis – as it happens, by capturing the immediate thrill of discovery and the fascinating challenge of interpretation. Some of the sites have been selected because they represent the kinds of locations that archaeologists deal with on a daily basis – not all archaeological work is spent digging pyramids or rich burial sites – and it is hoped that these sites will inspire you to join a local archaeology society and to visit similar sites in your area.

FIRST IMPRESSIONS
The 3.8-million-year-old Laetoli footprints in Tanzania symbolize the mystery and romance of humankind's obsession with its collective past.

gained can be set in context with a range of other evidence from the area. In this book, we have chosen a wide geographic and chronological range of sites – from a hunter-gatherer site, moments after the death of a mammoth, through the glories of the Classical period to the inventions of the industrial age.

ARCHAEOLOGICAL THEMES

Although we have presented this book in chronological order, it can be useful for archaeologists to approach the analysis of sites by grouping together those with similar cultural or physical themes. Some of the most dramatic sites are those that have the bulk of their evidence visible on the surface, such as buildings, tombs, and monuments. However, in many cases the clues to the exact use of the monument or structure may be missing. For example, at Avebury in England, archaeologists had the physical evidence of the henge standing before them but were still dependent on some small but vital clues found in the ground to suggest the actual use of the site.

Sometimes traces of whole towns and villages are visible, and these present a particular challenge for archaeologists. The temporary encampments of the Paleolithic age leave little evidence, but the complex settlements of the first cities are usually rich in deposits, and modern techniques can be applied very successfully. Geophysical surveying is a vital technique for discovering the foundations of buildings, and it can be used in conjunction with more traditional techniques, such as excavation. On these complex sites you will see how each piece of evidence can provide a vital clue – a piece of

READING THE PAST

Archaeology is simply a way of looking at the world through the analysis of material evidence left by past cultures. An archaeologist finding the ubiquitous Coca-Cola bottle a thousand years from now will be able to draw certain conclusions about our technical skills, our attitude to the environment, and our diet, from that single object. Other typical objects can also be "read" in this way, and the information

PREHISTORIC LIFE
Some sites produce finds that can be dated to over 200,000 years ago, like this bear tooth found in England.

MYSTERIOUS MONUMENTS
Some of the most famous archaeological sites in the world are mysterious stone monuments, such as Carnac in France and Avebury in England (see right), both of which had a ritual purpose.

carved ivory found in Palaikastro, Greece, points to trading contacts with the outside world, and traces of bone from a Neolithic settlement at Çatal Hüyük, Turkey, enable us to populate the fields around the town with domesticated animals.

These finds allow us to build up a more detailed picture of a site and its inhabitants. Each piece of evidence is an indicator of the sophistication and cultural inclinations of the society. Perhaps the most striking information that emerges from these more complex societies is the documentary evidence. From the historic period onward (after the advent of the written word), archaeologists can call on documentary evidence to supplement, or indeed initiate, the process of investigation and interpretation. Documents such as building plans or inventories can prove critical at industrial sites, as is

demonstrated at the Soho Works site in this book. Since the full history of a site cannot be learned just from documentary sources, archaeologists must also rely on excavated evidence. This type of evidence is extremely variable and can include animal remains, minute environmental materials, pottery, stone, glass, metal, and human remains. Burial sites are often very satisfying for archaeologists, because they provide a "closed-context" site (all the objects relate to one specific moment in time), full of cultural, religious, and scientific evidence. The discovery of skeletons, such as those found at Winterbourne Gunner in England, brings the reality of the past into sharp focus, as well as providing a subject for scientific analysis. Burial sites usually offer rich seams of evidence, including the body itself, clothing, ritual

deposits, and the whole cemetery context. These burials are often found as part of a ritual landscape, where monuments, earthworks, or buildings connected with the burial process are found. The ritualistic element of human development is complex and often difficult to interpret. At Navan in Northern Ireland, a mound site revealed evidence of a structure built during the Iron Age apparently for a single ritual event. This kind of site is as rare as it is fascinating, and it raises a number of questions that can only be speculated upon.

LOOKING AT THE SITES

In this book, each chapter begins with an illustrated map showing a wide range of sites from the period under consideration. It is standard practice for archaeologists to approach new sites with similar or relevant locations in mind or at least available for reference. This body of knowledge represents the painstaking work of previous generations of archaeologists, whose slow accumulation of evidence forms the foundation for all future archaeological work.

This world view is followed by a detailed look at evidence from one fairly typical site. This evidence comes partly from excavation, but also from a range of other techniques, including geophysical surveying and ground radar. We then show how the range of evidence discovered can be assembled and analyzed in order to suggest a visual interpretation of the original site. Naturally, the interpretation of evidence, particularly from the prehistoric period, can never be based on total certainty, but it is as near as we can get, given the techniques and knowledge currently available.

The wide variety of evidence frequently unearthed requires archaeologists to have a range of interpretative skills at their disposal. You will be able to see how the careful analysis of environmental evidence, such as the seeds and plants found on a Paleolithic site, enables archaeologists to build up a picture of the landscape as it existed 200,000 years ago. You will also see how magnetometry provides archaeologists with the exact positions of wall foundations and how GPS (Global Positioning by Satellite) can produce a map of the surrounding landscape.

PUTTING THE FINDS IN CONTEXT

As we travel forward in history, the nature of the evidence changes and the range of artifacts is increasingly complex. The impact of human ingenuity and technology requires the

THE SIGNS OF CIVILIZATION
Often people's interest in archaeology is inspired by a monument left by a great civilization, such as the Celsus Library at Ephesus (see left). These sites form an excellent starting point, since they are usually well documented.

UNEARTHING ARTIFACTS
*Most objects found in the ground will
show signs of decay. Metal objects, like
this dagger, disintegrate most easily.*

archaeologist to cross-reference with complex typologies against which specific objects can be compared. You will see this process at work as pieces of broken pottery are compared with a dated typology of pottery styles and types, enabling the site to be dated from a small shard. At the Saxon burial site Winterbourne Gunner, the study of personal artifacts, such as jewelry, enabled the burial to be dated by reference to changes in style that were known to have occurred at specific times. Interpreting archaeological finds often requires an understanding of the kind of technology used to create a particular object. Knowing the distinctive signs of a handmade pot or a mold-made bronze, being

ARCHAEOLOGISTS AT WORK
Archaeological excavation necessarily involves destruction, so it is vital that all finds are recorded in detail. Skilled archaeologists are meticulous and amazingly patient.

able to tell the way a flint object was retouched, or understanding the process of turning raw materials into glass can add immeasurably to an archaeologist's ability to interpret the evidence. After the detailed analysis of one site in each chapter, we look at a range of sites from around the world. This enables you to see how the main sites in the book compare with contemporary sites from different cultures across the globe.

Having introduced you to a specific site and a number of other sites in detail, we then provide a comprehensive gazetteer of sites from around the world, listed geographically. This will help you to locate sites that are near you, or in areas that you plan to visit, and to use this as a resource to carry out your own research.

Throughout the pages of this book we hope to lead you on a fascinating journey of archaeological discovery – from geophysical survey, excavated object, and comparative analysis, to a detailed reconstructed view of our shared past. Many of the techniques and skills you will see in use in this book could be applied to an area of the historical landscape near you, and we hope this book will inspire you to look at the present, of which you are a part, and the past, which helped to form you, with equal fascination.

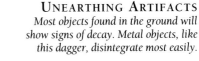

MICK ASTON & TIM TAYLOR

TECHNOLOGY IN USE
Much of archaeological work is done before and after excavation. At Avebury discoveries are being made using noninvasive geophysical surveys.

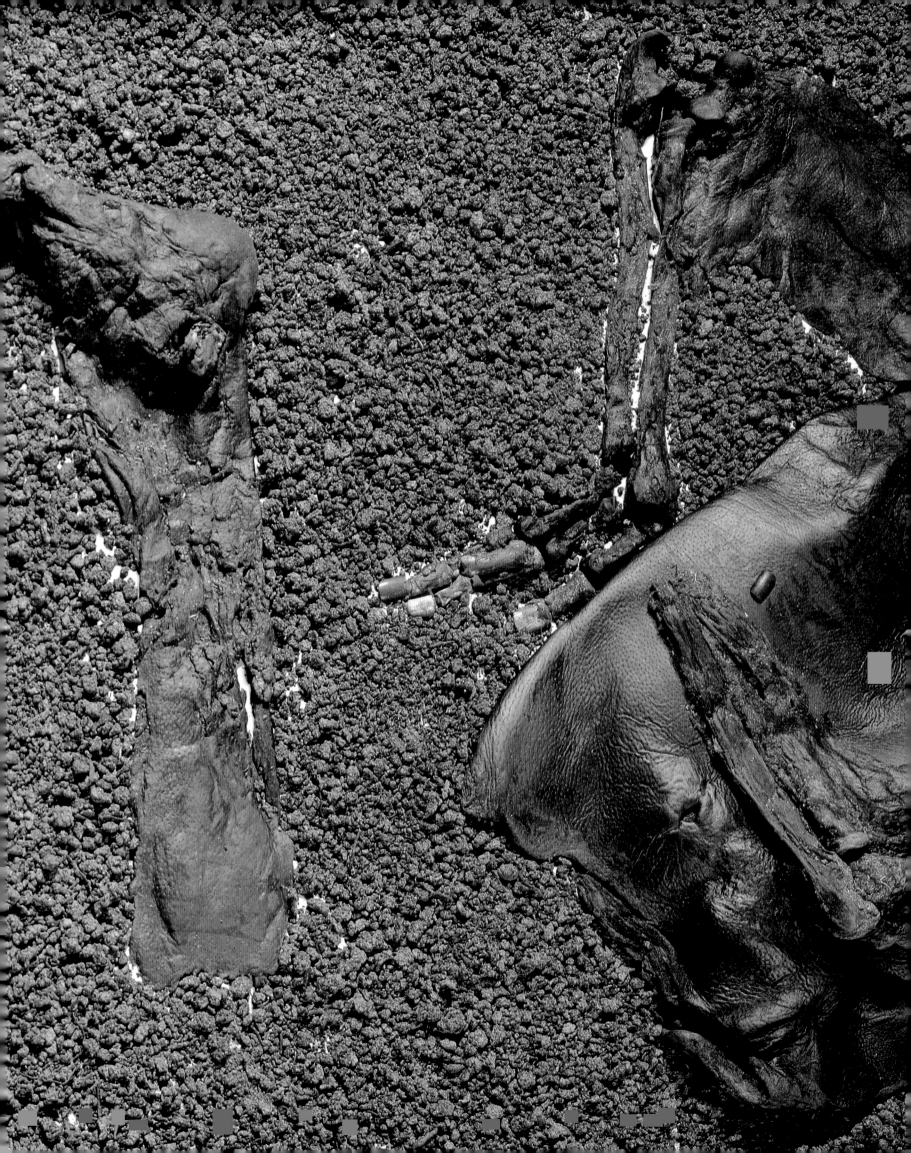

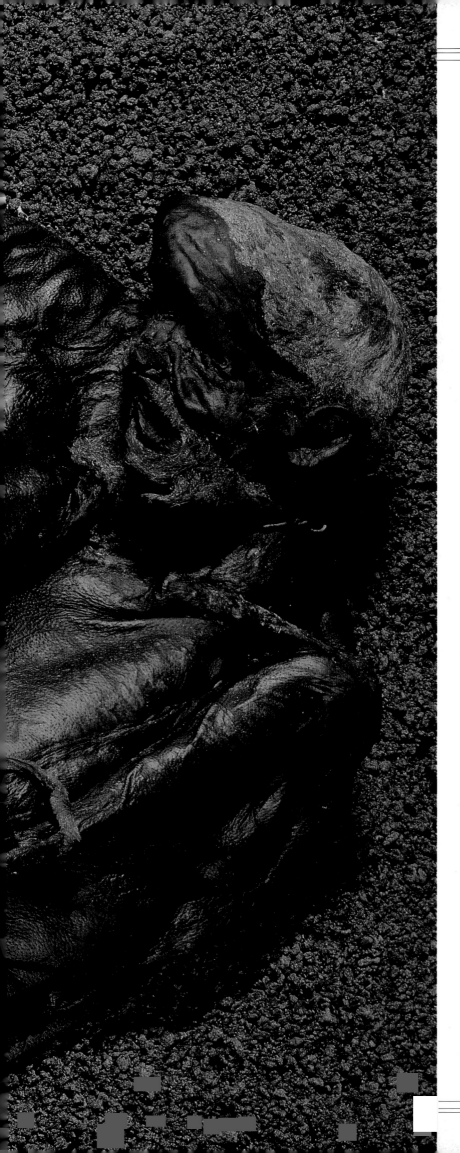

PART ONE
ARCHAEOLOGICAL SITES

THE WORLD'S KEY ARCHAEOLOGICAL SITES
HAVE ALL BEEN BROUGHT TO LIGHT THROUGH
THE SKILL, DEDICATION, AND PATIENCE OF
GENERATIONS OF ARCHAEOLOGISTS. THE SITES
CHOSEN TO APPEAR IN THIS BOOK REFLECT THE
AMAZING DIVERSITY OF HUMAN LIFE ACROSS
THE GLOBE, AND WILL GIVE YOU A SENSE OF
THE EXCITING POWERS OF DISCOVERY THAT
LIE AT THE HEART OF ALL ARCHAEOLOGICAL
INVESTIGATION. ARCHAEOLOGY IS ESSENTIALLY A
QUEST FOR KNOWLEDGE ABOUT OURSELVES AND
OUR SHARED PAST. ALTHOUGH EXCAVATION
REQUIRES A METICULOUS, SCIENTIFIC APPROACH,
IT CANNOT SUCCEED WITHOUT THE CREATIVE
IMAGINATION AND GENUINE ENTHUSIASM OF
THE ARCHAEOLOGISTS IN THE FIELD. IT IS THIS
ENTHUSIASM THAT WE HOPE TO COMMUNICATE
THROUGH THE PAGES OF THIS BOOK — BY LEADING
YOU ON A JOURNEY OF DISCOVERY THROUGH THE
KEY MOMENTS OF OUR SHARED EVOLUTION.

THE LINDOW MAN
This body of a young man was found in a bog in Cheshire,
England, in 1984. The man appears to have been struck
violently around the head, garroted, and had his throat cut.
Radiocarbon dating has shown that the man met his death
in the 1st or 2nd century AD and that, like the bodies found
at Grauballe and Tollund in Denmark, he had
probably been ritually sacrificed.

HUNTER-GATHERERS

FOR MOST OF THE TIME since humans first evolved, they have hunted or scavenged animals and birds, fished, and collected eggs and shellfish. Early peoples also gathered a wide range of plant foods – berries, roots, tubers, nuts, and fruit – when available. The lifestyle of the hunter-gatherers was mainly nomadic, with groups moving around as they followed migrating animals and exploited plant resources. Although we tend to think of these people as cave dwellers, they used many open-air and temporary campsites as well.

Only a small amount of archaeological evidence from these very early periods has survived, but it ranges from the stone hand-axes found in Africa to the harpoon points of the Arctic hunters. Several types of early art have also been uncovered from this period, including stunning cave paintings and portable art, such as bones carved into animal shapes. These finds reflect the essential, practical, and spiritual elements of our human experience that have lasted for more than one million years.

GREENLAND

① HEALY LAKE
The lowest layer of this site dates back 11,000 years and contains many microblades (small stone blades).

CANADA

NORTH AMERICA

② L'ANSE AMOUR
Antler bones, such as the one above, were found at this site. They were used to make harpoons for hunting land and sea animals.

③ DANGER CAVE
The tools found in this well-preserved cave site tell us much about the activities of the local people 11,000 years ago.

NORTH ATLANTIC OCEAN

④ BLACKWATER DRAW
These tools from 12,000 years ago provide the first evidence of humans in America.

MEXICO

⑤ GUILÁ NAQUITZ
By 6400 BC, the people of this community had changed from hunters to farmers.

⑥ TAIMA-TAIMA
A butchered elephant carcass indicates that people hunted large game here from 15,000 years ago.

VENEZUELA

Orinoco

PACIFIC OCEAN

Amazon

⑦ TAPERINHA
The shell middens (piles of refuse) found here date to more than 7,000 years ago and indicate that hunter-gatherers were utilizing marine resources.

Andes

PERU

Andes

BRAZIL

SOUTH AMERICA

CHILE

⑧ MONTE VERDE
Around 13,000 years ago, people in this area of Chile were living in wooden huts and using wood, bone, and hand-shaped stone tools.

STONE AGE CAVE PAINTINGS
There are several caves in the Valltorta Gorge in Spain that contain Stone Age art. Originally painted in white, the pictures were repainted many times in red and black, using pigments and ash, probably mixed with vegetable juices or animal fats. The Cueva del Civil (see left) shows stylized people holding weapons, and is likely to have been painted during the transition from a hunting to a farming society.

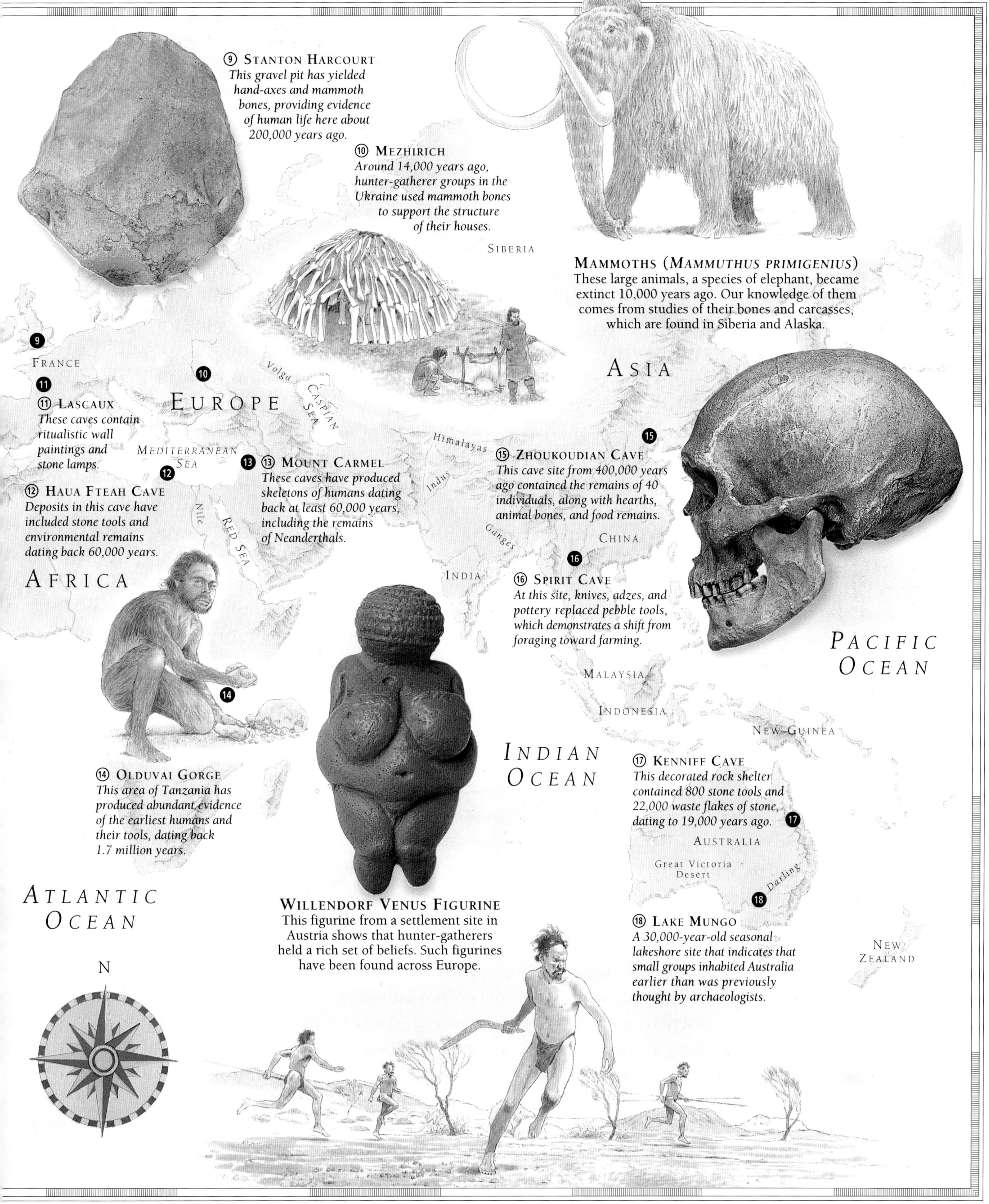

⑨ STANTON HARCOURT
This gravel pit has yielded hand-axes and mammoth bones, providing evidence of human life here about 200,000 years ago.

⑩ MEZHIRICH
Around 14,000 years ago, hunter-gatherer groups in the Ukraine used mammoth bones to support the structure of their houses.

SIBERIA

MAMMOTHS (MAMMUTHUS PRIMIGENIUS)
These large animals, a species of elephant, became extinct 10,000 years ago. Our knowledge of them comes from studies of their bones and carcasses, which are found in Siberia and Alaska.

ASIA

FRANCE

EUROPE

Volga

CASPIAN SEA

⑪ LASCAUX
These caves contain ritualistic wall paintings and stone lamps.

Himalayas

MEDITERRANEAN SEA

⑬ MOUNT CARMEL
These caves have produced skeletons of humans dating back at least 60,000 years, including the remains of Neanderthals.

Indus

⑮ ZHOUKOUDIAN CAVE
This cave site from 400,000 years ago contained the remains of 40 individuals, along with hearths, animal bones, and food remains.

⑫ HAUA FTEAH CAVE
Deposits in this cave have included stone tools and environmental remains dating back 60,000 years.

Nile

RED SEA

Ganges

CHINA

INDIA

⑯ SPIRIT CAVE
At this site, knives, adzes, and pottery replaced pebble tools, which demonstrates a shift from foraging toward farming.

AFRICA

PACIFIC OCEAN

⑭ OLDUVAI GORGE
This area of Tanzania has produced abundant evidence of the earliest humans and their tools, dating back 1.7 million years.

MALAYSIA

INDONESIA

NEW-GUINEA

INDIAN OCEAN

⑰ KENNIFF CAVE
This decorated rock shelter contained 800 stone tools and 22,000 waste flakes of stone, dating to 19,000 years ago.

AUSTRALIA

Great Victoria Desert

Darling

ATLANTIC OCEAN

WILLENDORF VENUS FIGURINE
This figurine from a settlement site in Austria shows that hunter-gatherers held a rich set of beliefs. Such figurines have been found across Europe.

⑱ LAKE MUNGO
A 30,000-year-old seasonal lakeshore site that indicates that small groups inhabited Australia earlier than was previously thought by archaeologists.

NEW ZEALAND

N

13

A PALEOLITHIC VALLEY

STANTON HARCOURT, UNITED KINGDOM

GAZETTEER P. 144 – MAP REF. G7

SITE EVIDENCE

○ Large numbers of animal bones, teeth, and tusks have been found during gravel extraction in the valley, suggesting occupation by prehistoric animals.

○ Environmental evidence, including insects, pollen, mollusks, and plant remains have been retrieved from the riverbed.

○ Geophysical survey and excavations have uncovered the course of a prehistoric river channel.

DURING THE PALEOLITHIC AGE, advancing and retreating ice sheets produced widespread devastation across the northern hemisphere. During some of the retreats, known as interglacial periods, there was enough time for topsoil to be formed and for vegetation to grow in the warmer climate. This lush environment supported a variety of animals, which were in turn utilized by the human population. The actions of the fluctuating ice sheets make the search for evidence of this period very difficult. Remains are often limited, and it is difficult to know whether the evidence was created by nomadic people passing through an area or by a period of settlement, and how much the action of ice and water affected the site. The excavation at Stanton Harcourt raised one key archaeological question: Were the finds a result of a flash flood following a glacial meltdown, or did the various humans, plants, animals, and insects live around the wide river channel?

Area of excavation

Low resistance produced by moist clays

KEY

|⟞—————⟝| 33 ft (10 m)

——— Area of excavation

Areas of high resistivity

Areas of low resistivity

THE SITE
Gravel extraction at Stanton Harcourt had created a rich by-product for archaeologists, and investigations had been under way for a number of years. The site was due to be filled in, so the excavators had only a limited time to complete their work. Excavation concentrated on the lowest and, therefore, earliest phase of gravel layers, immediately above the ancient river clays.

High electrical resistance produced by gravels

Plant debris

GEOPHYSICAL MAP
Geophysical surveying was used to locate "Paleo-channels" – the hidden traces of the early river systems – where datable material would have accumulated. The channel edges had a higher electrical resistance than the former riverbeds, which had a higher moisture content.

SOIL SAMPLE
Samples of the soil from the bottom of ditches and channels were taken and processed to see what organic (plant or animal) remains were contained within them. These remains were examined and the species identified. Using this information, the climate and landscape of the area could be fairly accurately identified.

Width: 4in (10cm)

Eroded ivory

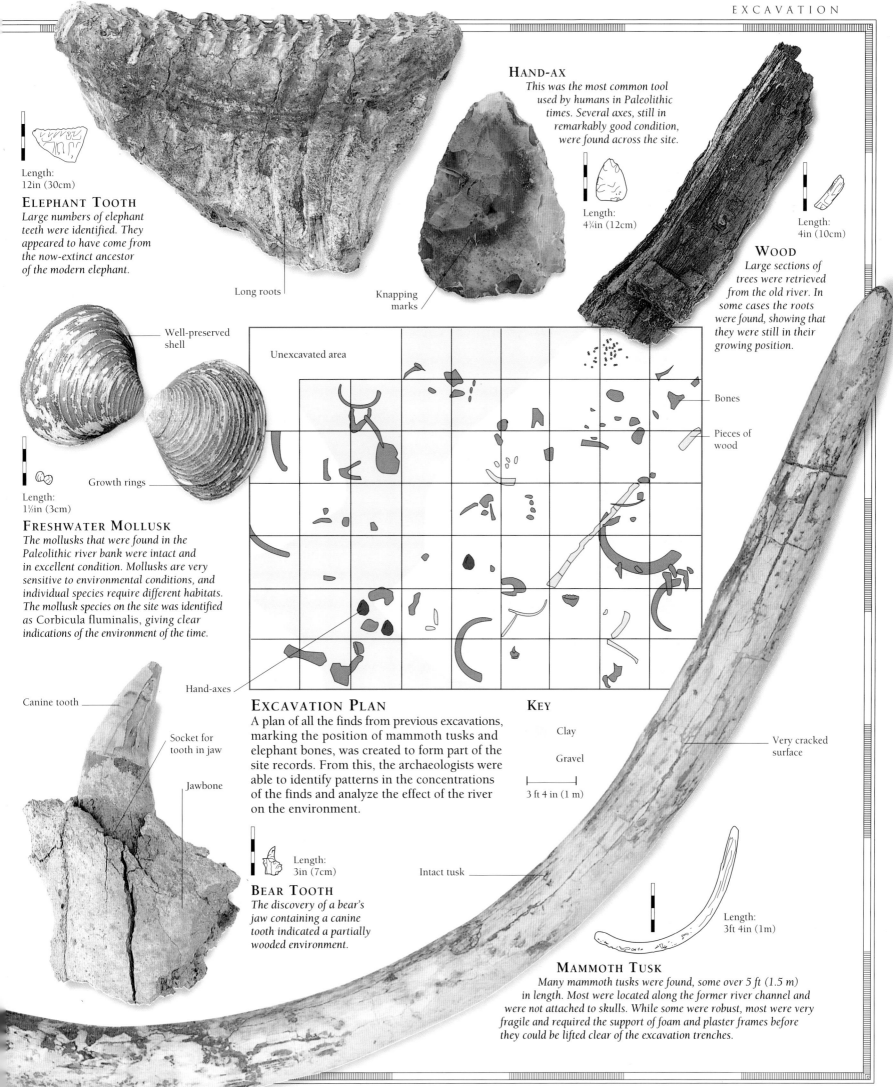

ELEPHANT TOOTH
Large numbers of elephant teeth were identified. They appeared to have come from the now-extinct ancestor of the modern elephant.

Length: 12in (30cm)

Long roots

HAND-AX
This was the most common tool used by humans in Paleolithic times. Several axes, still in remarkably good condition, were found across the site.

Length: 4¾in (12cm)

Knapping marks

WOOD
Large sections of trees were retrieved from the old river. In some cases the roots were found, showing that they were still in their growing position.

Length: 4in (10cm)

Well-preserved shell

Growth rings

Length: 1½in (3cm)

FRESHWATER MOLLUSK
The mollusks that were found in the Paleolithic river bank were intact and in excellent condition. Mollusks are very sensitive to environmental conditions, and individual species require different habitats. The mollusk species on the site was identified as Corbicula fluminalis, giving clear indications of the environment of the time.

Unexcavated area

Bones

Pieces of wood

Hand-axes

EXCAVATION PLAN
A plan of all the finds from previous excavations, marking the position of mammoth tusks and elephant bones, was created to form part of the site records. From this, the archaeologists were able to identify patterns in the concentrations of the finds and analyze the effect of the river on the environment.

KEY
Clay

Gravel

3 ft 4 in (1 m)

Canine tooth

Socket for tooth in jaw

Jawbone

Length: 3in (7cm)

BEAR TOOTH
The discovery of a bear's jaw containing a canine tooth indicated a partially wooded environment.

Intact tusk

Very cracked surface

Length: 3ft 4 in (1m)

MAMMOTH TUSK
Many mammoth tusks were found, some over 5 ft (1.5 m) in length. Most were located along the former river channel and were not attached to skulls. While some were robust, most were very fragile and required the support of foam and plaster frames before they could be lifted clear of the excavation trenches.

A PALEOLITHIC VALLEY

INTERPRETING THE FINDS

THE MAIN ARTIFACTS left to us by our Stone Age ancestors are the flint tools they used and the remains of the flora and fauna that made up their environment and diet. Remains of the people themselves or their campsites are rare. There were some key questions that the archaeologists had to ask themselves when faced with this Paleolithic site. Were the objects *in situ* or had they been washed into the site by glacial or river action, and, therefore, how near was the site to a human settlement? Do the animal bones and other evidence display any signs of human activity?

The wide range of environmental material enabled archaeologists to compose a picture of Stanton Harcourt. The climatic, vegetational, and faunal conditions of the ancient landscape inhabited by the hunter-gatherers were reconstructed by assuming that the various animal and plant species identified from the site required the same environmental conditions in the past as they do now.

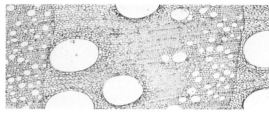

WATERLOGGED WOOD
By sectioning a sample of the wood and examining it under a microscope, the archaeologists identified the species as oak. Different species have different types of patterns and sizes of vessels.

Partially wooded landscape

Surrounding countryside containing caves

ARTIST'S IMPRESSION
From the evidence, we can imagine a wide, reed-fringed river flowing through a partly wooded and open grassland landscape. Remains of mammoths and other animals drift down the river and get caught in tree roots and fallen trees, especially at points where the river meanders.

BROWN BEAR
Brown bears eat mainly berries and fruit, but also scavenge meat from mammal carcasses. Their presence indicates a partially wooded landscape and an abundance of food.

APHODIAS FASCIATUS   **HISTER QUADRINACULATUS**   **DERMESTES MURINUS**

BEETLES
Several different species of beetle were identified from the soil sample, suggesting a specific environment. These included Hister quadrinaculatus, *which lived off dead bodies infested with flies, and* Aphodias fasciatus, *which fed on the dung of large herbivores.*

Distinctive leaves

VALERIAN
One of the plant species found at the site was Valerian (Valeriana sp.). This is a widespread, flowering herb that grows across Europe. The various plants discovered suggest a varied, flowering grassland in the valley.

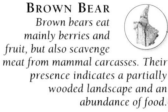

Seeds of various types and sizes

SEEDS
By sifting samples, a wide variety of plant seeds were identified. Each species has seeds of different shapes and sizes that are distinctive and identifiable.

Plant and animal debris trapped in reeds

Reeds fringing the river channels

TOOTH COMPARISON

Both elephants and mammoths are represented at Stanton Harcourt, and they can be distinguished by differences in the structure of their teeth. The plates that make up the teeth are more compressed in mammoths, giving more ridges on the grinding surfaces. This was an adaptation for crushing the coarse vegetation that characterized cold climates. In contrast, elephant teeth have widely spaced ridges for soft vegetation. So the teeth suggest that mammoths, normally associated with, and adapted for, very cold climates, were trapped in the increasingly warm climate of Britain as it was made into an island by the rising sea level at the end of the Ice Age.

ELEPHANT TOOTH

Compressed ridges

Widely spaced ridges

(TOP VIEW)

Widely spaced plates

Pointed roots

(SIDE VIEW)

MAMMOTH TOOTH

Grinding surface

(TOP VIEW)

Compressed plates

(SIDE VIEW)

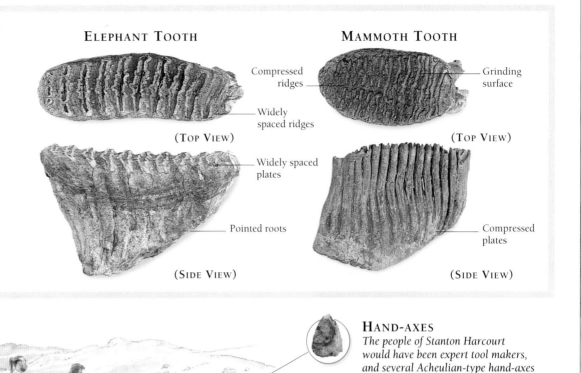

HAND-AXES

The people of Stanton Harcourt would have been expert tool makers, and several Acheulian-type hand-axes (see pp. 18–19) were found near the site. No waste from the manufacture of axes was found, and their good condition suggests that they were brought here from another area. Had the axes been more damaged, it would have suggested that they had been carried by either ice movement or rivers.

People butchering the mammoth carcass for meat and skin

MAMMOTH CARCASS

Some evidence suggests that dying or dead animals would have been washed downriver by flood waters. The remains of these animals would then be an easy source of food for the people living near the river's edge. They would have butchered the carcasses using stone tools.

Large tusks

CONCLUSIONS

🔍 The environmental evidence indicates the changing climate in the valley in Paleolithic times.

🔍 The plant species suggest the vegetational habitats along the river bank, surrounding plains, and woodland.

🔍 A wide range of animals grazed and hunted in the valley and could be identified from their bones, teeth, and tusks.

🔍 The hand-axes suggest that humans were using the area, but had not settled there.

FRESHWATER MOLLUSKS

These small shells (Corbicula fluminalis) were found intact in small family groups, which indicated that the area uncovered was an undisturbed prehistoric land surface adjacent to the river. Only very small undisturbed areas remained.

Dead mammoth caught in the shallows

SOFT ANTLER HAMMER

STONE PEBBLE

AX-MAKING TOOLS

In some cases, the hammers used to make tools are found with the tools themselves. Stone hammers survive well, whereas antler, which was often used to "pressure-flake" – remove flakes by pushing instead of striking the stone – survives less well.

STONE TOOLS

BY COMPARING A RANGE OF TOOLS from different time periods, archaeologists can begin to understand more about the technological capabilities of different societies. In the case of stone tools, there is often a period of more than 100,000 years between distinctive changes in types of tool. Each new stage of development tends to be characterized by a more sophisticated technique of production, particularly with regard to the relative length of usable cutting edge produced from a single piece of flint. Another way of analyzing a group of tools would be to imagine the amount of work taken to create each implement. It takes only a few strikes of stone against stone to create an Oldowan chopper, whereas to produce a flaked and ground ax takes considerably longer. One of the most significant stone-tool developments is the change from using just a core, with the flakes removed, to using the flakes themselves, reworked into a variety of implements.

OLDOWAN CHOPPER
—— 1.9 MILLION YEARS AGO ——

Some of the earliest stone tools discovered to date are those found in the Olduvai Gorge in Tanzania, dating to 1.9 million years ago. These types of tools were simple pebbles with occasional flakes knocked off them to create a cutting edge. The smaller pieces that had been chipped off were probably used for cutting and scraping. These tools were used by *Homo habilis*, one of the ancestors of modern humans. The simple nature of these tools often makes it difficult to distinguish them from naturally broken and cracked pebbles in the field. A wide variety of types of stone were used to make these tools, including quartzite and volcanic rocks.

PALEOLITHIC HAND-AX
—— 1.5 MILLION YEARS AGO ——

Around this time a newly evolved species of human called *Homo erectus* spread from Africa to the rest of the world, excluding the Americas. These people carried a new kind of stone tool with them that is now described as being of the "Acheulian" type. This style of hand-ax is characterized by being flaked on both sides.

LEVALLOIS AX
—— 200,000 YEARS AGO ——

Homo sapiens emerged about 200,000 years ago, and a new form of tool (named after the Levallois region of France, where it was first recognized) is associated with this period. This new technique created a tortoise-shaped core surrounded by a series of sharp facets all around the edge. From the end of this core, one large flake is skillfully detached. This flake is shell-shaped – bulbous in the center, with sharp edges – which gives it great strength. This technique, still in use until about 35,000 years ago, enabled the maker to predict and control the exact shape of the final flake.

1.9 MYA

Roughly chipped working surface

Smooth part sits in the palm

Flakes removed from all over

Defined cutting edge

Flake removed for tool

Tortoise-shaped core

—— KEY CHARACTERISTICS ——
Pebbles used as the raw material.
Smooth part to sit in the palm.
Chipped and flaked cutting edge.

—— KEY CHARACTERISTICS ——
Defined cutting edge.
Flaked all over to produce two faces.
Distinct and elongated ovate shape.

—— KEY CHARACTERISTICS ——
Tortoiselike shape from core preparation.
Ax is one large flake struck from core.
Cutting edge is chipped for strength.

KNAPPING A PALEOLITHIC AX

Making (or knapping) a stone ax is a job that requires great skill and patience. The skill lies in striking the core (a specifically selected stone of the correct weight and dimensions) at exactly the right place each time, so that the fracture produced creates a flake of the size and shape required. Even after thousands of years, the results of this process are often still visible on individual axes. This is because the place of impact, called the point of percussion, and the ripple fractures, created by the blow, are often well preserved.

1 A fine-grained, faultless stone that will fracture easily is selected. The stone core is prepared using a hard hammer stone. Major flakes are removed by striking the core to fracture it, and a rough ax shape is produced.

2 Once the stone is roughly the correct shape, a soft bone hammer is used to chip off specific areas of the stone in a controlled manner. This stage, which is known as "secondary flaking," requires patience and precision.

3 Once the stone is ax-shaped, the knapper can begin to shape the ax further by "pressure flaking" – pushing not striking – the ax with a chisel. This task requires a leather hand cover to protect the palm.

UPPER PALEOLITHIC BLADES
40,000 YEARS AGO

Around 40,000 years ago, an advanced form of the core technique was developed, in which a series of blades were struck from one core. After they had been secondary flaked (the removal of flakes from each edge) the blades were used to make knives and scrapers. This method created the maximum length cutting edge from one core.

CORE

Evenly sized blades

BLADE   BLADE

Pointed end

KEY CHARACTERISTICS
Parallel-sided blades.
Pointed or chisel ends.
One core produces many blades.

MICROLITHS
10,000 YEARS AGO

Microliths were heavily used from about 10,000 years ago. They were much smaller blades, struck from a core and used in composite tools such as harpoons, sickles, and spears, where several small blades could be set together as barbs and serrations. The larger blades with the chisel-like, thicker ends are known as burins and were probably used for engraving.

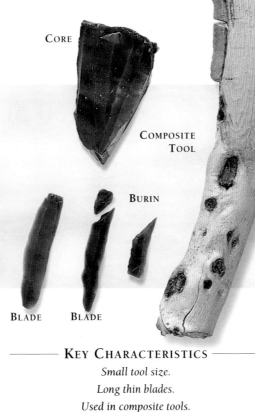

CORE

COMPOSITE TOOL

BURIN

BLADE   BLADE

KEY CHARACTERISTICS
Small tool size.
Long thin blades.
Used in composite tools.

NEOLITHIC AX
4000 BC

Ground and polished stone axes appear in the Neolithic period (see pp. 24–33). These larger axes, strengthened by grinding the cutting edge with another stone, represent a large expenditure of human time and effort. As a result, they were often symbolic as well as utilitarian objects.

Hard-wearing tool

Smooth cutting edge

Flaking scars

4000 BC

KEY CHARACTERISTICS
Regular shape.
Sharp, ground edge.
Smooth, polished surface.

COMPARATIVE SITES

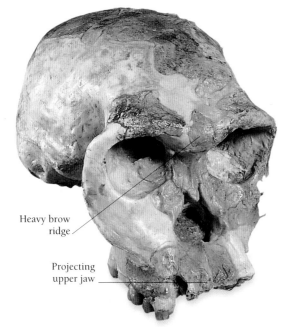

AN EARLY HUMAN SKULL
This is a skull of Homo habilis, *the earliest human species known to have made stone tools. Unlike in modern human skulls, the brain case is small in relation to the facial area, the top jaw projects, and there are heavy brow ridges.*

Heavy brow ridge

Projecting upper jaw

OLDUVAI GORGE
SERENGETI PLAIN, TANZANIA
GAZETTEER P. 142 – MAP REF. F5

OLDUVAI GORGE WAS CREATED around 500,000 years ago, when a geologic rift occurred and a river cut its way into the Serengeti Plain. As the river made its way through the layers of sediment and volcanic debris, evidence of human occupation around the shores of a now-extinct lake was revealed. By dating the volcanic strata in the gorge using the technique of potassium–argon dating, scientists have proved that the archaeological remains date back over 1.7 million years and that the site was used until 100,000 years ago. The first ancient hominid bones were found at Olduvai in 1959, and ongoing work has uncovered pieces of more than 60 skeletons and huge quantities of stone tools and animal bones. The earliest fossil remains identified belong to *Australopithecus boisei* and *Homo habilis*, the first hominids to make and use stone tools in the form of very simple pebble choppers (see below). Around 1.5 million years ago, the more advanced *Homo erectus* appeared, and their arrival brought about the development of more sophisticated hand axes and cleavers.

BIRTHPLACE OF HOMO HABILIS

Olduvai Gorge is important to archaeologists studying hunter-gatherers because its long sequence of stratified deposits enables us to follow the biological development of early humans and the changes in stone-tool technology associated with human development. The first *Homo habilis* finds in the gorge were made in 1964, and Louis Leakey (archaeologist, 1903–72) and his team soon uncovered other remains dating back 1.7 million years. Leakey quickly attributed the many tools found in the area (dating back 1.9 million years) to this new genus of "handy" humans. The tools showed clear evidence of butchery, but the meat itself was probably obtained through scavenging rather than hunting. Since their discovery, the finds from Olduvai Gorge have caused great controversy because the scatters of broken stones, animal bones, and hominid fossils are very difficult to interpret.

VIEW OF THE GORGE
Olduvai is a Y-shaped gorge, and this view shows the area where most of the excavations have taken place. The steep sides are the result of water erosion through the soft deposits, and it is here that the finds are made.

Signs of flake removal

PEBBLE TOOLS
The characteristic tool is a flaked pebble, from which pieces have been removed to create sharp edges.

MEZHIRICH

RIVER DNIEPER, UKRAINE

GAZETTEER P. 147 – MAP REF. M5

THE MAMMOTH-BONE HOUSES discovered at Mezhirich, in the Ukraine, were located on an extensive open-air site next to the Dnieper River and probably represent the world's oldest architecture. The site is believed to have been the base camp of a small group of hunter-gatherers, who would have traveled around the region, using different sites throughout the year, while still returning to Mezhirich on a regular basis. The site has been dated, using radiocarbon dating, to a period over 18,000 years ago. Archaeologists have found evidence for five houses at the site, each one built of mammoth bones and containing hearths, work areas, and large amounts of debris. The bones were used to provide a solid frame, which probably would have been covered with hides. Essentially, they were built as strong tents rather than as houses, and the careful arrangement of bones would not have been visible from the outside.

MAMMOTH-BONE EXCAVATIONS
The excavations at the site revealed each house as a heap of mammoth bones, as shown above. Systematic removal of the bones one by one, and careful recording of the position of each bone in relation to the others, have enabled archaeologists to reconstruct the structure of each house before it collapsed.

SEASONAL SHELTERS

Each house at Mezhirich had a unique architecture, utilizing different patterns of bones to create the dome shape. The large quantities of bones used to build them and the labor involved indicate that they were important structures. One house alone used the mandibles of 95 mammoths, and archaeologists have estimated that each house would have taken 10 people 5 or 6 days to build.

Many objects were also found, reinforcing the idea that it was an important site or base camp.

Decorated objects and figurines indicate a rich material culture and possible ritual or symbolic beliefs. Jewelry was also found, including fossilized shells, perforated for wearing, that came from 187 mi (300 km) away, and amber from the Baltic, over 62 mi (100 km) away. In addition to this, analysis of the tools found has shown that the stone used to make them was imported from some distance.

From the structure of the houses and the artifacts found, we can build up a much richer picture of the lifestyle of the people who

used them than is usually possible for hunter-gatherers. The houses suggest socially organized groups, capable of remarkable architectural achievements, while the objects provide us with a glimpse of a society with ritual and artistic facets. The artifacts transported from great distances suggest wide-ranging contacts, although archaeologists can only guess about their nature and significance.

CARVED FIGURINE
This ivory female figurine was found at Mezhirich. Similar figurines of this period have been found all across Europe.

MAMMOTH-BONE HOUSE
This reconstruction shows the mammoth bones carefully and systematically stacked together. They probably would have been covered with animal hides.

Large tusk doorway

MOUNT CARMEL

NEAR HAIFA, ISRAEL

GAZETTEER P. 148 – MAP REF. E4

SINCE THE DISCOVERY OF HUMAN REMAINS in the caves of Skūl and Tabūn at Mount Carmel in the 1920s, new and controversial evidence concerning the evolution of hominids and early modern humans has been discovered. Neanderthals were a species of early hominid, with specialized characteristics adapted for the cold climates of Europe and western Asia. There has been much debate about their relationship to the evolution of modern humans – some scholars see them as our ancestors, while others regard them as a parallel development who were replaced by modern humans. The caves at Mount Carmel are central to this debate because they have produced skeletal and stone-tool evidence for both Neanderthals and early modern humans, with dates that suggest that they coexisted for a substantial period of time. The skulls found in the cave of Tabūn are all of a Neanderthal type; those at Skūl are of the modern type.

NATUFIAN BURIAL
During the epipaleolithic period (final phase of the Paleolithic) the Natufian culture with more formal burials flourished in Israel. This nearby burial contained typical Natufian goods, including a seashell headband.

CAVE DWELLERS

The smallest of the caves is at Skūl, where at least ten bodies appear to have been buried in a single area of deposit. There has been more excavation at Tabūn, but its material – bodies and sophisticated stone tools – is less extensive and more widely scattered. Initially it was not known whether the Neanderthals and modern humans had lived at the same time or whether the Neanderthals had evolved into modern humans.

This debate could not be solved until the application of new dating techniques – ESR (electron spin resonance) to date teeth, and thermoluminescence to date burned flint and rocks – clarified the relationship between the inhabitants of the caves and showed the individuals to be much older than previously thought. At Skūl, dates of 100,000 to 80,000 years ago were obtained, whereas the Tabūn evidence was dated by ESR to be 200,000 to 90,000 years old. A Tabūn skeleton was dated to 110,000 years ago, suggesting that she had been an early Neanderthal living at the same time as early modern humans, who began using the caves about 90,000 years ago, while the Neanderthals disappeared about 45,000 years ago. The finds in the caves on Mount Carmel shows how crucial dating can be in understanding archaeological sites.

CAVE FINDS
A number of objects were found inside the caves alongside the bones. These included Paleolithic art, such as this bone carved into the shape of an animal. Finds like these are common across Europe and western Asia.

DANGER CAVE

UTAH, UNITED STATES

—— GAZETTEER P. 137 – MAP REF. J7 ——

LOCATED IN NORTHERN UTAH, Danger Cave can be found on slopes 112 ft (34 m) above Great Salt Lake, which would have been an important focus of activity for the prehistoric hunter-gatherers in the Great Basin area. It is one of a series of caves in the area that are crucial to reconstructing the hunter-gatherer lifestyle in the American West. The cave has very deep deposits and was occupied intermittently for many thousands of years, from about 9000 BC until the 1st millennium AD, so it provides a long sequence of datable artifacts and environmental remains. These remains allow archaeologists to look at how the Archaic groups changed and adapted their lifestyle, through time, to the harsh desert environment and varying availability of resources. Because the cave is so dry, a wide range of objects has survived – in addition to the usual stone tools – demonstrating how important plant and animal products were to early communities.

EXCAVATION OF THE CAVE
Detailed three-dimensional recording techniques are used to excavate caves. This is because the deep deposits, over 10 ft (3 m) at Danger Cave, have complex stratigraphy, with many layers of built-up earth and trash. Here, the vertical depth of each layer is being recorded.

USING THE ENVIRONMENT

Excavations at Danger Cave have revolutionized our understanding of Archaic hunter-gatherer life in the Great Basin because radiocarbon dating has been used to establish its long sequence of human occupation. This showed that the local people had developed their own culture much earlier than previously thought, maturing independently of groups farther south that had always seemed more dominant to archaeologists.

The first human occupation dates from 9000 BC, when sparse remains of waste stone flakes suggest that groups following big-game herds used the cave as a temporary shelter. By 8300 BC, the hunters were making hearths in the cave.

The first major changes in lifestyle occurred in about 7500 BC, when the increasingly warm and dry climate forced the hunters to adapt to the different plants and animals becoming available. They became more mobile, hunting smaller animals like deer and antelope, and collecting marsh and scrub plants. Because of the dry environment, many artifacts made from these plants and animals survive, including wooden and stone tools, animal hides, fur, basketwork, matting, bones, and debris, including human feces, which can be microanalyzed to reveal diet and health.

Small family groups continued to use the cave during the dry period and the wetter one that followed. They developed new tool styles, and gathered seeds that they began to process.

CAVE EXTERIOR
The cave is located in the hills to the west of Great Salt Lake. The landscape is very arid and harsh, and the cave would have provided convenient shelter.

MOCCASIN
The dry environment has preserved artifacts like this shoe (6000 BC), which would normally have perished, for thousands of years. It was made of hide and sewn together with animal sinew.

THE NEOLITHIC

ABOUT 10,000 YEARS AGO, some hunting communities began to cultivate plants and domesticate animals to support themselves, while persisting with traditional hunter-gatherer activities, relying on the use of stone for their tools and weapons. This process of management and selection took place over many generations. The crops grown and animals domesticated by early farmers varied from region to region, just as staple foods do today. By tracing the biological development of different species, we can map the origins and spread of early agriculture across the world. Once they had established a constant food source, many Neolithic communities began to build permanent settlements. This allowed a wider range of objects to be used, including pottery for storage and cooking, and stimulated trade between cultures. These durable structures and objects survive in much better condition than the ephemeral traces of their hunter-gatherer predecessors. As a result, we can readily see the impact of farming on the physical landscape, as well as on developing urban communities.

GREENLAND

CORN
Corn was first cultivated in Central America and became the staple crop of the region. These early corncobs were much smaller than those eaten today.

NORTH AMERICA   ATLANTIC OCEAN

PACIFIC OCEAN

① **TAMAULIPAS**
Evidence for plant domestication has been found at this dry cave site, along with some stone tools. A variety of plants were also found here, including corn, pumpkins, beans, and bottle gourds.

② **TEHUACÁN VALLEY**
A valley containing some of the first evidence of corn domestication in Central America, from 7000 BC.

CARIBBEAN SEA

③ **VALDIVIA**
This site gives its name to the earliest form of pottery known in the Americas, and dates from around 3200 BC.

LLAMAS
Llamas were an important resource in South America. They were domesticated by 3000 BC and were used as beasts of burden, as well as for wool and meat.

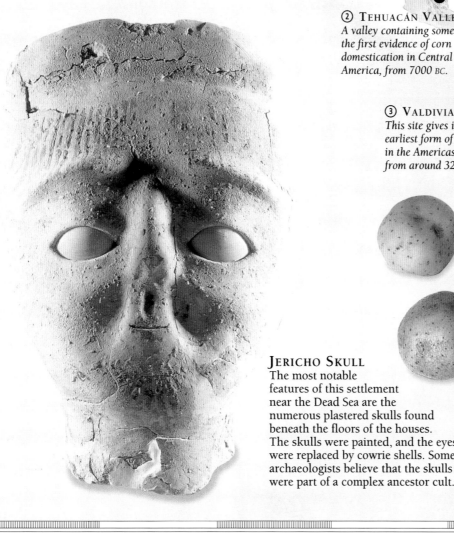

ANDEAN POTATOES
The Andean peoples cultivated many plant forms, including the potato, which became a staple crop.

JERICHO SKULL
The most notable features of this settlement near the Dead Sea are the numerous plastered skulls found beneath the floors of the houses. The skulls were painted, and the eyes were replaced by cowrie shells. Some archaeologists believe that the skulls were part of a complex ancestor cult.

N

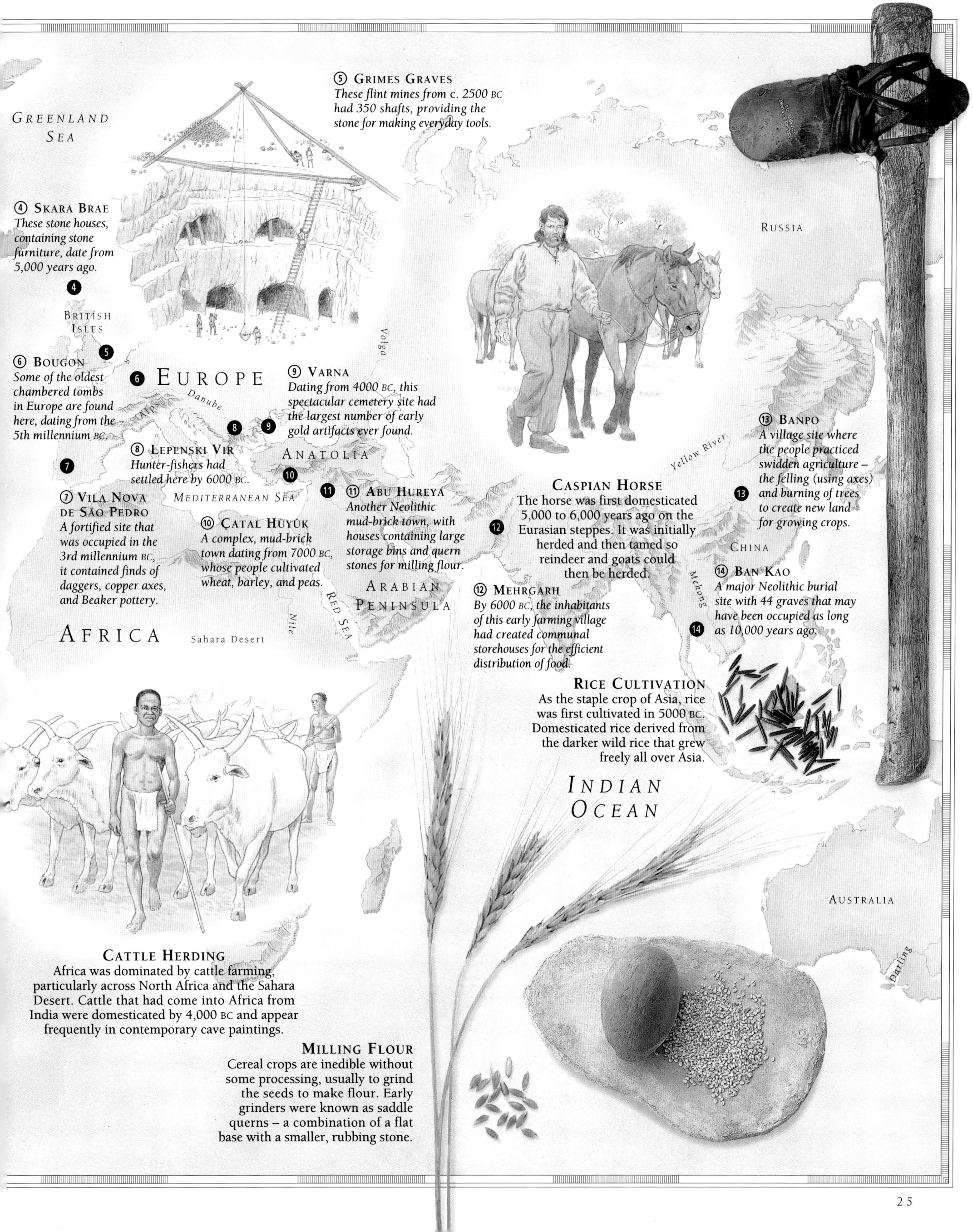

GREENLAND
SEA

⑤ GRIMES GRAVES
These flint mines from c. 2500 BC had 350 shafts, providing the stone for making everyday tools.

④ SKARA BRAE
These stone houses, containing stone furniture, date from 5,000 years ago.

4

BRITISH
ISLES

5

⑥ BOUGON
Some of the oldest chambered tombs in Europe are found here, dating from the 5th millennium BC.

6 EUROPE

RUSSIA

Volga

Danube

Alps

7

⑨ VARNA
Dating from 4000 BC, this spectacular cemetery site had the largest number of early gold artifacts ever found.

⑧ LEPENSKI VIR
Hunter-fishers had settled here by 6000 BC.

ANATOLIA

8 **9**

10

11

⑦ VILA NOVA DE SÃO PEDRO
A fortified site that was occupied in the 3rd millennium BC, it contained finds of daggers, copper axes, and Beaker pottery.

⑩ ÇATAL HÜYÜK
A complex, mud-brick town dating from 7000 BC, whose people cultivated wheat, barley, and peas.

MEDITERRANEAN SEA

⑪ ABU HUREYA
Another Neolithic mud-brick town, with houses containing large storage bins and quern stones for milling flour.

ARABIAN
PENINSULA

AFRICA

Nile

RED SEA

Sahara Desert

Yellow River

⑬ BANPO
A village site where the people practiced swidden agriculture – the felling (using axes) and burning of trees to create new land for growing crops.

13

CHINA

CASPIAN HORSE
The horse was first domesticated 5,000 to 6,000 years ago on the Eurasian steppes. It was initially herded and then tamed so reindeer and goats could then be herded.

12

⑭ BAN KAO
A major Neolithic burial site with 44 graves that may have been occupied as long as 10,000 years ago.

Mekong

14

⑫ MEHRGARH
By 6000 BC, the inhabitants of this early farming village had created communal storehouses for the efficient distribution of food.

RICE CULTIVATION
As the staple crop of Asia, rice was first cultivated in 5000 BC. Domesticated rice derived from the darker wild rice that grew freely all over Asia.

INDIAN
OCEAN

AUSTRALIA

Darling

CATTLE HERDING
Africa was dominated by cattle farming, particularly across North Africa and the Sahara Desert. Cattle that had come into Africa from India were domesticated by 4,000 BC and appear frequently in contemporary cave paintings.

MILLING FLOUR
Cereal crops are inedible without some processing, usually to grind the seeds to make flour. Early grinders were known as saddle querns – a combination of a flat base with a smaller, rubbing stone.

A NEOLITHIC TOWN

ÇATAL HÜYÜK, TURKEY

GAZETTEER P. 148 – MAP REF. D3

SITE EVIDENCE

🔍 Ongoing excavations suggest a town with a complex grid of rooms and buildings, with mud-brick walls covered with plaster and unique religious adornments.

🔍 The rich collection of artifacts and decorative fittings in the buildings suggests a complex and sophisticated lifestyle.

🔍 There is plentiful evidence for the economic basis of the settlement and for understanding farming practices within it.

ON THE FERTILE PLAINS OF CENTRAL TURKEY, in about 7000 BC, a farming town grew up, with densely packed houses built around small courtyards. All around the settlement, crops of wheat, barley, lentils, and peas were grown, and wild fruits and nuts were harvested. Herds of cattle were also kept to supplement the diet of wild-animal meat. The town was built up over successive periods into a tell, a mound of earth created by continual collapse and repair of the mud-brick buildings. The tell has been investigated by archaeologists, and although only about three percent of the site has been excavated, the richness of the archaeological evidence teaches us a great deal about the life of this early farming community, from diet and economy to the rich spiritual life of the inhabitants. Çatal Hüyük is situated in one of the world's foremost areas for the development of farming, and the evidence from this settlement is crucial to our understanding of this process.

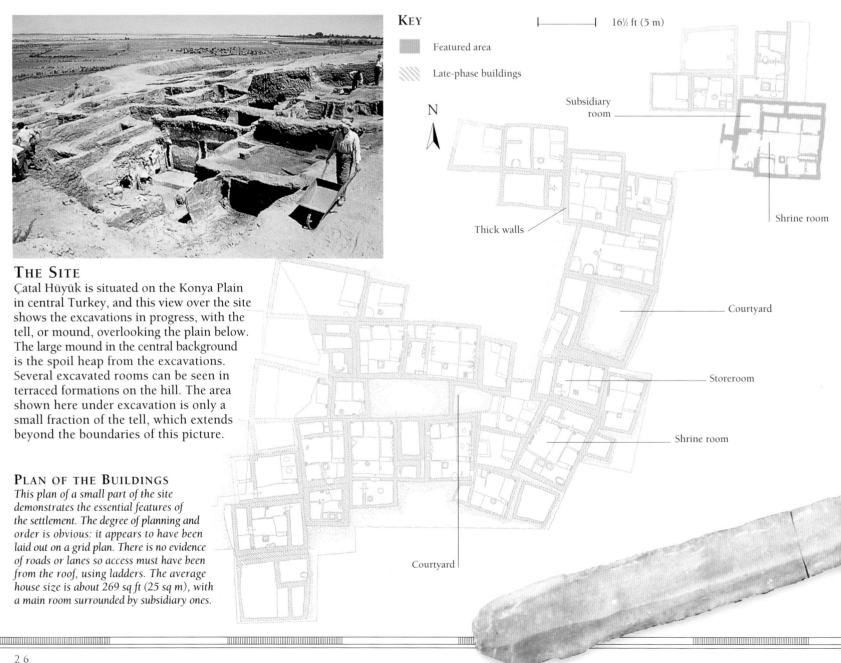

KEY

| 16½ ft (5 m)

▨ Featured area

▨ Late-phase buildings

N

Subsidiary room

Thick walls

Shrine room

Courtyard

Storeroom

Shrine room

Courtyard

THE SITE

Çatal Hüyük is situated on the Konya Plain in central Turkey, and this view over the site shows the excavations in progress, with the tell, or mound, overlooking the plain below. The large mound in the central background is the spoil heap from the excavations. Several excavated rooms can be seen in terraced formations on the hill. The area shown here under excavation is only a small fraction of the tell, which extends beyond the boundaries of this picture.

PLAN OF THE BUILDINGS

This plan of a small part of the site demonstrates the essential features of the settlement. The degree of planning and order is obvious: it appears to have been laid out on a grid plan. There is no evidence of roads or lanes so access must have been from the roof, using ladders. The average house size is about 269 sq ft (25 sq m), with a main room surrounded by subsidiary ones.

BULL SHRINE ROOM

Over 40 rooms in the settlement are decorated with plaster bull's heads and actual bull's horns. These, and other rooms spread throughout the site, are also decorated with rich wall paintings, showing leopards, rams, and goddesses.

HOUSE PLAN

Each unit in the settlement contained defined areas for different activities. This plan shows one of the houses that contained a bull shrine. There is also a hearth and an oven for cooking, and separate living and storage spaces. The town was composed of hundreds of similar small family units.

HUMAN REMAINS

The people of this settlement did not bury their dead in a formal cemetery. Instead, the bodies were exposed in the air, to be picked clean by vultures. The skeletons were then buried under the sleeping platforms in the houses. Ancestor worship was probably practiced, with successive generations being buried in the same rooms.

Unique pattern

Length: 2in (5cm)

CLAY SEALS

Many baked clay seals, carved with various designs, have been found. They were probably used for stamping paint patterns onto cloth or walls. Each design is unique, perhaps indicating that they represent individual people.

Horned bull's head

Bull-horn shrine

Raised platform

Open hearth

Shrine platform

Oven

Throne

Carbonized cloth

Height: 5in (12cm)

MOTHER GODDESS

About 50 pregnant female figurines made from terra-cotta, chalk, alabaster, marble, limestone, and volcanic rocks have been found at the site. These may have been used as part of a fertility ritual.

Simple weave

Length: 7in (17cm)

PIECE OF FABRIC

Many skeletons were buried wrapped in cloth, and where the bodies have been burned, the wrappings survive as carbonized material. Several fine-quality fabrics have been identified, including wool and probably linen.

Irregular edge

Length: 6in (15cm)

Length: ¼in (0.5cm)

FLINT KNIFE

Many basic weapons made of flint and obsidian have been excavated at the site, including daggers of high quality. The blades would have been mounted onto wooden handles and kept in leather sheaths.

CARBONIZED CROPS

Grains and seeds that have been burned survive well and provide useful evidence for the food used in the settlement. Carbonized crops can also be used for radiocarbon dating (see p. 126).

A NEOLITHIC TOWN

INTERPRETING THE FINDS

THE EXCAVATIONS CARRIED OUT at Çatal Hüyük show that the mound is artificial and the result of a succession of settlements on the same site, from before 7000 BC until more than 1,000 years later. From the buildings and rich assembly of artifacts, it is possible to build up a detailed picture of everyday life. The abundance of carbonized organic material reminds us that our farming ancestors would have used a wider variety of materials and objects than archaeological methods may uncover. The miraculous survival of wall paintings and three-dimensional sculptures vividly suggests the complex spiritual life of the community, and the many skeletons found reveal the health of the people and the dynamics of family life in the town. From the range and quality of the finds, it appears that Çatal Hüyük was a highly sophisticated farming and trading community with a complex range of rituals.

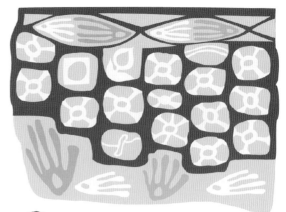

CLAY SEAL PATTERN
Walls could have been decorated using clay stamps to create panels of colorful geometric patterns. Many designs were found, including parallel horizontal or vertical lines, concentric circles, and checkerboard patterns.

WALL PAINTINGS

There are few wall paintings surviving in the world from the Neolithic period. Most wall paintings of this period were painted using mineral pigments and soot, which was combined with egg or oil and then applied to plastered walls with brushes or rags. The colors used were mainly earth tones – red, brown, yellow, and black. The wall paintings at Çatal Hüyük probably held deeply symbolic significance for the inhabitants, and were not just for decoration.

WALL-PAINTING REMAINS
Many of the wall paintings were damaged, but it was still possible to make out their original designs before conservation work was done.

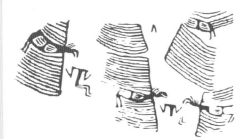

DRAWING OF WHOLE PAINTING
The picture clearly reveals headless figures and sweeping birds of prey. This may be depicting the exposure of dead bodies to vultures.

ARTIST'S IMPRESSION

Many of the walls in the settlement survived to a substantial height, so plenty of architectural detail is still present, with only the upper areas and roof needing reconstruction. On the left, the living area includes bed platforms and storage areas for crops, and on the right there is a shrine, with rows of bull's-head sculptures, bull's horns, and a burial. The roof has been reconstructed from fallen fragments, suggesting it was built of wooden poles covered in mud and reeds. Access to the rooms was by ladder (or maybe a more permanent staircase).

Storeroom entrance

Clay oven

EINKORN

EMMER

CEREAL CROPS
The discovery of carbonized grains indicate the cultivation of cereals, such as wheat and barley. Evidence of legumes, lentils, and vetches (a plant with a beanlike fruit) was also found. There were also traces of wild nuts and fruits, such as apples, pistachios, hackberries, almonds, and acorns. The crops shown left are two early varieties of domesticated wheat – emmer and einkorn.

Upper-story room

FABRICS
The inhabitants of the town appear to have been skilled weavers, so their clothing probably included sophisticated dyed garments. Only simple pieces of carbonized fabric were found and they probably had been used to wrap the bodies of the deceased.

MOTHER GODDESS

These mother-goddess figurines, with exaggerated female features, are often depicted with animals and found in storage vessels. It is because of these associations that they are thought to have represented a fertility cult, associated with hunting game, keeping domesticated animals, and growing crops.

Woman giving birth on a leopard-throne

BURIAL PRACTICES
After exposure, the articulated, fleshless, and desiccated skeletons were probably kept in a mortuary. They were then wrapped in cloth, skins, baskets, or matting; bound with rope or leather thongs and placed beneath the platforms in the shrine rooms or houses. This practice appears to have been carried out in the spring, when the houses were decorated.

BULL SHRINES

The shrines consisted of many varied details, including bulls' heads modeled in relief on the walls and benches with rows of horns (as shown left). There were also pillars with horns, known as bucrania, two of which are shown here. The bulls' heads often carry many layers of paint, suggesting continual renewal of the patterns and frequent redecoration of the inside of the shrine rooms.

CONCLUSIONS

◉ The town was obviously a densely occupied, enduring settlement of farmers, set in a rich natural and agriculturally exploited hinterland. The organic materials discovered reinforce this hypothesis.

◉ The well-preserved mud-brick buildings provide plenty of information about their construction, decoration, and use.

◉ The spiritual beliefs of the inhabitants are exceptionally well represented in the wall paintings, figurines, burials, and sculpture.

Ladder providing sole access to the upper stories

Recessed wall cabinet for fuel storage

Rectangular hearth with raised curb

Layers of mud-brick debris and walls from earlier buildings

WEAPONS AND TOOLS
Some of the knives, particularly those made of tough, imported flint, were high-status weapons, probably for ceremonial use. Obsidian, obtainable locally, was used for other functional implements, such as scrapers, cutting tools, spears, and arrowheads.

COMPARATIVE SITES

SKARA BRAE

ORKNEY ISLANDS, UNITED KINGDOM

GAZETTEER P. 144 - MAP REF. F1

THE ORKNEY ISLANDS, where the remains of the Neolithic settlement of Skara Brae are situated, have a cold and windswept climate. However, in about 3000 BC the climate was much milder, and wheat was cultivated. Cattle and sheep were also kept by farmers, who lived in small settlements. Despite their agricultural lifestyle, fishing and gathering were still very important to the inhabitants of the village, to supplement their diet and provide a wide range of tools and household goods.

The houses at Skara Brae, buried under sand for thousands of years, were first revealed in 1850 during a heavy storm. Several attempts to excavate them were made, before Gordon Childe (1892–1957), one of the most famous archaeologists of the 20th century, carried out a full excavation in 1928. He recorded the whole site in detail, but found it impossible to date it satisfactorily. Skara Brae is a very unique and isolated settlement, so, without modern dating techniques, Childe could not find enough parallels with similar sites of known dates to assign one to the village. He was, however, convinced that its inhabitants were Neolithic in technology and lifestyle, and modern dating methods have since proved him right.

Dry-stone walls made of local flagstone

THE SETTLEMENT

Six complete houses survive at Skara Brae, but, because the site is on the coast, other houses may have been eroded by the sea. They are all very similar in size and design and were built by cutting dry-stone walled rooms into piles of earth. As a result of the bleak and windy climate, there were no trees on the islands, so the inhabitants used the large quantities of flagstone found in the surrounding area. The stone broke easily into large, flat slabs, which were used in the walls. The walls were then plastered with clay, some fragments of which survive, to make them more insulated. The rooms of the houses are open today, but would originally have been roofed with thatch or turf, possibly supported on whalebone rafters. Each house consists of a single square room with rounded corners, having several stone cells connected to it by passages. At some point in the history of the community, the alleyways between the houses became tunnels, and they were roofed over with stone slabs. The main furniture was built of stone, with a large hearth in the center, a bureau or shelves for storage, and beds – each one consisting of a stone box that could be filled with plant material. There is also a series of small, stone, clay-lined tanks, about three or four per house. These were used either to store water or to hold shellfish for bait. There are also cabinets in the walls, which still stand to

THE INTERIOR OF A ROOM

The dry-stone walling of the house can be seen, as well as the remaining furniture. Originally, this area would have been totally covered by a turfed or thatched roof, which would have started where the turf line now is.

10 ft (3 m) tall. The beds have pillars to support a canopy, possibly to provide shelter or prevent dripping through the roof. The inhabitants of Skara Brae may have burned dung and whale and seal products, as well as wood. The cells may have been used for storage, although some have drains, which run away into a central drainage system outside the settlement, suggesting that there may have been toilets. One such cell contained hundreds of beads and other decorative and more functional objects.

Most early settlements in the Northern Isles, such as Skara Brae, have one unique building that is clearly different from the other houses. At Skara Brae, this building is slightly to the west of the

COASTAL SETTLEMENT

This view of the site shows the close proximity of the houses to the seashore. It can be seen that all the houses are linked together by a passage system, suggesting that the community was closely knit.

Passageway leading to rest of settlement

Remains of stone furniture

Turf line suggests original roof level

and the majority are associated with animal husbandry. There were also finds of carbonized cereals, flint and chert tools and implements, and wooden artifacts, some made from driftwood, including North American tree species. The bones of seals and whales have been found, as well as those of small sheep and large cattle. The artifacts found on the site reflect the rich and diverse environment that the inhabitants exploited in an efficient and variable fashion. The islands contained rich pasture and easily plowed sandy soils, and there were abundant marine food resources to compensate for the lack of land animals.

Fishing was probably done from small boats close to the shore, but larger sea animals would have been scavenged or captured when they were washed up or beached on the shoreline. Small fish and mollusks – especially

STONE FURNISHINGS

The bureau is the boxlike structure on the back wall of the house. The bed is situated on the left, while the hearth can be seen in the center. There are stone-lined pits in front of the bureau, whose exact use is unknown.

limpets (probably used as fish bait), oysters, crab, and other small coastal fish – would have been common, as would killer whales and walruses.

Most of the artifacts were made of one of three types of material – stone, pottery, or bone. Flint use was restricted because of its rarity, and flint tools would have been used only to make bone and wood tools and objects. From the artifacts found, it appears that the people of Skara Brae wore pendants and beads made from the marrow bones of sheep and the roots of cows' teeth and those of killer whales. Knucklebones may have been used as gamepieces. The most puzzling stone artifacts are small carved balls, known also from other sites in Scotland, whose function is not known. Other finds include pins, beads, pendants, carved stone objects, and bone and stone pots.

A SELF-SUFFICIENT COMMUNITY

Skara Brae was excavated in great detail by Childe, and because later archaeologists have reexamined the site and its artifacts, we now know a great deal about the Neolithic inhabitants of Skara Brae. The settlement lasted around 600 years, with small, stable family groups living in the network of houses and passages. These groups were resourceful and proficient, exploiting the varied environment to support themselves, without exhausting their natural surroundings. Nevertheless, Skara Brae was abandoned in around 2500 BC and was soon swallowed up by sand. The fierce storms that forced the abandonment of the site can, ironically, be thanked for eventually revealing one of the most complete Neolithic sites in northern Europe.

BONE TOOLS

Bone is the most common material excavated at Skara Brae. It was commonly made into tools and pins such as these, shown left. Bones of birds, cattle, and whales were all used, and were probably worked with flint tools.

village and is more oval in shape than the other houses, with a porch built on the front. It has a hearth, cabinets, and cells, but no bureaus or beds, and the walls are recessed and sections of the building partitioned off with slabs. This building also produced a different set of artifacts from the other houses, particularly burned materials and waste from flint production, which suggests that it was a workshop rather than a house. However, the fact that the room was almost clear of garbage and other artifacts also suggests that it may have been a ritual area, and that the tool production was associated with religion or magic.

The houses found at the site were all linked together by a network of passages, and were built into one interlocking system within a great mound of earth and debris, built up of dung, ash, clay, and animal refuse. Many of the artifacts come from these middens of earth and refuse,

Possible needle

Clothes fastener

Hairpin

BOUGON

DEUX-SEVRES, FRANCE

GAZETTEER P. 146 – MAP REF. G6

CHAMBERED TOMB
Here is one of the chambered tombs, showing the entrance to the passage that leads to the burial chamber. The regular, dry-stone walls and corbeled roof can be seen. These features are typical of the megalithic monuments so prevalent in the west of France. Most of the chambers at Bougon were reused as late as the Early Bronze Age (c. 3200–1950 BC).

ONE OF A SERIES OF MEGALITHIC BURIAL SITES in western France, Bougon was discovered and partially excavated in the 1840s by local antiquarians, making it one of the first Neolithic sites to be archaeologically excavated in the region. Since 1972, it has been systematically studied as part of a long-term project. There are eight main burial tombs, known as passage graves, in five mounds at the cemetery, which were in use throughout the Neolithic period. Recent excavations have included studies of soil and pollen, in an attempt to reconstruct the environment and landscape of the area when the tombs were built. Radiocarbon dating has also provided information for two of the cairns at Bougon, giving them a date of around 3800 BC. Examination of the pottery and lithic material has shown that the cemetery was also in use in the later Neolithic, through to the 3rd millennium BC. Megalithic cemeteries, due to their substantial stone construction, are essential to archaeologists in understanding the Neolithic cultures of western Europe.

BURIAL OF THE DEAD

The six cairns at Bougon are situated in an approximate east–west line. Each is different in its construction, and most were built in a number of phases. Some are what is known as "Atlantic" type, where a large, circular or oval mound is constructed of dry-stone walling and contains one or more chambers inside, which are reached from the entrance by small passages. Bougon also has very good examples of megalithic tombs, where the cairn is more rectangular and built with large stone blocks rather than with dry-stone walling.

The finds from Bougon are typical of many Neolithic sites, where pottery is used as a key for dating and understanding the site. Shards from round-based, undecorated bowls come from the early phase of burials at Bougon. Some objects called "vase-supports" have also been excavated. These are highly decorated artifacts, incised with lines, dots, and triangles, and are thought to be plinths of some sort.

These objects are found in chambered tombs throughout France and are thought to have had some ritual purpose. Later, straight-sided vessels of a different style are also found. Small beads and pendants have also been discovered in some of the tombs, as have flint tools, arrowheads, and small stone axes.

Huge numbers of human remains have been discovered at Bougon from the later Neolithic period. In one of the tombs, 200 individuals had been buried in two phases separated by over 1,000 years, while, in another, 38 skeletons were discovered. Two of the skulls showed evidence of trepanation, in which holes were cut in the bone for medical or ritual reasons. A large number of these bones have since disappeared, so they cannot be studied using modern techniques. There are many other chambered tombs and burial sites in western France similar to Bougon, but there are virtually no settlement sites to accompany them.

TREPANATION
Disks of bone have been removed from this skull, a process known as trepanation. It was practiced in many early societies for medical or spiritual relief. Remarkably, some individuals survived the repeated operations.

THE BOWL
This round, gourd-shaped bowl, nearly 6 in (15 cm) high, was recovered during the early excavations at Bougon. Two circular depressions are clearly visible on its body.

Archaeologists know far more about how these early farmers buried their dead than how they cultivated their crops or built their settlements. Their villages and homesteads were probably built of wood and thus have left little trace on the ground, while the tombs and cairns have survived because of their substantial stone construction and their prominence in the landscape. In addition to being burial sites, the tombs probably provided a visible territorial landmark for the local Neolithic communities and an important ancestral focus for emerging farming societies. Megalithic structures provide essential information for archaeologists because their differing construction from area to area reveals the emergence of disparate regional cultures.

Varna

BLACK SEA, BULGARIA

GAZETTEER P. 147 – MAP REF. L7

THE FIRST METALS USED in the Neolithic economy were copper and gold, probably because both occur naturally and are easily hammered without needing smelting or heating. The earliest-known metallurgy in Europe was discovered in the Balkan region, dating to around 5000 BC. Archaeologists usually refer to this transitional period as the Copper, or Chalcolithic, Age, rather than the Bronze Age, because the primary material used was still stone, and the small quantities of metals used were probably for status and symbolic purposes only, rather than serving practical purposes. Varna, dating to the 5th millennium BC, is one of the most spectacular cemetery sites in eastern Europe. Over 2,000 gold artifacts have been recovered from Varna since its discovery in 1972, revolutionizing archaeological thought about the end of the Neolithic period.

Double-band bracelet

Simple, hammered gold bracelet

GOLD BRACELETS
Bracelets of gold and shell found in the graves mostly lay around the elbow, suggesting that they were worn high up on the arm. Great skill was needed to make such regular, geometric forms to a standard size.

THE TRANSITIONAL ERA

The large cemetery at Varna was discovered by accident in 1972, near the shores of the Black Sea. So far, 204 graves have been excavated, and they have yielded huge numbers of artifacts, from human skeletons to gold ornaments, pottery, stone and copper tools, and clay figurines and masks. The discovery of such large quantities of metalwork dating back to such early times completely altered what archaeologists had believed about the development from Neolithic communities to Bronze Age societies.

Metallurgy is important to archaeologists because it often indicates the beginning of class differentiation and status in society, becoming more apparent in the more sophisticated societies of the Bronze Age. At Varna, such a wide-ranging and abundant collection of artifacts has been found that archaeologists can suggest how societies may have been organized from the layout of the burials. Several different types of grave have been identified, based on the way the body was laid, lying either on its back or on its side, the types of decoration,

and the amount of metalwork found in each one. For example, only a few of the burials contained gold scepters, which are interpreted as symbols of power and high status, while some of the graves contained no gold at all.

Osteological (bone) examination of the skeletons has provided information about burial practices and the health of those interred. Many of the supine burials were male, while females were buried flexed on their side. Different burial practices were used for people with conditions such as epilepsy or broken limbs. Although waterside settlements were also excavated, the conclusions of the archaeologists – that Varna was a socially structured, complex society, with a rich ritual life and elaborate funerary practices – came only from the burials.

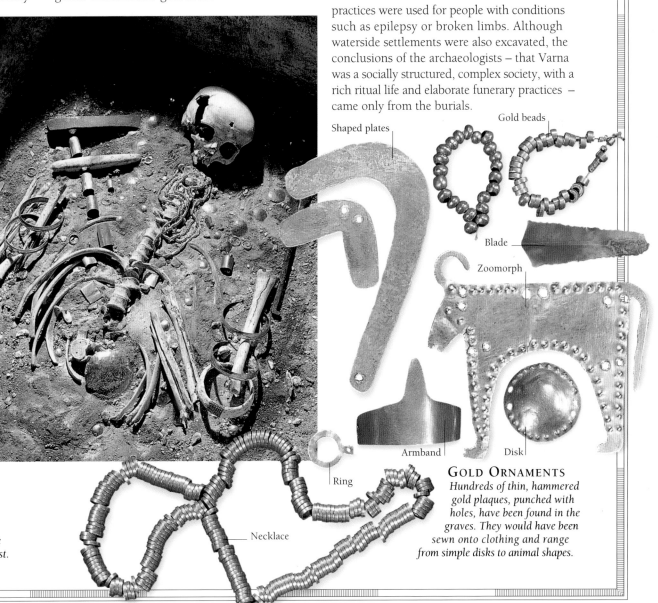

Shaped plates

Gold beads

Blade

Zoomorph

Armband

Disk

Ring

Necklace

GOLD ORNAMENTS
Hundreds of thin, hammered gold plaques, punched with holes, have been found in the graves. They would have been sewn onto clothing and range from simple disks to animal shapes.

THE GRAVE GOODS
The picture (above right) shows one of the graves with the artifacts laid out in a typical fashion around the body. There are stone axes above the right shoulder of the body, bracelets around the elbows, and beads and ornaments across the chest.

THE MONUMENTAL ERA

FROM THE PYRAMIDS of ancient Egypt to the stone circles and burial mounds of Europe, monuments are the largest and most impressive structures surviving from earlier times. Dating from the 4th millennium BC, they probably had a variety of religious, funerary, and public uses, although together they represent a common phase in the development of many different societies. The building of such large structures signifies the growing ability of communities to mobilize large quantities of resources, including labor. The structures also reflect the power of the community leaders and the cultural traditions of each society. In some areas, the monuments accompany the development of the first cities of the world, but for those in other areas we still know very little about the everyday life and settlements of the people who built them.

Archaeologists studying these structures and the artifacts associated with them can usually learn a great deal about their construction and the technology and materials that were used to build them. The actual function of such monuments, however, has proved far more difficult to discover and comprehend.

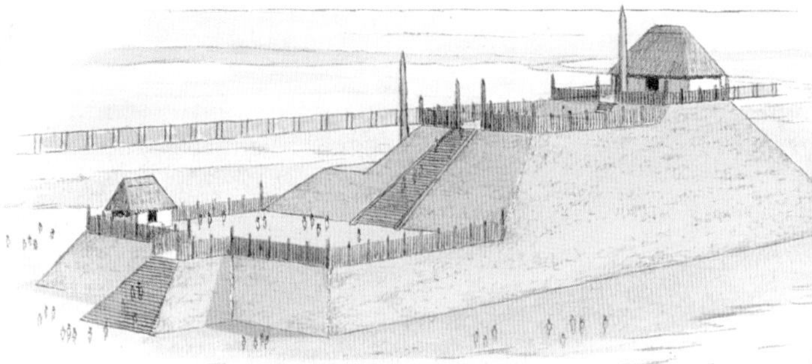

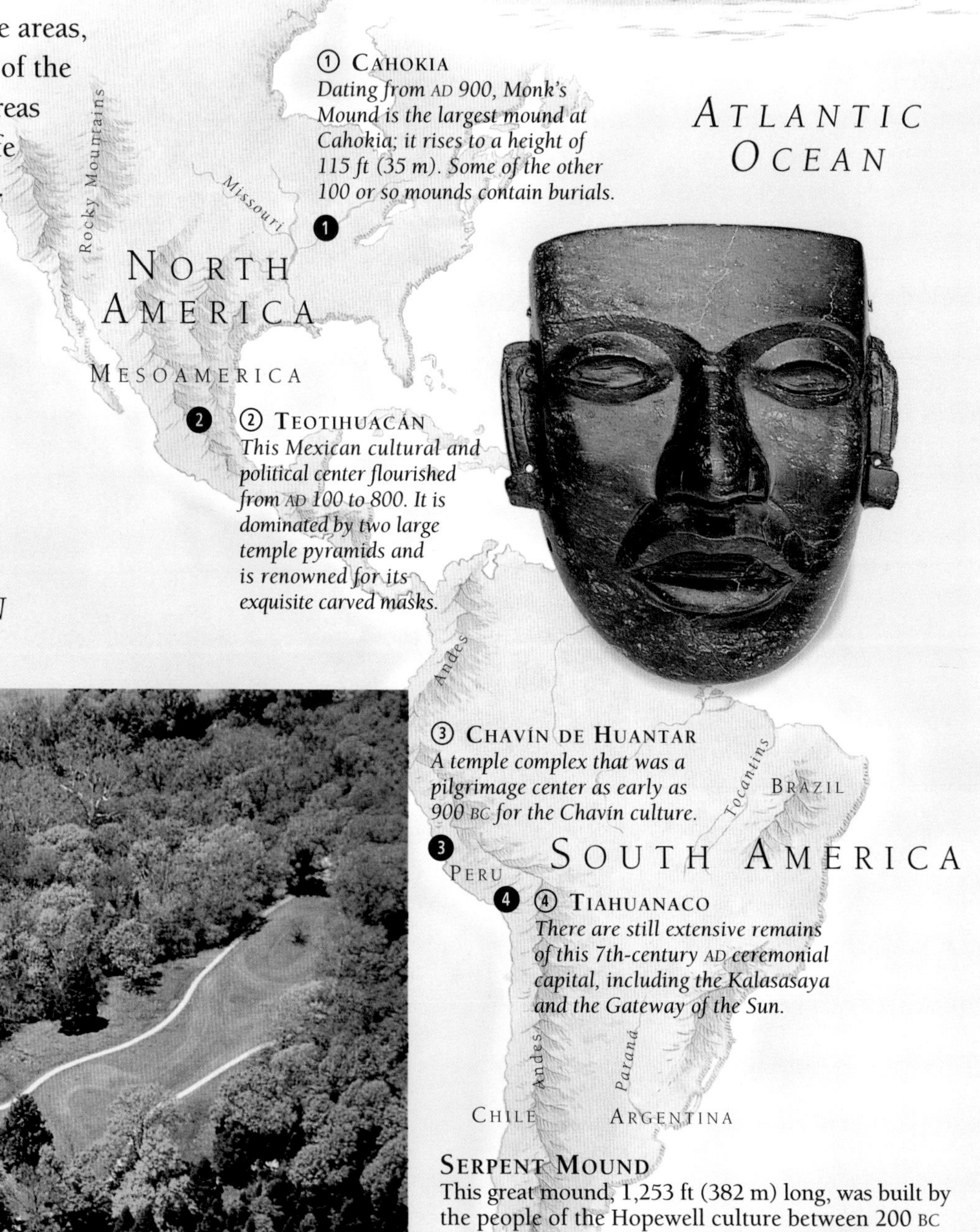

Rocky Mountains
Missouri
NORTH AMERICA
MESOAMERICA
ATLANTIC OCEAN
PACIFIC OCEAN
Andes
Tocantins
BRAZIL
SOUTH AMERICA
PERU
Andes
Parana
CHILE
ARGENTINA

① CAHOKIA
Dating from AD 900, Monk's Mound is the largest mound at Cahokia; it rises to a height of 115 ft (35 m). Some of the other 100 or so mounds contain burials.

② TEOTIHUACAN
This Mexican cultural and political center flourished from AD 100 to 800. It is dominated by two large temple pyramids and is renowned for its exquisite carved masks.

③ CHAVÍN DE HUANTAR
A temple complex that was a pilgrimage center as early as 900 BC for the Chavín culture.

④ TIAHUANACO
There are still extensive remains of this 7th-century AD ceremonial capital, including the Kalasasaya and the Gateway of the Sun.

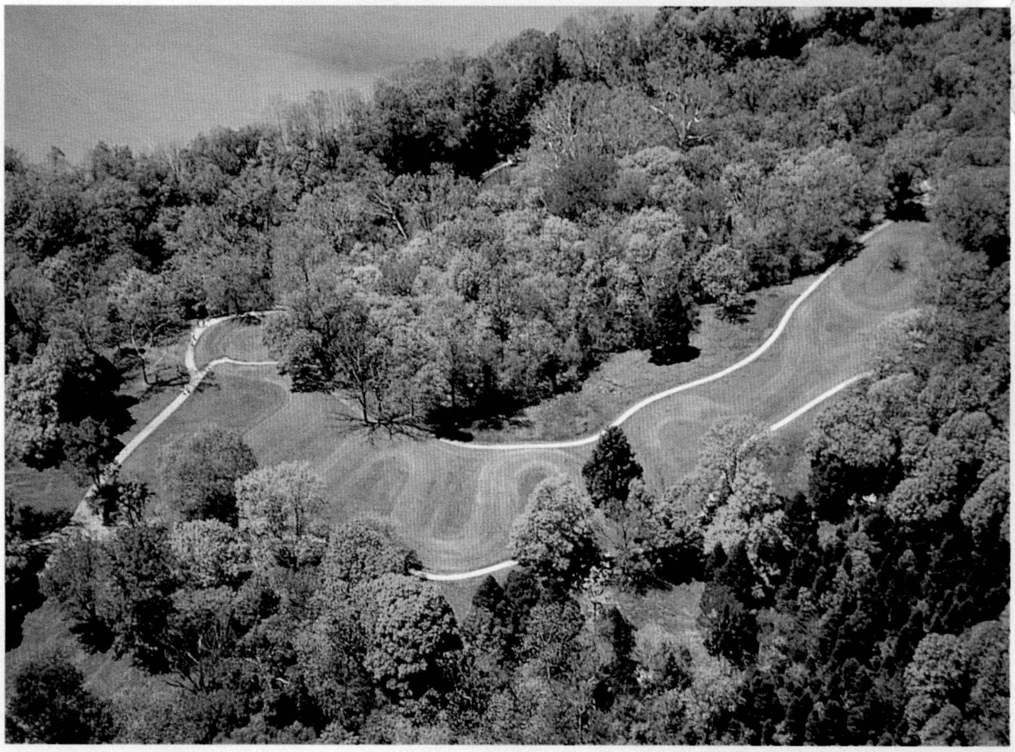

SERPENT MOUND
This great mound, 1,253 ft (382 m) long, was built by the people of the Hopewell culture between 200 BC and AD 400. It was built as a funerary monument, with an egg-shaped burial mound in the jaws of the serpent. There are many mounds in the Ohio valley, each of which would have required a huge investment of labor, up to 200,000 hours' work. Many objects were found in the mounds, including shark teeth and silver ornaments.

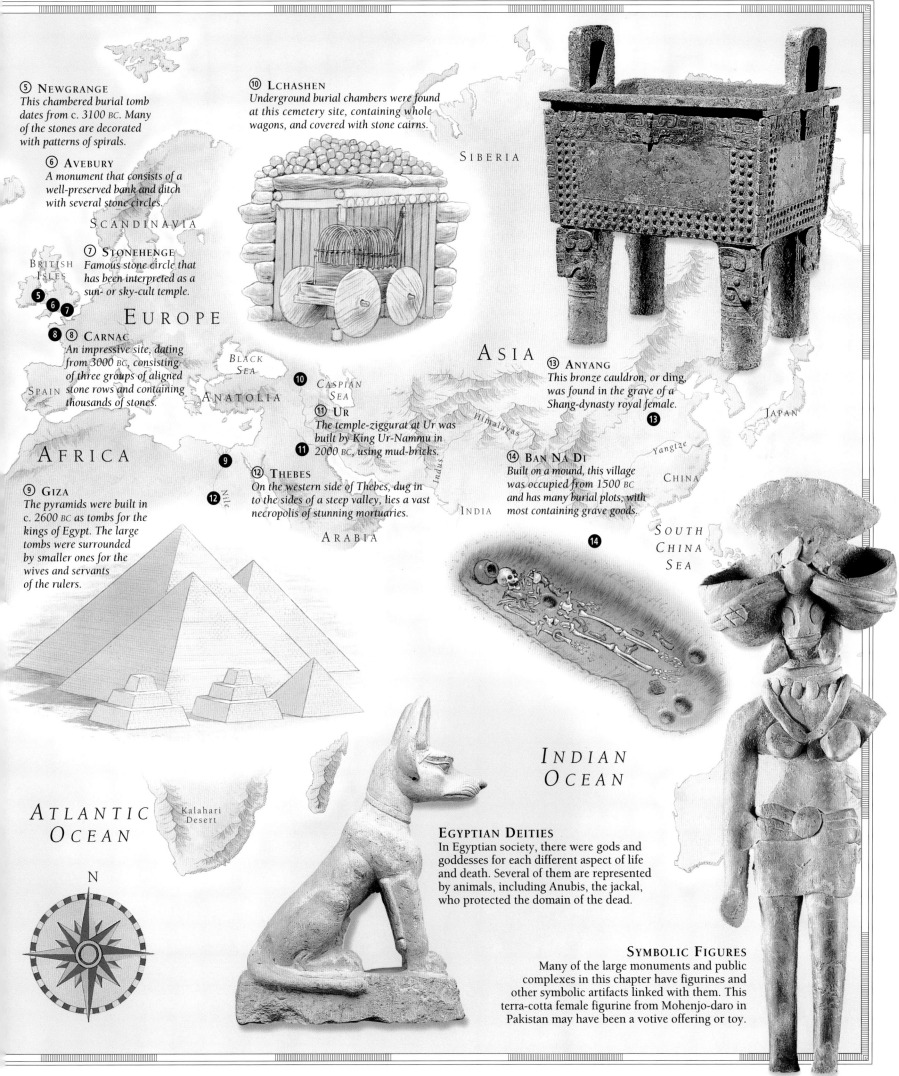

⑤ NEWGRANGE
This chambered burial tomb dates from c. 3100 BC. Many of the stones are decorated with patterns of spirals.

⑥ AVEBURY
A monument that consists of a well-preserved bank and ditch with several stone circles.

SCANDINAVIA

⑦ STONEHENGE
Famous stone circle that has been interpreted as a sun- or sky-cult temple.

BRITISH ISLES

EUROPE

⑧ CARNAC
An impressive site, dating from 3000 BC, consisting of three groups of aligned stone rows and containing thousands of stones.

SPAIN

⑩ LCHASHEN
Underground burial chambers were found at this cemetery site, containing whole wagons, and covered with stone cairns.

SIBERIA

BLACK SEA

ANATOLIA

CASPIAN SEA

⑪ UR
The temple-ziggurat at Ur was built by King Ur-Nammu in 2000 BC, using mud-bricks.

⑫ THEBES
On the western side of Thebes, dug into the sides of a steep valley, lies a vast necropolis of stunning mortuaries.

ARABIA

ASIA

Himalayas

Indus

INDIA

Nile

AFRICA

⑨ GIZA
The pyramids were built in c. 2600 BC as tombs for the kings of Egypt. The large tombs were surrounded by smaller ones for the wives and servants of the rulers.

⑬ ANYANG
This bronze cauldron, or ding, was found in the grave of a Shang-dynasty royal female.

⑭ BAN NA DI
Built on a mound, this village was occupied from 1500 BC and has many burial plots, with most containing grave goods.

Yangtze

CHINA

JAPAN

SOUTH CHINA SEA

INDIAN OCEAN

ATLANTIC OCEAN

Kalahari Desert

N

EGYPTIAN DEITIES
In Egyptian society, there were gods and goddesses for each different aspect of life and death. Several of them are represented by animals, including Anubis, the jackal, who protected the domain of the dead.

SYMBOLIC FIGURES
Many of the large monuments and public complexes in this chapter have figurines and other symbolic artifacts linked with them. This terra-cotta female figurine from Mohenjo-daro in Pakistan may have been a votive offering or toy.

HENGE MONUMENT

AVEBURY, UNITED KINGDOM

GAZETTEER P. 134 – MAP REF. F8

SITE EVIDENCE

○ Both surviving earthworks and standing stones provided physical evidence of large-scale prehistoric engineering.

○ Previous excavations of the site had produced a number of objects, mostly relating to the construction of the henge.

○ New archaeological methods, such as geophysical survey and aerial photography, had recently brought new information on the enclosure to light.

THE SWEEPING CHALK DOWNLANDS of southern Britain contain a number of enigmatic monuments that are testament to the outstanding engineering feats of the Neolithic and Bronze Age peoples. At Avebury, in Wiltshire, there is a large henge monument – a deep ditch and bank enclosing an area containing stone circles – that is known to date back to *c.* 2600 BC. The ditch was first excavated in 1910, and the stones were examined again in the 1930s, when many fallen and buried stones were re-erected. Today's archaeologists believe that there are still discoveries to be made at Avebury and have recently applied nonintrusive techniques, including geophysical surveying, to study each quadrant of the enclosure inside the ditch. This has already revealed new features, as has recent aerial photography, taken when the grassed area was parched. Finds from earlier excavations, together with these new discoveries, have enabled us to learn even more about the henge.

THE SITE

Excavations at the beginning of the 20th century, and recent surveys of the four quadrants of the enclosure, have attempted to discover the positions of all the 247 original standing stones. These studies have also shown the original profile of the ditch, as well as details about the construction of the bank.

Area of ditch excavation

GEOPHYSICAL SURVEY

Geophysics proved extremely valuable in locating the original positions of the stones that had disappeared or been dislodged. A geophysical survey of the northern quadrants revealed the exact position of the northern inner circle.

THE STONES

All of the remaining stones have been identified as being made of local sarsen – a quartzite sandstone. They also appear to have been erected in their naturally quarried shapes without alterations. There do, however, seem to be only two distinctive shapes of stone – long, column-shaped stones and broad, lozenge-shaped stones that are almost triangular.

Lozenge-shaped stones

Outer circle

NORTHWEST QUADRANT

SOUTHWEST QUADRANT

Natural surface with no evidence of carving or shaping

Column-shaped stone

N

KEY

— Area of ditch excavation

▦ Area of geophysical survey

├———┤ 148 ft (45 m)

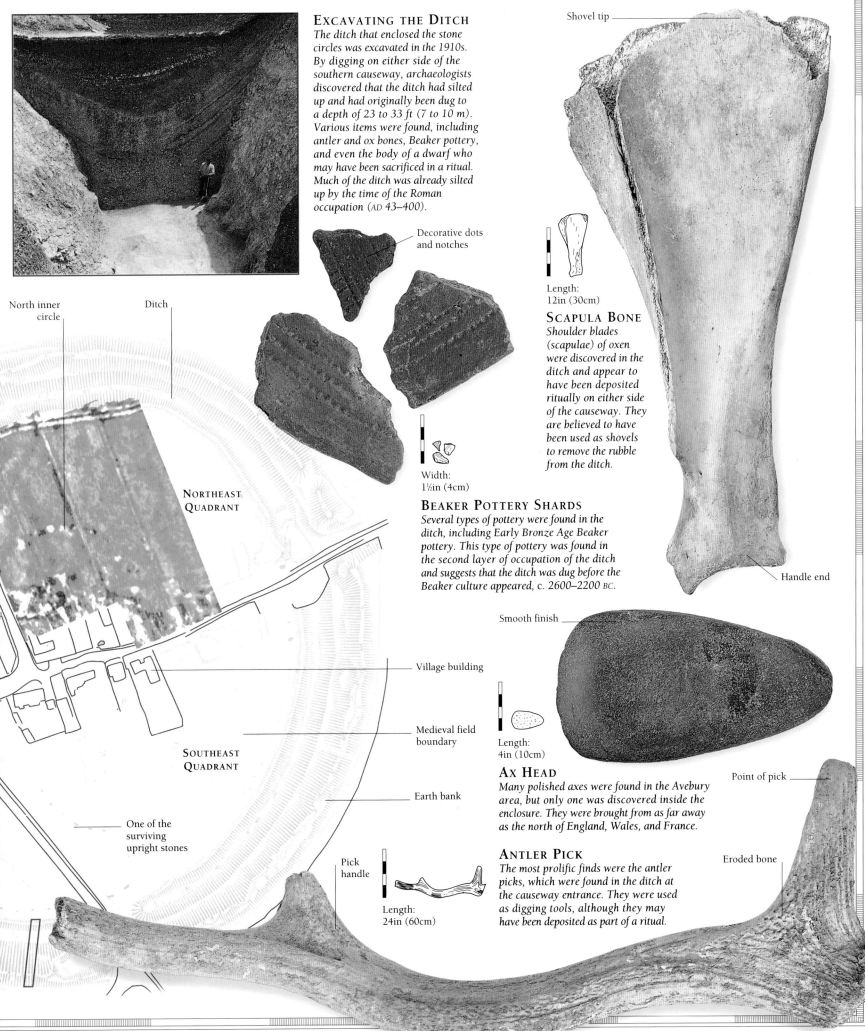

EXCAVATING THE DITCH

The ditch that enclosed the stone circles was excavated in the 1910s. By digging on either side of the southern causeway, archaeologists discovered that the ditch had silted up and had originally been dug to a depth of 23 to 33 ft (7 to 10 m). Various items were found, including antler and ox bones, Beaker pottery, and even the body of a dwarf who may have been sacrificed in a ritual. Much of the ditch was already silted up by the time of the Roman occupation (AD 43–400).

Shovel tip

Length: 12in (30cm)

SCAPULA BONE

Shoulder blades (scapulae) of oxen were discovered in the ditch and appear to have been deposited ritually on either side of the causeway. They are believed to have been used as shovels to remove the rubble from the ditch.

Decorative dots and notches

Width: 1½in (4cm)

Handle end

BEAKER POTTERY SHARDS

Several types of pottery were found in the ditch, including Early Bronze Age Beaker pottery. This type of pottery was found in the second layer of occupation of the ditch and suggests that the ditch was dug before the Beaker culture appeared, c. 2600–2200 BC.

North inner circle

Ditch

NORTHEAST QUADRANT

Smooth finish

Length: 4in (10cm)

Village building

AX HEAD

Many polished axes were found in the Avebury area, but only one was discovered inside the enclosure. They were brought from as far away as the north of England, Wales, and France.

Medieval field boundary

SOUTHEAST QUADRANT

Earth bank

Point of pick

ANTLER PICK

The most prolific finds were the antler picks, which were found in the ditch at the causeway entrance. They were used as digging tools, although they may have been deposited as part of a ritual.

One of the surviving upright stones

Pick handle

Length: 24in (60cm)

Eroded bone

HENGE MONUMENT

INTERPRETING THE FINDS

FROM THE EXCAVATIONS AND SURVEYS, as well as from the physical remains themselves, it is possible to imagine how the henge at Avebury was constructed – from the digging of the ditch and the transportation of the stones to the erecting of the stone circle. The huge amount of labor and organization that would have been required to build the henge can be estimated using experimental archaeology, and its relationship to the surrounding landscape can then be assessed. However, there is little direct evidence to answer the second key question asked by archaeologists: what was it used for? There are few finds from the henge, apart from the tools used to build the monument, and archaeologists must use other methods to determine its function. There is no doubt, however, that Avebury was a prominent landscape feature and a focus for the local people over many generations.

BRONZE AGE BEAKER
When assembled correctly, the shards make up a typical Bronze Age beaker with a distinctive shape, pattern, and hematite slip coating. It is assumed that these were drinking vessels, probably for beer or water.

GEOPHYSICS INTERPRETATION

The geophysical survey was useful because it confirmed many features already known about Avebury, as well as revealing some new ones. As many of the original stones are missing, it is helpful to be able to see the marks of the original stone holes and to complete the circle. The survey also discovered a circular feature in the ground to the northeast of the stone circle that is not apparent to the naked eye. This feature is unusual, and archaeologists are unsure of its original use. It probably existed before the building of the henge.

Area of disturbance caused by stone restoration

Linear feature of unknown date

Recently discovered enclosure

NORTHWEST QUADRANT

Unknown feature, possibly geologic

Circular feature of unknown origin

Evidence of stone holes

NORTHEAST QUADRANT

ARTIST'S IMPRESSION

All the documentary and physical evidence enabled the archaeologists to hypothesize about the building of the monument. Evidence suggests that the ditch was dug and a bank constructed before the stones were erected. However, this illustration shows all stages of the work happening simultaneously. Wooden rollers, ropes, and timbers would have been used to position the stones.

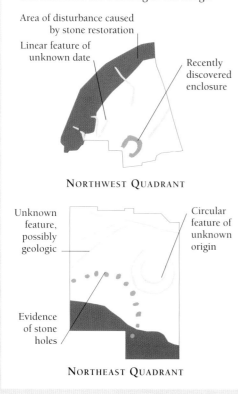

Northern inner circle

Outer circle of stones

Chalk causeway at the western entrance

Supporting wooden struts

Stones being erected using ropes and timbers

THE DITCH AND THE BANK
The excavation of 1910 revealed that the bank had originally been 55 ft (16.7 m) above the ditch. It has been estimated that to produce the ditch and bank it would have taken 100 people three years.

THE RITUAL LANDSCAPE

Looking at the landscape surrounding Avebury, we can see how it is part of a wider group of stone circles, burial mounds, and enclosures connected by stone avenues. From the south, an avenue of stone ran for several miles to the Sanctuary, another stone circle. There may also have been an avenue to the west. Silbury Hill, the largest artificial mound in prehistoric Europe, can be found nearby. Long barrows of the Neolithic period are situated all around the site, including the West Kennet barrow, and many more mound barrows date to the Bronze Age. The high investment in labor and materials indicates that Avebury was a special site, probably a ritual complex or stone temple.

THE STONE CIRCLES

There were originally 98 stones in the outer circle, set at 36-ft (11-m) intervals. The north and south circles contained 27 and 29 stones, respectively. The south circle centered on an obelisk and the north around a cove (three vast stones in an open square shape).

Moving stones on wooden rollers

Eastern causeway entrance

PRIZED POSSESSION

Only one ax head was found in the circle itself, indicating that the people building the henge kept domestic goods outside the enclosure. Ax heads were traded from far afield, making them valuable items that would have been carefully looked after.

Southern inner circle

CONSTRUCTION TOOLS

Most of the tools found were located at the causeway entrances, where the excavations took place. Lumps of chalk would have been dislodged with the antler picks, and then shoveled with bone scapulas, probably into baskets. The rubble was then carried or raised to the surface and dumped to form the bank.

CONCLUSIONS

🔎 The artifacts from the site tell us how the monument was built, but provide very little information to help with the interpretation of its use.

🔎 The surrounding countryside, as well as the size and nature of the henge itself, suggest that it was a ritual site and part of a wider monumental landscape.

🔎 New archaeological techniques have shown that there are still many questions unanswered about Avebury, and that there is much information yet to be gained.

Larger entrance stones

RITUAL DEPOSITS

It appears from the nature of the finds adjacent to the southern entrance that the people had made ritual deposits in the ditch during construction.

POTTERY

POTTERY IS THE MOST COMMON MATERIAL found on archaeological sites across the world and is linked with many cultures. Because of its weight, bulk, and fragility, it is usually associated with settlements and agricultural societies, when the accumulation of containers and stored produce was first made possible. There are three key stages in pottery making; building and shaping the pot, decorating it (using relief, slips, and glazes), and firing it. The technology of each of these stages has progressed over time, but the development of better kilns with higher firing temperatures is the most significant. Glazing and more advanced firing techniques made pottery more durable, so that vessels could be used for a wider variety of purposes, including cooking and storage. Pottery-making techniques and styles developed at different rates in different areas, so the easiest way for archaeologists to date sites is by placing any shards they find within a typology.

TEMPERING THE CLAY
To make sure that pottery fires successfully, materials called temper are added to the clay. These range from vegetable matter, such as chaff or grasses, to crushed quartz and flint (shown above). These inclusions can be seen in the cross-section of the pottery.

NEOLITHIC BOWL
— 3700–3000 BC —

Pottery appears at different dates throughout the world, but the earliest type was being made in Japan as long as 12,000 years ago. Very early pottery, like this bowl from the Ukraine dating from 5,700 years ago, was made in a very simple way by puddling mud to form a lump of clay, and then gouging out and working the lump into a bowl shape. This produced an irregular shape and texture. The bowl would have been fired at quite a low temperature, probably in a bonfire, so the vessel would not have been particularly hard or durable.

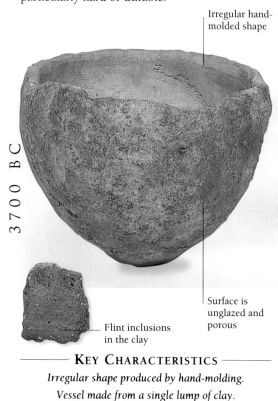

3700 BC

Irregular hand-molded shape

Surface is unglazed and porous

Flint inclusions in the clay

KEY CHARACTERISTICS
Irregular shape produced by hand-molding.
Vessel made from a single lump of clay.
No glaze or decoration is applied.

BEAKER
— 2500–1800 BC —

The development of coiling for making pots meant that more regular and evenly shaped vessels could be produced. This Early Bronze Age beaker from England was made in this way, with the design impressed on the hardening surface with a twisted cord. The surface has a hematite slip, and many beakers look polished or burnished.

Regular shape

Pattern created by lines of depression

KEY CHARACTERISTICS
More regular, even shapes produced.
Many styles of decoration used.
Ridges left by coils sometimes feature.

GREEK POT
— 900–700 BC —

Greek pottery was made on a wheel, allowing much more even and complex shapes to be made. Large vessels were often made in pieces and joined together. The geometric patterns were painted using a slip – a very thin layer of clay, which gives the pot its distinctive colors.

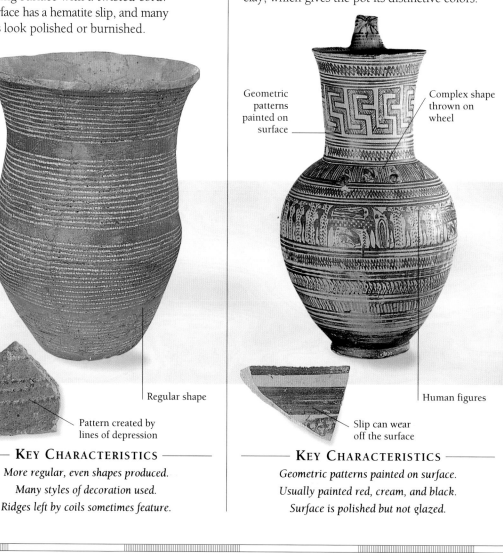

Geometric patterns painted on surface

Complex shape thrown on wheel

Human figures

Slip can wear off the surface

KEY CHARACTERISTICS
Geometric patterns painted on surface.
Usually painted red, cream, and black.
Surface is polished but not glazed.

MAKING A COILED POT

Before the invention of the potter's wheel, making a pot was often done using the coiling technique. Fairly sophisticated and regular-shaped vessels could be made in this way without the potter needing much equipment, and a wide variety of pot types, such as pitchers and beakers, could be made. Even when wheels were in use, coiling might still have been used to make the basic vessel shape, which would then be finished off on a wheel. This was especially useful for large pots that were too heavy to throw on a wheel.

1 *The clay is dug up and mixed with water in a pit. As a result, coarse particles separate out and leave a fine mixture that can be dried and worked into a smooth lump of clay. The clay is then rolled into long sausage shapes.*

2 *The lengths of clay are coiled around and built up in circles to produce the shape of pot required. While still damp, the coils are worked together to create a smooth-sided pot and to prevent them from cracking apart later.*

3 *When the clay becomes "leather" hard, the pot is rubbed and polished with a cloth or piece of leather. This leaves a smooth surface that can be decorated by stamping, incising, or glazing. The pot is then fired.*

SAMIAN WARE
— 1ST CENTURY BC —

During the Roman period, high-quality red tableware was manufactured in Italy and northern Europe and traded around the empire. This pottery, called Samian ware, was made in molds and mass-produced in huge numbers. A thin slip was applied to the surface to make it glossy and to produce the distinctive color. Samian bowls and dishes are found across the whole of the former Roman Empire. After production ceased in the 3rd century AD, many factories produced copies.

Shiny slipped surface

Molded pattern

— KEY CHARACTERISTICS —
Distinctive red clay and slip used.
Standardized shapes and patterns from molds.
Sometimes stamped with manufacturer's mark.

MEDIEVAL "GREEN WARE"
— 13TH CENTURY AD —

The introduction of glaze, a glassy mineral layer used to cover vessels, made it possible to apply new colors and decorations to pots, making them less porous. Medieval European glazes were normally lead-based, with copper added to create green colors.

Handle made separately and attached

Spout formed for pouring

Green glaze clearly visible

Incised-line pattern

— KEY CHARACTERISTICS —
Green color, often mottled and uneven.
Jugs and pitchers commonly found.
Coarse, heavy-duty wares common.

WEDGWOOD
— 19TH CENTURY AD —

From the late 18th century onward, new styles of pottery emerged in Europe, influenced by Chinese imports. Blue and white patterns, often applied with transfers, were used, as on this distinctive Wedgwood jug. Harder and finer vessels, such as porcelain, could be produced using the new firing techniques and clays.

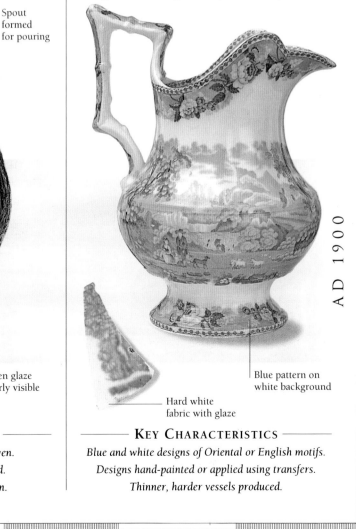

Blue pattern on white background

Hard white fabric with glaze

— KEY CHARACTERISTICS —
Blue and white designs of Oriental or English motifs.
Designs hand-painted or applied using transfers.
Thinner, harder vessels produced.

AD 1900

COMPARATIVE SITES

CARNAC

BRITTANY, FRANCE

———— GAZETTEER P. 146 – MAP REF. F6 ————

MEGALITHIC STRUCTURES CAN BE FOUND all over northwestern Europe, the best known being Stonehenge in Wiltshire, England. The complex of ritual Neolithic monuments at Carnac evolved over hundreds of years, with changes in monument type, shape, and form mirroring the development of the local culture. As part of a major ritual-monument group situated around the Gulf of Morbihan in Brittany, Carnac stretches over several square miles and is composed of many different sites and structures. There are single standing stones, massive alignments of stones, long mounds containing burial chambers, and passage graves adorned with megalithic art. Many theories have been put forward about the stones and their meaning, but the exact use of the monuments will always remain unknown.

THE STANDING STONES

Stones used for megalithic monuments are invariably "natural" – they are not cut or shaped, but deliberately selected for their size, shape, and type. There are almost 3,000 remaining standing stones at Carnac.

MEGALITHS AND MOUNDS

The megaliths appear as single stones (*menhirs*), groups of single stones (rows or circles), and megalithic complexes that form chambers (passage graves, chambered tombs, or *tumuli*). The huge, ovoid burial mounds were often built against a passage grave and were associated with great wealth. These burial mounds include the massive Tumulus de Saint-Michel, 33 ft (10 m) high and 410 ft (125 m) long, which was excavated in the early 20th century. It consisted of a number of small cists underneath a huge pile of stones, all of which were encapsulated in a thick layer of clay and a further covering of stones.

The largest class of megalith found at Carnac is the single standing stone. The Grand Menhir Brisé weighs more than 330 tons (300 tonnes). It may well have been a *menhir* that broke up in transit and never reached its destination. Suggestions that its purpose was as part of a lunar foresight for worship or as an astronomical calendar may well be correct.

THE MENEC ALIGNMENT

Many stones at Carnac are aligned in rows. One of the best examples is that of Le Menec, over ⅔ mi (1 km) long. It consists of 11 parallel lines of stones, graded in size, and culminating in an oval ring.

JADEITE AXES

The Tumulus de Saint-Michel contained a chamber in which a number of finely polished axes, including some made of jadeite, were found. These appear to have been made purely for ceremonial use because they are in excellent condition.

TEOTIHUACAN

VALLEY OF MEXICO, MEXICO

GAZETTEER P. 138 – MAP REF. D5

THE AVENUE OF THE DEAD
This view is taken from the top of the Pyramid of the Moon, looking south along the Avenue of the Dead, the main axis of the city. The huge Pyramid of the Sun can be seen on the left. The street was deliberately aligned north–south, and aerial photography has shown that there was a major east–west street with which it intersects.

REACHING ITS PEAK BETWEEN AD 200 AND 650, Teotihuacán was the greatest city of Mesoamerica, indeed probably of the preindustrial world, with a population that may have reached 250,000 people. Despite its eventual collapse in around AD 750, it was later venerated by the Aztec religion as the origin of the cosmos. Founded on the fertile land of the Valley of Mexico, where intensive agriculture was practiced, the city became the center of a huge political state. The excellent strategic position of Teotihuacán allowed it to have a strong domination over trade, especially in obsidian, and artifacts from the city can be found across all of Mesoamerica. The state was bound together by a ruthless and efficient system of tight military control and strong religious cults, and violent practices, such as human sacrifice, were used to demonstrate the power of the rulers. Yet it is the ceremonial art and architecture of the Teotihuacános that has left the most lasting impression.

THE CITY OF MONUMENTS

An archaeological project begun by the University of Rochester, New York, using low-level aerial reconnaissance and surface survey, has produced detailed topographical maps of the ancient city. It has also shown that the huge monuments and streets of the ceremonial area in the center of the city were accompanied by a large area of administrative and residential buildings.

The street plan of this massive urban center was laid out on a grid pattern, and was based on astronomical observations. The 3-mi- (5-km-) long Avenue of the Dead runs through the center of the city and is lined with religious and ceremonial complexes, the most impressive structures being the Pyramid of the Sun – at 246 ft (75 m) high, the largest artificial structure in precolonial America – and the Pyramid of the Moon. These structures were truncated pyramids with temples on their flattened summits and were sites of great ritual importance. The size of all these buildings underlines the power of the rulers of Teotihuacán, and their ability to mobilize vast amounts of labor for monumental construction.

The city rulers lived in elaborately frescoed single-story palace complexes, while other parts of the city consisted of rectangular compounds belonging to merchants, of which there were many; Teotihuacán had a monopoly over the sources and manufacture of obsidian – there were over 300 workshops in the city alone. The city contained a number of other districts, including an area where foreigners gathered and worshiped their own gods.

It is not known exactly why Teotihuacán was destroyed, but it was most probably the result of an invasion or internal political unrest. Nonetheless, the great city and its huge empire acquired a central position in Aztec mythology in later years, and its ruins have fascinated archaeologists for over a century.

THE PYRAMID OF THE SUN
The third-largest pyramid in the world was built with the effort of thousands of laborers, who brought in millions of baskets of rubble to build the structure. A natural cave was found beneath it, which had been enlarged for ritual use.

ONYX MARBLE VESSEL
The geometric form and planar faces of this vessel are reminiscent of the stone masks that decorate many of the façades of the site's buildings. This unique sculpture is in the form of an ocelot – a leopardlike wild cat.

THEBES

UPPER EGYPT, EGYPT

GAZETTEER P. 143 – MAP REF. E7

THE ANCIENT CITY OF WASET, renamed Thebes by the ancient Greeks, lies on the Nile River about 400 mi (640 km) south of Cairo. It began as a minor trading post in the Old Kingdom (*c.* 2700–2160 BC), surviving into the New Kingdom (*c.* 1550–1070 BC), when it became a new capital. The city developed in two parts – the residential area, on the east bank of the Nile at Luxor, and the huge ceremonial and funerary complexes of the pharaohs on the west bank. Much of the ancient city has been covered by modern development, and the collection of temples and monuments that survives is scattered over a wide area. Napoleon Bonaparte was one of the first people to become interested in the monuments of ancient Egypt during his military campaigns in the area, and there has been a continuous tradition of historical and archaeological investigation carried out since then by scholars of all nationalities. Today, archaeological work continues in tandem with conserving and preserving the monuments from damage caused by pollution and tourism.

THE MORTUARY TEMPLES
The temples at Thebes were the most important and wealthy in dynastic Egypt. Even when Avaris replaced Thebes as the capital, the pharaohs were still buried there, and their status encouraged cults to continue after their death through worship in the temples built in their name.

RITUAL CITY OF THE PHARAOHS

Thebes became famous throughout the ancient world for the magnificence of its ritual structures. The city really began its great expansion during the reign of Thutmose I (*c.* 1500 BC) and his descendants, who spent lavishly on the construction and decoration of temples, shrines, and palaces. The vast necropolis – a huge collection of temples and tombs – lies on the west bank of the Nile. In the Valleys of the Kings and Queens, royal tombs were cut into the cliff faces and mortuary complexes were built on the narrow desert plains.

The most famous tomb is that of Tutankhamen, a minor king whose undisturbed burial produced many rich artifacts when excavated in the 1920s by Howard Carter. In fact, amazing collections of artifacts have been unearthed from many tombs, due to the ancient Egyptian practice of leaving burial goods to join the deceased in the afterlife. Unfortunately, most tombs were looted many centuries ago, and early archaeologists, more interested in the artifacts than the sites themselves, have also destroyed valuable information.

Today, archaeologists have been concentrating on burials away from the main area of the west bank. Increasing concern about the fragility of the monuments, particularly the delicate wall paintings inside many of the tombs, has made the conservation of previously excavated sites as important as uncovering new evidence.

HATSHEPSUT'S FUNERARY TEMPLE
This magnificent mortuary temple at Deir el-Bahri contained the body of Queen Hatshepsut. When her husband (her half-brother Thutmose II) died, Hatshepsut took power rather than allowing it to pass to her stepson, Thutmose III. She decided to become pharaoh rather than queen, assuming the usual masculine titles and clothing.

THE COLOSSI OF MEMNON
This pair of massive, seated statues of Amenhotep III once marked the entrance to his 18th Dynasty mortuary temple. At 69 ft (21 m) high, they dominate the surrounding landscape. After suffering earthquake damage, one of the statues was reputed to make musical sounds at dawn.

Carved pharaonic headdress

The figures are seated on thrones

Hieroglyphic inscriptions

Heavily eroded stonework

Repairs carried out by Romans

THE FIRST CITIES

DURING THE BRONZE AGE in the Old World (Europe, Asia, and Africa) and later in the New World (the Americas), large-scale urban centers developed almost exclusively around fertile river valleys, such as those of the Nile, the Indus, the Tigris, and the Euphrates. Stable agricultural systems, often based on irrigation and water management, allowed large populations to congregate in specific areas, where they produced food more efficiently and developed specialized crafts. These early urban centers had certain features in common. Most were fortified, with planned streets, elaborate public facilities, and great temple complexes (see previous chapter), which provided a central focus for the city dwellers. The first evidence of writing is associated with these civilizations, probably because it became essential for the administration of long-distance trading networks. Political and administrative power soon became centralized, and complex social hierarchies developed, leading to the creation of independent city-states. Headed by powerful rulers, the city-states managed their surrounding areas and, after swift unification and expansion, soon controlled huge empires.

HUDSON BAY

CUNEIFORM WRITING
This early form of writing developed in Mesopotamia in around 3000 BC as a method of accounting, recording crops, and marking property. The marks were created by pressing a reed into wet clay tablets.

PACIFIC OCEAN

NORTH AMERICA

ATLANTIC OCEAN

Mississippi

MESOAMERICA

① LA VENTA
The Olmecs were one of the early city-based peoples of the Americas in the early 1st millennium BC. The civilization was characterized by monumental architecture, as shown by this colossal basalt depiction of a ruler's head.

PACIFIC OCEAN

SOUTH AMERICA

Amazon

② CAHUACHI
Renowned for their unusual ceramics, the Nazca people lived in coastal settlements, such as Cahuachi, and built large ceremonial centers.

CHILE

ARGENTINA

DOLPHIN FRESCO
The Minoan culture, based on the fertile valleys of Crete, was the first urban civilization to emerge in Europe. Royal palaces, consisting of hundreds of rooms around large courtyards, were built and also acted as industrial, religious, and administrative centers. The palace walls were extensively decorated with paintings.

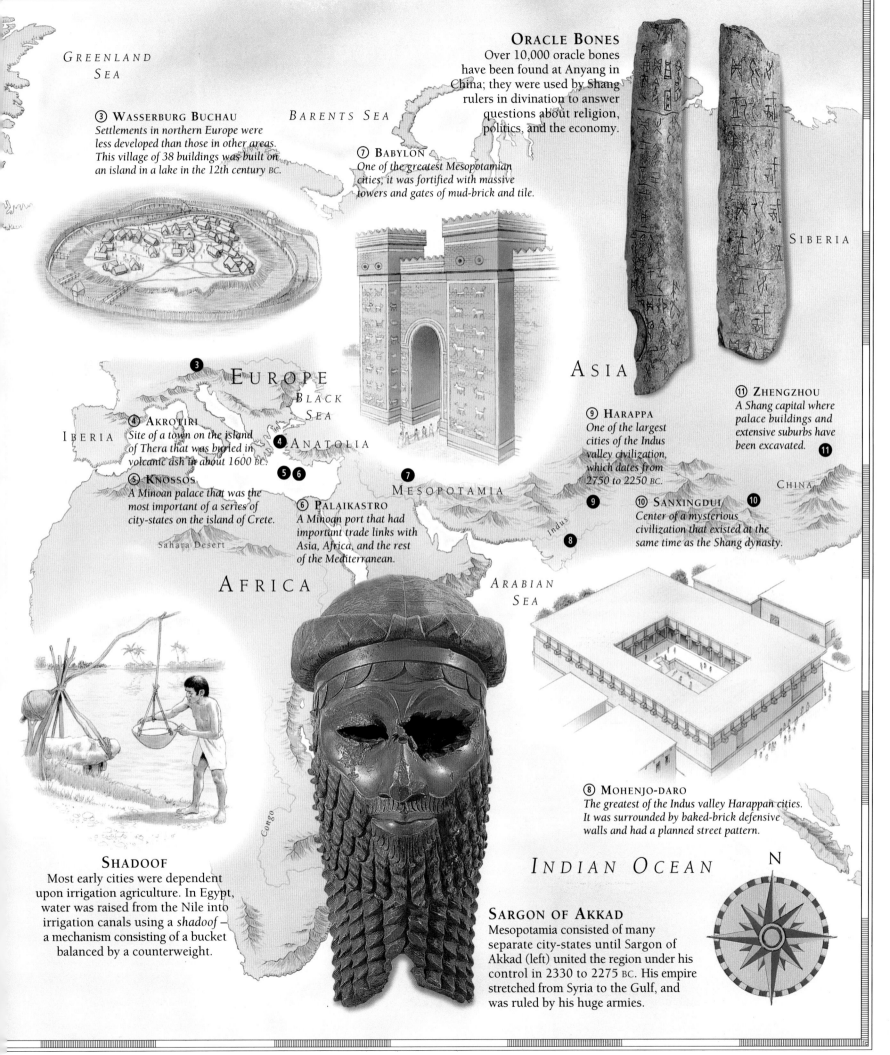

ORACLE BONES
Over 10,000 oracle bones have been found at Anyang in China; they were used by Shang rulers in divination to answer questions about religion, politics, and the economy.

③ **WASSERBURG BUCHAU**
Settlements in northern Europe were less developed than those in other areas. This village of 38 buildings was built on an island in a lake in the 12th century BC.

⑦ **BABYLON**
One of the greatest Mesopotamian cities; it was fortified with massive towers and gates of mud-brick and tile.

GREENLAND SEA

BARENTS SEA

SIBERIA

EUROPE

BLACK SEA

ASIA

⑪ **ZHENGZHOU**
A Shang capital where palace buildings and extensive suburbs have been excavated.

④ **AKROTIRI**
Site of a town on the island of Thera that was buried in volcanic ash in about 1600 BC.

⑤ **KNOSSOS**
A Minoan palace that was the most important of a series of city-states on the island of Crete.

⑥ **PALAIKASTRO**
A Minoan port that had important trade links with Asia, Africa, and the rest of the Mediterranean.

IBERIA

ANATOLIA

MESOPOTAMIA

Sahara Desert

⑨ **HARAPPA**
One of the largest cities of the Indus valley civilization, which dates from 2750 to 2250 BC.

⑩ **SANXINGDUI**
Center of a mysterious civilization that existed at the same time as the Shang dynasty.

CHINA

Indus

AFRICA

ARABIAN SEA

Congo

SHADOOF
Most early cities were dependent upon irrigation agriculture. In Egypt, water was raised from the Nile into irrigation canals using a *shadoof* — a mechanism consisting of a bucket balanced by a counterweight.

⑧ **MOHENJO-DARO**
The greatest of the Indus valley Harappan cities. It was surrounded by baked-brick defensive walls and had a planned street pattern.

INDIAN OCEAN

N

SARGON OF AKKAD
Mesopotamia consisted of many separate city-states until Sargon of Akkad (left) united the region under his control in 2330 to 2275 BC. His empire stretched from Syria to the Gulf, and was ruled by his huge armies.

47

A BRONZE AGE TOWN

PALAIKASTRO, GREECE

GAZETTEER P. 137 – MAP REF. L9

SITE EVIDENCE

🔍 Surface finds and surveys have identified the area covered by the ancient town.

🔍 Excavations have revealed the layout of the streets and houses.

🔍 Artifacts have been found that suggest the daily life of the inhabitants of the city.

🔍 Exotic items, such as ivory, have been found, which imply a long-distance trading network centered on the town.

IN A CRESCENT-SHAPED BAY on the eastern shore of Crete, overlooked by high mountains, the Bronze Age town of Palaikastro grew up, surrounded by farms, vineyards, and olive groves. Busy roads led to and from the town and linked it to other settlements inland, such as the great palace at Knossos. Other roads also ran down to the harbor, from where cargo ships sailed across the Mediterranean Sea.

Archaeologists have been working on the site since 1901, but recent excavations have concentrated on specific buildings and have uncovered a wide range of finds. These artifacts have enabled the archaeologists to date Palaikastro as far back as 4,000 years ago. The town was probably the capital of eastern Crete in the Minoan period of the Bronze Age, but when this civilization fell into decline so did Palaikastro, and by 1200 BC, it was deserted. After this there was only an isolated sanctuary of Dictaen Zeus at the site, and, as a result, the town survived without modification.

THE SITE

The old town of Palaikastro, known as Roussolakkos (red field) because of the mud-bricks used to build it, has been known about for many years. The first excavations took place in 1901, and since 1986 more digs have been undertaken. Surveys of the surrounding plains have identified fortifications and farms contemporaneous with the town.

Area of excavation

Visible surface remains

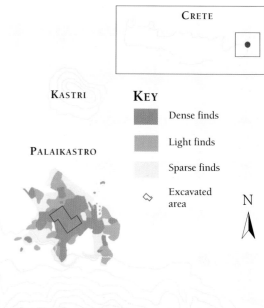

CRETE

KASTRI

PALAIKASTRO

KEY

⬛ Dense finds

⬛ Light finds

⬜ Sparse finds

◇ Excavated area

N

⅗₀ mi (1 km)

TOWN BOUNDARY PLAN

Only a small part of the town has been excavated, but its approximate limits have been identified from collections of pottery shards and stone from the surface. The varying densities (from dense to sparse) of the finds show where the central focus of the town was situated. This shows a concentration of finds inland, about 1,640 ft (500 m) from the coast. The town covered at least ⅗₀ sq mi (1 sq km).

REMAINING STRUCTURES

Some of the walls of the buildings survive up to 3 ft (1 m) high, and so they can be quite accurately reconstructed. Many were built of mud-brick on a stone foundation, while some, like the one shown left, were constructed of large, cut-sandstone blocks.

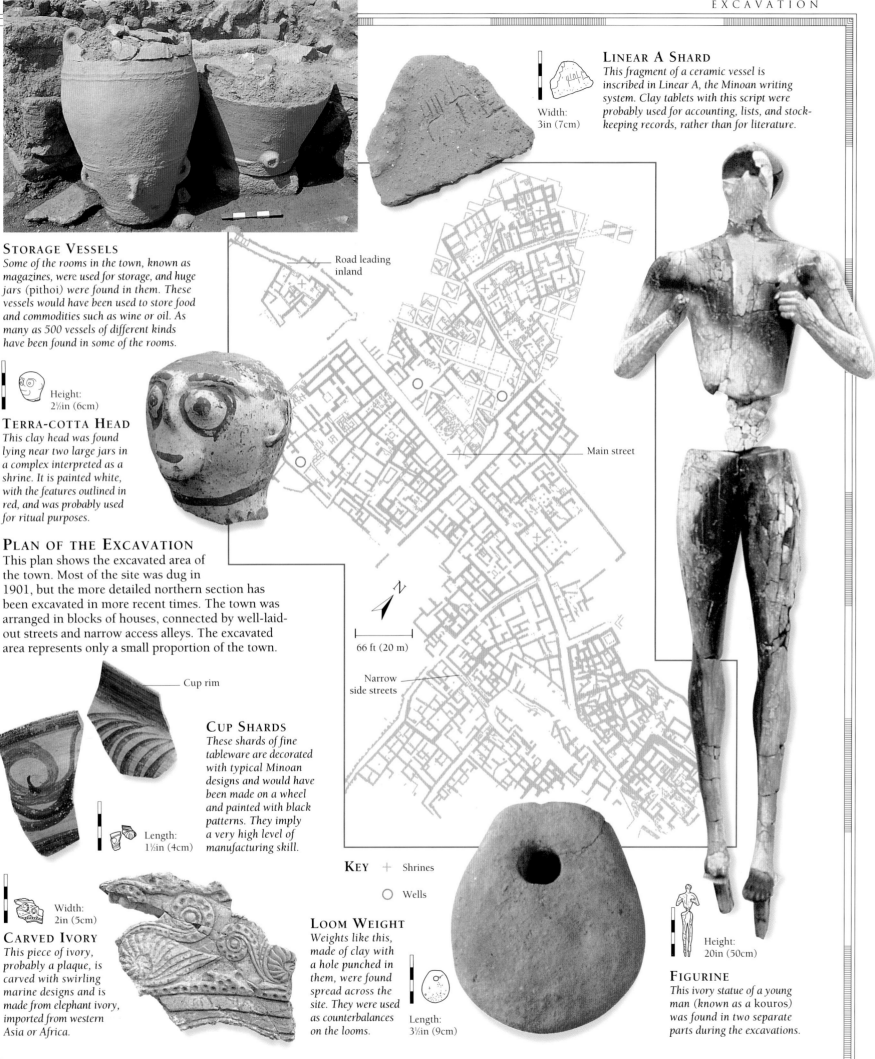

STORAGE VESSELS
Some of the rooms in the town, known as magazines, were used for storage, and huge jars (pithoi) were found in them. These vessels would have been used to store food and commodities such as wine or oil. As many as 500 vessels of different kinds have been found in some of the rooms.

Height:
2⅜in (6cm)

TERRA-COTTA HEAD
This clay head was found lying near two large jars in a complex interpreted as a shrine. It is painted white, with the features outlined in red, and was probably used for ritual purposes.

PLAN OF THE EXCAVATION
This plan shows the excavated area of the town. Most of the site was dug in 1901, but the more detailed northern section has been excavated in more recent times. The town was arranged in blocks of houses, connected by well-laid-out streets and narrow access alleys. The excavated area represents only a small proportion of the town.

Cup rim

CUP SHARDS
These shards of fine tableware are decorated with typical Minoan designs and would have been made on a wheel and painted with black patterns. They imply a very high level of manufacturing skill.

Length:
1½in (4cm)

Width:
2in (5cm)

CARVED IVORY
This piece of ivory, probably a plaque, is carved with swirling marine designs and is made from elephant ivory, imported from western Asia or Africa.

Road leading inland

Main street

66 ft (20 m)

Narrow side streets

KEY + Shrines

○ Wells

LOOM WEIGHT
Weights like this, made of clay with a hole punched in them, were found spread across the site. They were used as counterbalances on the looms.

Length:
3½in (9cm)

LINEAR A SHARD
This fragment of a ceramic vessel is inscribed in Linear A, the Minoan writing system. Clay tablets with this script were probably used for accounting, lists, and stock-keeping records, rather than for literature.

Width:
3in (7cm)

Height:
20in (50cm)

FIGURINE
This ivory statue of a young man (known as a kouros) was found in two separate parts during the excavations.

A BRONZE AGE TOWN

INTERPRETING THE FINDS

ARCHAEOLOGICAL INVESTIGATIONS have revealed a great deal about the lifestyle of the inhabitants in this Bronze Age town. While a detailed picture of the development of the streets and houses can be built up using existing architectural evidence, other finds provide clues to the insides of the buildings, which contained plastered floors and walls, cabinets, and hearths. Other excavated evidence gives us a glimpse of the ritual life of the inhabitants, the produce and commodities they used for food, and the luxury goods and bronze objects that they traded. The variety of goods – along with the unexcavated harbor and the road network – indicates that Palaikastro was a key part of the wider Bronze Age society that existed on Crete and across the Mediterranean region. This early town was obviously a sophisticated, hierarchical, and spiritual society with complex trading networks.

SHRINE BUILDING
Some of the buildings have been interpreted as part of a cult center for the town. A kouros figurine, probably used in cult ceremonies, was found in a courtyard, near a stone building that had a balcony decorated with sandstone horns.

ARTIST'S IMPRESSION
This reconstruction of the town shows how it might have looked to a traveler on the hill of Petsofa looking down on the Bay of Palaikastro in 1500 BC. It has been created using the excavated evidence, the surface surveys, and comparative information about architecture from other well-preserved Minoan sites, such as Knossos. The open, spacious districts are inland, and the more crowded, older ones are nearer the harbor. The streets that cut through the town lead out into the countryside, to nearby farms, and down to the sandy harbor.

STORAGE BUILDINGS
Some of the remains of the storage buildings in the town suggest that they were left suddenly by the inhabitants. The buildup of debris around the storage jars indicates a slow accumulation of garbage, followed by the collapse of the roof.

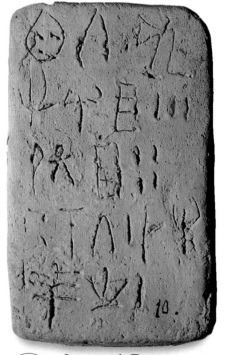

Fields around the town

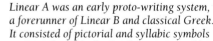

LINEAR A TABLET
Linear A was an early proto-writing system, a forerunner of Linear B and classical Greek. It consisted of pictorial and syllabic symbols and has not yet been deciphered. The characters would have been drawn into the wet clay tablet and then dried in the sun. They survive intact only if they have been accidentally burned at some later time.

WOVEN MATERIAL
The Minoans would probably have produced cloth made out of wool and linen. The loom weights found across the site indicate widespread domestic, not commercial, production.

HOUSES
Variation in status can be recognized in the range of houses that have been excavated. Spacious houses built around courtyards were found inland. Nearer the harbor, in the older, more crowded districts, the houses and workshops were closely packed together.

TRADING BOATS

The town developed in this specific location because of the natural harbor. Bronze Age ships would have carried goods around the Mediterranean and would have brought metals and exotic goods, such as ivory, from western Asia and Africa. A variety of foodstuffs, including oil and wine, were exported from Crete. These advanced trading arrangements enabled Palaikastro to develop into a rich and vital community.

12th-century BC refuge site

Early cemeteries

Sheltered bay

Trading boats

Promontory

Densely packed buildings

Road leading along the coast

THE KOUROS

From the pieces of decoration found, and by comparison with similar statues, the appearance of the *kouros* figurine can be accurately reconstructed. The figure had an elaborate hairstyle and clothing and wore gold shoes and a scabbard.

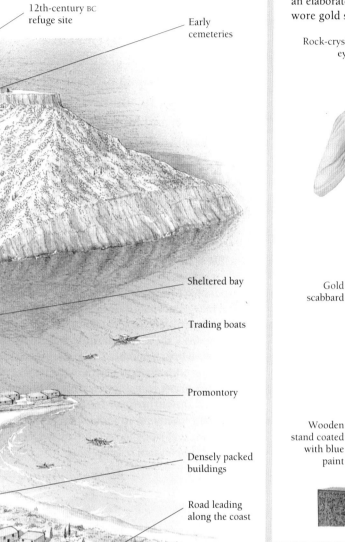

Rock-crystal eyes

Details of veins and muscles carved on limbs

Gold scabbard

Wooden stand coated with blue paint

Gold sandals

CONCLUSIONS

 Excavated remains imply a well-developed, densely populated town.

The variety of objects uncovered show an overall high standard of living.

Exotic objects reveal extensive contacts with the eastern Mediterranean and Africa.

Pottery and carved figures show very high levels of skilled production.

The two types of stone construction suggest different domestic and cult architecture.

POTTERY VESSELS

In addition to potshards, many complete vessels have been found at Palaikastro, some still stacked in cabinets and storerooms. Pottery production on Crete during the Bronze Age would have been a well-organized industry, with specialized workshops turning out high-quality wheel-made pots. Libation vessels found at the site include conical cups, pithoi, and a rhyton.

COMPARATIVE SITES

MOHENJO-DARO

INDUS VALLEY, PAKISTAN

GAZETTEER P. 151 – MAP REF. 14

ONE OF THE LARGE CITIES of the Indus Valley civilization, Mohenjo-daro flourished during the 3rd and 2nd millennia BC. At this time, the Indus Valley was one of the most quickly developing areas in the world, as farming settlements grew into urban centers, based on efficient irrigation agriculture. Mohenjo-daro was built of mud-brick and laid out in a grid pattern, with alleys connecting the buildings. There were well-developed public facilities, such as wells, drainage, and public toilets, to serve the estimated population of 40,000. A range of housing can be identified, from merchant houses with spacious courtyards to single-roomed tenements, lending support to the assertion that the urban civilizations of the Indus Valley were comparable in scale to those of Mesopotamia and ancient Egypt.

PRIEST-KING STATUE
One of the few surviving sculptures from the city, this soapstone statue of one of the city's ruling priest-kings stands 3 in (7.6 cm) high and was found in one of the houses of the lower town. It would probably have been decorated with red paint.

AERIAL VIEW OF THE REMAINS
This view shows some of the excavated remains of the city, revealing how densely packed the buildings were and how well preserved they were by the dry conditions.

HARAPPAN CIVILIZATION

Mohenjo-daro was discovered in 1922 by archaeologists excavating a Buddhist temple, 400 mi (640 km) south of Harappa (which itself had been discovered in 1856). Like other Harappan sites, Mohenjo-daro was centered on a citadel, which had been built on raised ground so that it towered 50 ft (15 m) above its surroundings.

Houses and workshops were situated to the east. The streets were laid out on a grid pattern, and the main ones were wide. Excavations have revealed that the Indus Valley cities were generally part of a carefully planned and ordered world. The Harappan civilizations were involved in long-distance trade, mainly along the Persian Gulf and central Asia, from Mesopotamia, where Harappan seals have been found, through to India. Rare goods, such as gold, precious stones, ivory, and spices, were exchanged for oil, figs, cloth, pottery, and beads. Standard sets of weights and measures suggest that the economy was well regulated. Further standardization is found in the mud-bricks used to build the city, which were of a uniform size throughout the Indus Valley.

One of the first major structures to be excavated was the Great Bath, which was 39 ft (12 m) long and 23 ft (7 m) wide. Built of brick and lined with asphalt, it was extremely likely to have been used for ritual purposes. Its central position suggests its importance, and, along with the extensive drainage system (every house in Mohenjo-daro included a bathroom and a well), shows how important bathing and sanitation were to the city's inhabitants. In fact, the excavations revealed a sense of civic responsibility that was not matched until the Roman Empire 2,000 years later.

THE CITADEL

One of the most famous British archaeologists of all time, Sir Mortimer Wheeler (1890–1976), undertook an excavation at Mohenjo-daro in 1950 using local villagers as assistants. They began with an excavation of the great brick mound. At first, it was thought that the site had been just a fortified citadel, but soon the excavations produced vast quantities of finds. Four weeks after the project started, 12 wagonloads of selected pottery were

sent to the headquarters. As the walls were excavated, it became clear that the structure, with its numerous passages, might have been a great civic granary. A grid of airducts was discovered below the floor of what would have been a great wooden barn. The ducts would have been used to provide ventilation, to keep the grain from rotting. Loading platforms and ramps for wagons were identified. The citadel represented the economic focus of the city – a sort of city treasury. Unfortunately, the high level of the water table made excavations of lower levels impossible, although archaeological evidence uncovered does supply us with a detailed picture of the diet of the inhabitants. Wheat and barley were the staple crops, with peas, dates, melons, and cotton also grown. Cattle were used for their milk, meat, and for work, while their dung was used for fuel and fertilizer. There is also evidence that the inhabitants used buffalo, sheep, dogs, pigs, and even elephants.

Mohenjo-daro went into decline in about 1900 BC, but the exact nature of the city's decline,

DETAILED VIEW OF THE SITE
The carefully built brick buildings can be seen here with their thick walls and alleys in between. While the fronts of the houses each presented a blank façade to the streets, the interiors opened out onto central courtyards.

probably through natural causes, is not yet known. Despite the wealth of archaeological information, the Harappan civilization presents an enigma for archaeologists. While there is much evidence of urban life, the diet, and trade links (all of which point to a sophisticated civilization), without deciphering Harappan script or finding more evidence about their beliefs, art, and administration, we can only speculate about their culture, government, and religious beliefs.

Buddhist stupa above the remains of the earlier city

THE CITADEL MOUND
The citadel acted as the administrative and ceremonial center of the city and included the Great Bath. The "College," which can be seen in the foreground, may have been used to accommodate the city's priests.

HARAPPAN SEALS
Over 2,000 of these seals were found. They were probably used by civic officials for stamping wax. The language on the seals, consisting of about 400 different pictorial signs, has not yet been deciphered.

SANXINGDUI

SICHUAN PROVINCE, CHINA

GAZETTEER P. 152 – MAP REF. F6

THE SICHUAN PROVINCE has long been recognized as a potential area of archaeological interest. Recent excavations and radiocarbon dating at Sanxingdui have revealed a large city dating to 2800–1000 BC. It appears to have been the center of an ancient state, tentatively identified as the Shu state by Chinese archaeologists, and it is believed to have extended over much of the Sichuan province. Little is known about the culture, because there is no direct mention of the state in Shang or later writings. It is likely that Sanxingdui was part of an autonomous state but still had its own independent culture. Thus, all conclusions have to be based on the archaeological evidence, the richness of which demonstrates what an important center Sanxingdui must have been.

BRONZE HEAD

BRONZE OBJECTS IN PIT 2
Many objects were found together in large pits, including surreal masks and anthropomorphic figures, the largest weighing over 400 lb (180 kg). This picture shows the huge number of bronze objects deposited in Pit 2, many of which had been deliberately broken by their makers.

Scroll-shaped projection

THE SACRIFICIAL PITS

The Bronze Age fortified city of Sanxingdui covered an area of approximately 4½ sq mi (12 sq km) and was laid out on a north–south axis. Archaeological research has been carried out on the site since the early 1980s, although a previous excavation in 1929 yielded isolated bronze and stone finds, which gave an indication of the riches that might be found in the area.

Earthwork survey and excavation has shown that the city was surrounded by high walls built of earth and wood, indicating that the site was heavily fortified. The central axis was raised up, and it is from here that the richest of the archaeological finds have been discovered. The city contained many buildings of different sizes and shapes, which were built and rebuilt over hundreds of years, leaving many layers of foundations for archaeologists to uncover. The houses were built of wood and clay, and the closely packed buildings indicate that the city was densely populated. One building, 240 sq yd (200 sq m) in size, stood out from the rest and may have been a public building or meeting-place. Artisans' quarters, with evidence of large-scale specialized production, along with the quantity and quality of the artifacts found, suggest that, as well as being a large center of population, Sanxingdui was

Hole to attach to wall or object

Large hand to hold object

RITUAL MASK
This was one of three masks that were found in Pit 2. The holes on either side of the face indicate that it was connected to a wall or an object. The distorted features, such as the elongated nose and protruding eyeballs, suggest that the mask was some sort of mythical representation.

a ritual focus and perhaps also a place of pilgrimage. The abundance of materials such as gold, jade, and bronze, which were used to make elaborate tools, sculptures, and other enigmatic objects, seems to justify this conclusion.

Four sacrificial pits have been scientifically excavated at Sanxingdui – one in 1929 and three during the 1980s. Each pit was large and rectangular – up to 16 ft (5 m) in length – and contained a huge wealth of artifacts. All four had been filled in distinct layers, each of which contained different artifacts that had been deliberately broken or burned. Bronze items, ranging from ritual masks and sculptures of birds and animals to a life-size figure of a man, were discovered. Pit 2 (shown above right) contained an exceptionally large number of bronze items.

Jade items were also common finds in the pits, particularly small disks, scepters, and weapons. Burned animal

MYSTERIOUS FIGURE
This life-size bronze figure with strong features was found deliberately broken in one of the sacrificial pits. The huge hands and their peculiar positioning suggest that it once held a large, curved object, such as an elephant's tusk. Whatever the object was, it holds the key to the figure's interpretation.

bones, mainly those of elephants, are probably the remains of sacrifices. Pit 2 contained over 60 elephant tusks, indicating how important the animal must have been to the culture. Other objects were also placed in the pits, including seashells, gold objects, and ceramic vessels. However, there were no finds of human bones or everyday objects, and this seems to support the belief that the city had been used mainly for ritual purposes.

ANCIENT RITUAL CENTER

The excavation of the city (and especially of its amazing sacrificial pits) clearly shows what an important ritual center Sanxingdui must have been. A huge amount of expert craftsmanship would have been needed to produce the complex bronze and jade objects, and the scale and quality of the sacrificed items indicate great attention to detail and a very wealthy society. Many of the ritual objects can be interpreted by comparing them with documented Shang artifacts, but the many items unique to this site reveal a culture that reveled in its individuality and distinction from other contemporaneous states. Unfortunately, we can only guess at what elaborate ceremonies were performed to accompany the sacrifices to the deities by the inhabitants of Sanxingdui, but there is hope that the riches yet to be found in the region will shed more light on this mysterious culture.

AKROTIRI

IN THE MIDDLE OF THE 2ND MILLENNIUM BC, the volcano in the middle of the Aegean island of Thera (modern Santorini) erupted, burying the prosperous Minoan coastal town of Akrotiri beneath 13 ft (4 m) of pumice. The entire town was engulfed, and ongoing excavations have revealed over 11,960 sq yd (10,000 sq m) of uniquely preserved street plans and two- or three-story houses. This area is estimated to be less than half the town, and further work will ascertain whether the buildings uncovered are typical buildings or luxurious residences. No human remains were found – the inhabitants appear to have fled after an earthquake just before the volcanic eruption, taking portable goods with them, presumably to another Cycladic island or to neighboring Crete. The narrow, winding, paved streets of the site resemble those of the nearby modern town of Fira. The buildings or complexes were separate from one another, and had elaborate sewage and drainage systems. The various finds, rich frescoes, and surviving architecture have all allowed archaeologists to build up a detailed picture of life in Bronze Age Akrotiri.

FRESCO OF A FISHER-BOY
One of a pair of frescoes from a shrine complex in the upper story of the West House. The partially shaved hairstyle of the figures (represented by blue) and the locks of hair reflect their youthfulness.

THEORIES ON THE PAST

Akrotiri is most famous for the excavation of its many intact buildings. The West House (see right) appears to have been a public building, having a shrine on the upper story. It overlooks the triangular *plateia*, an area that probably held ritual gatherings, and crowds may well have witnessed rituals through the large, open windows. Wooden frames running through and along the walls, doors, and windows reinforced the buildings against earthquakes. Organic materials, such as wood, have not survived, but cavities in the pumice indicate the original locations. Although a limited number of personal possessions have been discovered, the beautiful frescoes found on the walls clearly illustrate the daily life and rituals of the inhabitants.

The greatest archaeological interest in the site, however, stems from the ongoing controversy over the absolute date of the Thera eruption. Dendrochronologists date it to 1628 BC, due to the unstable worldwide weather that year and from ash found in ice cores in Greenland. However, Minoan pottery suggests that the date is much later. The latest evidence comes from recent excavations in Egypt – pumice, which must have come from the Thera eruption, dates only as far back as 1550 BC. Whichever date proves to be correct, it is unlikely to reveal the reason for the end of the Minoan civilization.

POTTERY FINDS
Large quantities of pottery have been found in the excavations. They were probably too heavy and not valuable enough for the inhabitants to take with them when they left after the initial earthquake.

STANDING BUILDINGS
Some buildings up to three stories high survive at Akrotiri, with their windows and doors intact. These houses, which are similar to many structures on Crete, may well have belonged to rich merchants.

THE IRON AGE

EVIDENCE FROM THE IRON AGE allows us, for the first time, to construct a detailed picture of the daily life of the different types of people living in Iron Age societies, including farmers, warriors, craftsmen, and chieftains. Along with the many iron objects found throughout Europe and eastern Asia, numerous farmsteads, monuments, and settlements also remain in the modern landscape. Great fortified centers of power survive, particularly from the developing empires of western Asia and China, with monumental buildings and artifacts, and evidence for ritual worship, trade, and growing industry. In the west, this period is associated with the Celts – an assorted collection of peoples who lived throughout Europe and who produced a distinctive artistic style. Many of the best-preserved remains from this archaeological period are defensive structures, such as the hill forts of Europe, the fortified citadels of western Asia, and the monuments of China and Africa.

① NAVAN
Pagan ritual center that included a central round building, circular earthworks, and sacred pools.

② HILL OF TARA
An Iron Age complex that was the political and spiritual center of Celtic Ireland.

③ MONT LASSOIS
The 6th-century BC hill fort of a Celtic chieftain with the burial site of Vix nearby.

④ GRAUBALLE MAN
A male body, found in a bog, who appears to have been sacrificed around AD 310.

⑤ MANCHING
This large 2nd-century BC oppidum has yielded extensive evidence for trade and industry.

⑥ HALLSTATT
A multiphase cemetery site and salt mine dating from 1000 BC onward. It has given its name to a widespread style of Iron Age art and culture.

SCANDINAVIA

NORTH SEA

BRITISH ISLES

GAUL

Danube

ATLANTIC OCEAN

AFRICA

Sahara Desert

Niger

LAKE CHAD

SAHEL

Niger

GULF OF GUINEA

Congo

GOLD CELTIC BOAT
This gold model boat, dating to the 1st century BC, forms part of a hoard found at Broighter in Northern Ireland in 1896. The oars, mast, benches, and yardarm are clearly visible, and the model appears to be based accurately on contemporary vessels.

THE GUNDESTRUP CAULDRON
This huge gilded silver cauldron, dating to the 1st or 2nd century BC, was found in a peat bog in Denmark in 1891. The panels, hammered out in relief around the bowl, are decorated with pictures of gods and goddesses, human sacrifices, and animals. The cauldron is typical of an Iron Age ritual offering, where a valuable object is deposited in a pond or river. It also demonstrates the extensive number of deities and rituals that were significant in contemporary religions.

⑨ PAZYRYK
These frozen burials of steppe chieftains, dating to the 5th century BC, contain bodies, preserved textiles, and wooden objects, such as this carved bridle ornament.

SIBERIA

Steppes

⑨

SCYTHIAN HORSEMAN

During the Iron Age, the steppe region of central Russia was populated by pastoral warrior nomads who were skilled archers and horsemen and participated in the long-distance silk trade. The information about their clothes and horse trappings comes from numerous preserved burials.

BACTRIA

CHINA

⑦ BABYLON
This sandstone stela celebrates the rebuilding of this great city by the Assyrian Empire in 672 BC.

Euphrates

Tigris

⑦

⑧

⑧ TILLYA-DEPE
Ancient Bactrian site that is famous for its royal tombs containing finds of gold vessels and jewelry.

Yangtze

EGYPT

Nile

THE GULF

ARABIA

Indus

ARABIAN SEA

⑩

⑩ KAUSAMBI
A fortified city that was typical of the period, and several iron tools have been found here.

GREAT WALL OF CHINA

Covering over 994 mi (1,600 km), with forts placed along its length, the Great Wall was built by the first emperor of China between 221 and 200 BC, to protect the newly formed empire from invading nomads.

SOUTH CHINA SEA

Mekong

INDIAN OCEAN

INDIA

N

⑪ BUHUYA
This early iron-working site in Tanzania has clay furnaces with symbolic female features, reflecting the productive associations of metal production.

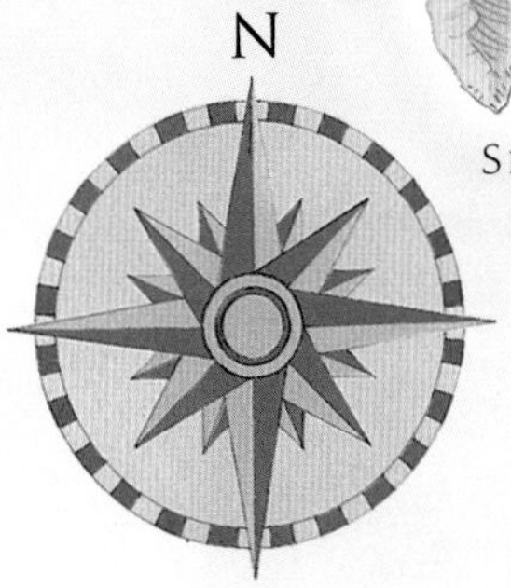

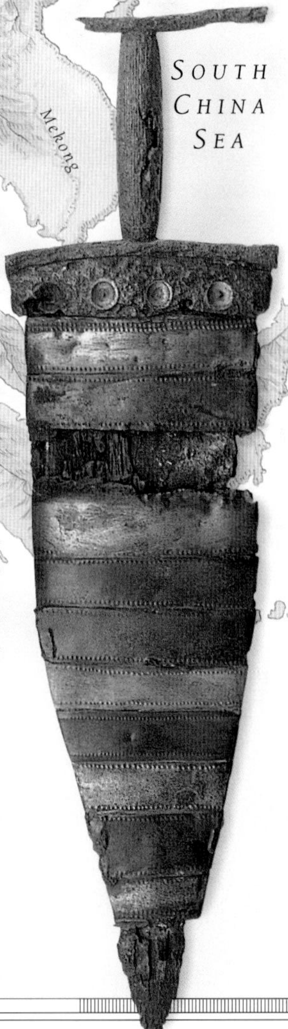

SRI LANKA

IRON DAGGER

Many Iron Age societies were warlike, paying great attention to defenses and weaponry. Typical of this period is this British dagger from 550 BC, made of iron and sheathed in bronze, which would probably have belonged to a tribal chieftain. The reasons for the rise of these violent societies is not clear, but increasing populations, pressure on resources, changing social aspirations, and the development of sophisticated weaponry may all have played a part.

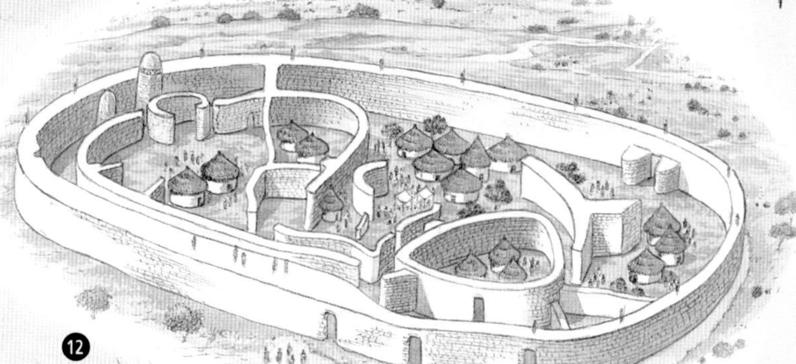

⑪

⑫

⑫ GREAT ZIMBABWE
Erected for the local chief and his family, this stone-built enclosure probably contained wooden structures. Many luxury objects have been found at the site, indicating long-distance trade in metals, shells, and pottery.

AN IRON AGE FORT

NAVAN, UNITED KINGDOM

GAZETTEER P. 144 – MAP REF. D5

SITE EVIDENCE

🔍 Documents written by Irish monks in the 7th and 8th centuries referred to a capital in the north of Ireland called Emain Macha. Geographical and other archaeological evidence equate this with Navan Fort, the capital of the Uiti people, who gave their name to the area that is now known as Ulster.

🔍 The mounds at Navan are clearly artificial and exist in a landscape that contains other Bronze and Iron Age features nearby.

IN THE 7TH AND 8TH CENTURIES, Irish monks wrote down the stories that were part of the oral tradition of the Celtic people of Ulster. The greatest of these was "The Ulster Cycle," which told of an Iron Age community whose capital, Emain Macha, was sited at Navan and was occupied by the great King Conchobor. Navan appears to be a typical Iron Age hill fort. Two mounds can be seen on top of a raised area, and there is some evidence of a large surrounding ditch. Archaeologists have discovered evidence of occupational activity at Navan dating back to 300 BC, and recent excavations have uncovered traces of a large Iron Age structure. It appears that in 94 BC, a huge, round, wooden building was erected, almost immediately filled in with stones, and then burned and turfed over. Using further excavation and analysis of objects, the archaeologists hoped to discover how and why the structure had been built and destroyed.

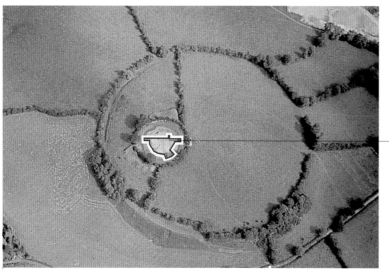

PLAN OF THE EXCAVATION

The remains found by the archaeologists were over 131 ft (40 m) across and consisted of numerous postholes set within a cairn of stones that appeared to radiate from a central post.

— Area of excavation

KEY

— Area excavated to greatest depth

○ Unexcavated molds in cairn

◉ Position of wooden post in posthole

|⊢————⊣| 10 ft (3 m)

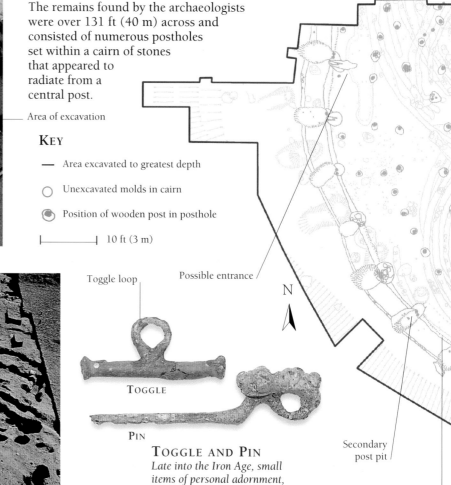

Possible entrance

N

Toggle loop

Secondary post pit

Horizontal wall structure

THE SITE

The mound site at Navan was first excavated by archaeologist Dudley Waterman between 1963 and 1971. The artificial cairn appears as a mound 9 ft (2.8 m) high, with the turf cover adding another 8 ft (2.5 m) to the height. Excavations on the top of the larger mound revealed remains of a round house compound that had been built over with a single large wooden structure.

POSTHOLE MOLDS

The most useful discovery at the site was the huge number of posthole molds forming the distinct shape of the structure. These molds are the holes left by disintegrating wooden posts. The posthole molds were of a regular size, suggesting a building that had even-sized posts supporting a shallow-sloping roof. This suggested a more ritualisic than practical use for the building.

TOGGLE

PIN

TOGGLE AND PIN

Late into the Iron Age, small items of personal adornment, such as toggles and pins, were still being made of bronze. A site such as Navan would have had skilled metalsmiths on hand to supply such items, which would have been worn by most people in society.

Length: 2in (5cm)

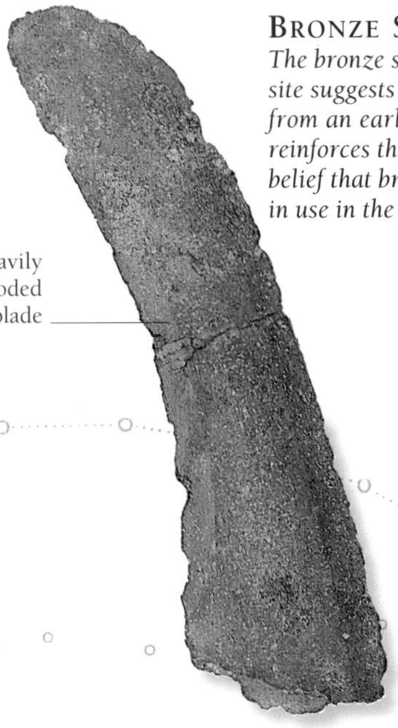

BRONZE SICKLE

The bronze sickle found at the site suggests continuity of use from an earlier period and reinforces the archaeologists' belief that bronze remained in use in the early Iron Age.

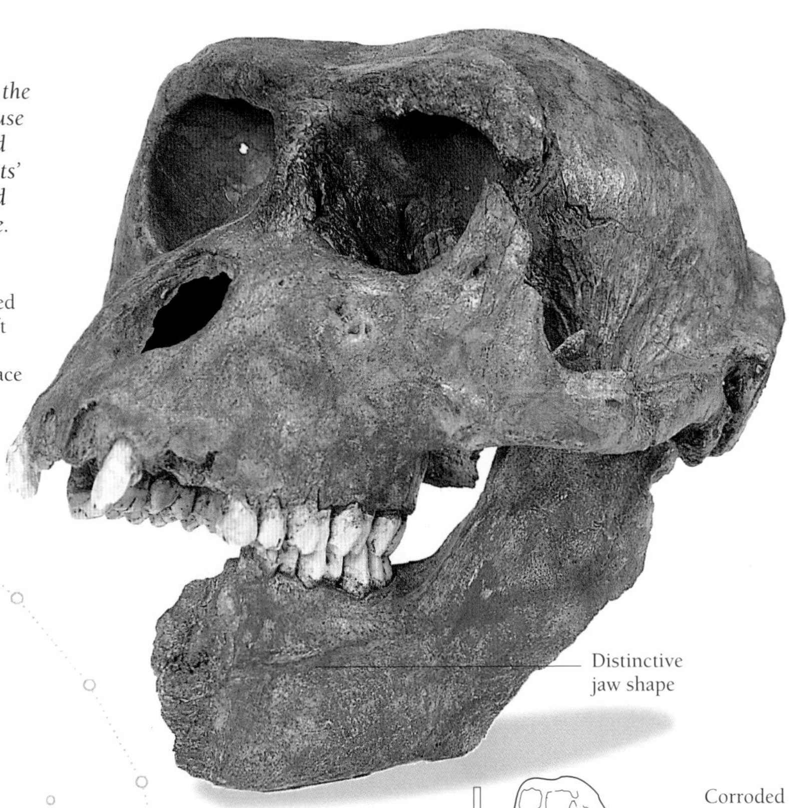

Length:
4⅞in (12.5cm)

Heavily corroded blade

Area above line excavated to within 3 ft (1 m) of old ground surface

Distinctive jaw shape

Corroded metal edge

Width:
4in (10cm)

BARBARY APE SKULL

Along with other animal bones, the skull of a Barbary ape was found in the area. Its existence in Navan implies the presence of a high-status chieftain, for whom such an animal would have been an appropriate gift.

Central posthole

Ramp down to central post

Stones to pack the posts in pit

Secondary posts for support

Coarse texture

POTSHERDS

Found just outside and to the east of the Navan enclosure, these fragments of coarse pottery date from the late Bronze Age or early Iron Age. They are from a typical flat-rimmed vessel and suggest that a settlement had been situated to the east of the structure.

Width:
2in (5cm)

SWORD OR DAGGER

This metal weapon, which is a cross between a sword and a dagger, is one of many metal items found at Navan. Swords made of iron became increasingly common around AD 100. Their design often replicated elements of earlier Bronze Age patterns.

Narrow section for fitting into handle

Length:
14in (35cm)

AN IRON AGE FORT

INTERPRETING THE FINDS

THE COMBINATION OF THE LARGE SCALE of the structure and the unusual nature of the finds suggested that the building had some ritualistic purpose. The uniform size of the postholes, the massive size of the structure, and the discovery of the ape skull all imply that the site had been some kind of ceremonial monument. Local archaeologists point to parallels in Caesar's account of the Gallic Wars, which referred to the Celtic practice of creating massive wooden structures – and how these would be burned down as an offering to the gods. Although the date is not known, because no excavations were carried out, the surviving earthworks also suggest a ritual, rather than a defensive, function for the site, since the bank is outside the ditch. Other ritual pools and structures can be found in the surrounding countryside.

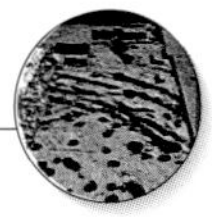

POSTHOLES
The posthole molds at Navan were of an even size, whereas in most Iron Age structures the supporting postswould have increased in size toward the center. This implies that the structure had a ritualistic rather than practical function.

Concentric rings of posts

Interior wooden post

Probable entrance

Outer post

BROOCH DESIGN

Brooches and fibulae (safety-pin-style brooches) are some of the most common artifacts found on Iron Age sites in the British Isles. Changes in taste and style make these types of decorative objects invaluable to archaeologists for dating sites. The Navan fibula (see below) has a distinctive ball-and-socket arrangement that demonstrates the high levels of skill required to produce such a brooch. Local craftsmen would have produced regional variants that can now easily be identified and put into a typology by archaeologists.

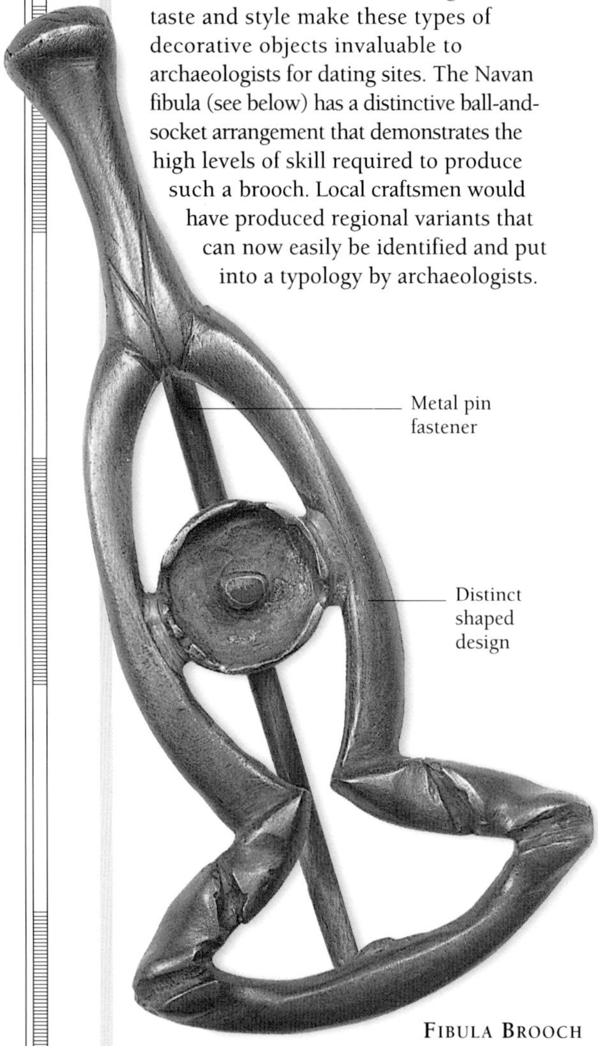

Metal pin fastener

Distinct shaped design

Trumpeter heralding the arrival of the chieftain

FIBULA BROOCH

ARTIST'S IMPRESSION
This structure would have been the focus for the whole community, a place where the chieftains would congregate and the Celtic rituals to mark the passing of the seasons and the celebrations of victories in warfare could take place. Other possible reasons for this structure and ceremony may have been to bring together local people to strengthen tribal bonds, such as a funeral for a king, or for the ritual preservation of a building associated with a powerful figurehead. This reconstruction shows a roofed structure, but some archaeologists believe it could have been an unroofed, wooden, henge-type monument.

BARBARY APE
Possibly brought as a gift by traders, a Barbary ape from Africa would have been presented to the ruler as a prestigous gift.

FLAT-RIMMED WARE

Flat-rimmed ware is typical of this period and would have formed the basis of a simple collection of pots owned by each family. The clay often contained additives, such as shell or grit, to prevent the pot from cracking when it was fired.

Thatch

Low-pitched roof

Double posts

Horizontal wooden slatting

SCABBARD

The new metal weapons were sharp and strong, and they conferred status on their owner. Decorated scabbards indicated even greater personal wealth.

People attending the ceremony

BRONZE SICKLE

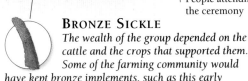

The wealth of the group depended on the cattle and the crops that supported them. Some of the farming community would have kept bronze implements, such as this early sickle, hanging from a leather belt.

LA TÈNE PIN

The La Tène-designed pin would have been a display of wealth for its owner. Possibly designed by local craftsmen, it reflected the influence of continental Europe.

LA TÈNE

The deposition of valuable objects in pools was an important element in the ritual life of the Celts. The discovery of hundreds of bronze and iron objects at the lakeshore site of La Tène in Switzerland gave rise to the term, La Tène, to describe both the time period and the style of art. The characteristic curvilinear style of La Tène has a more freehand appearance than the earlier Hallstatt style (see p. 59) and enables archaeologists to distinguish between the art of the early and late Iron Age. At Loughnashade, near Navan in Northern Ireland, four long horns were found in a lake, and one is displayed in the National Museum in Dublin. Dating to the 1st century AD, these long trumpets are decorated in typical Celtic style and have been made from sheets of bronze, hammered out and then riveted. The pattern on the trumpet end was created using the repoussé technique and bears the hallmarks of the later La Tène style.

Beautiful curvilinear pattern

TRUMPET END OF THE HORN

CONCLUSIONS

🔎 The site shows, from archaeological evidence, the background to the early heroic literature associated with the site.

🔎 Excavation of the main mound established that the site dated to the Iron Age, although occupation nearby extended over a much longer period.

🔎 The posthole arrangements, stone structures, and central post, together with the finds, all indicate a ritual rather than practical use for the site.

COMPARATIVE SITES

THE HILL COMPLEX
The granite outcrops that still dominate the landscape provided an excellent starting point for the first phase of building. The dry-stone walling followed the contours of the natural boulders, and so most of the walls are curved. Some doorways were topped by monolithic structures.

GREAT ZIMBABWE
NEAR FORT VICTORIA, ZIMBABWE
—— GAZETTEER P. 142 – MAP REF. F7 ——

THE 15TH AND 16TH CENTURIES saw the birth of extensive global commerce as European nations set up trading colonies around the world. In Africa, long-distance trading networks had been established long before the European arrival. Iron Age cultures in southern Africa had exploited local minerals, and a thriving trade could be found along the east coast of Africa. At its peak between the 13th and 15th centuries, the citadel of Great Zimbabwe controlled a large area of eastern Africa, and the extent of its ruins are testament to its former importance. Unfortunately, there is little historical evidence about the site, which makes its archaeology all the more significant. It is hoped that future investigations may explain even more about this civilization and its role in shaping the history of eastern and southern Africa.

THE RUINS

Lying in the heart of southern Africa, on a plain between the Zambezi River in the north and the Limpopo River in the south, Great Zimbabwe's (*dzimba dza babwe* means "houses of stone" in Shona) unparalleled monumental stone-building makes it possibly the most important Iron Age site in Africa. Archaeological excavations have revealed that the site was settled in four phases. Radiocarbon dating has shown that it was first settled between AD 100 and 300, but its second phase (AD 350–1050) lasted longer and has yielded a number of clay figurines of cattle. It was after the arrival of the Shona people, in about AD 900, that the stone phase of the site began, with most of the monuments built between 1270 and 1450. The final phase was between 1450 and 1850. But one archaeologist, Peter Garlake, argues that Great Zimbabwe may have been abandoned entirely by the late 16th century.

The ruins as they stand consist of three parts: stone-walled enclosures around a granite hill that included a royal residence; a central area in the valley below, with the royal wives' complex, grain silos, and the Great Enclosure; and the town where the rest of the population lived.

Over time, crudely built functional buildings gave way to tall, close-fitting walls and huge buildings made of rectangular blocks. One of the most impressive structures was the Great Enclosure, an elliptical building 787 ft (240 m) long and 33 ft (10 m) high, assumed to have also housed the huts of the king or chief. The top of its wall is decorated with a chevron pattern. Inside the enclosure, there is a solid conical tower, likely to have been the enclosure's focal point.

It is from the later periods, when the Rozwi people occupied Great Zimbabwe, that the largest number of artifacts have been found. Soapstone carvings, bowls, phalli, and gold, copper, and iron ornaments were all discovered. Though a few querns were found, most of the excavation shows that the economies of the communities at Great Zimbabwe were based on the production of copper, iron, and bronze.

RITUAL FIGURINE
This soapstone figure, thought to represent a mythical eagle carrying messages to the gods, was one of a number of votive objects found.

A THRIVING ECONOMY

Great Zimbabwe was situated on a direct route between the center of gold production in Matabeleland to the west and the East African coast. Its position meant that it was able to control trade between the interior and the coast, and, as a result, it could dominate the area economically. Gold and copper artifacts and imports, such as Chinese ceramics and glass beads, suggest far-reaching economic links. Most significantly, the geographic spread of similar stone-built enclosures indicates the scope of Great Zimbabwe's influence.

Great Zimbabwe became the capital of a huge Shona empire, stretching over much of southern Africa. Imported ceramics and comparisons with finds from Portuguese trading stations date Great Zimbabwe's peak to the 14th and 15th centuries, the same time as the Islamic ports dominated the east coast. Its economic decline started in about 1450. Portuguese accounts blame a severe shortage of salt, but this may just reflect centuries of over-exploitation of land and resources or a breakdown of trade. One Shona group migrated north to the area of the middle Zambezi and established

CONICAL TOWER
Inside the Great Enclosure stands a solid conical tower, reaching a height of 34 ft (10.5 m), with a diameter of 18 ft (5.5 m). A patterned frieze can be found around its top. The exact use of the tower is not known.

THE GREAT ENCLOSURE WALL
All the walls were built with flat slabs of granite without any mortar. The outer wall of the Great Enclosure was 38 ft (10 m) high, with a narrow corridor running between it and an inner wall.

themselves there. This would have lessened Great Zimbabwe's monopoly and the extent of its power, contributing further to its economic decline.

Cattle were kept in the Great Enclosure until the 19th century, when the site was finally abandoned. Soon after, Europeans looking for gold founded a prospecting company. Its activities were eventually terminated, but not before its employees looted many important objects and destroyed evidence during their searches.

THE POLITICS OF DISCOVERY

There was much controversy in the early 20th century about Great Zimbabwe. White prejudice questioned the possibility that these huge stone buildings could have been built by indigenous black Africans, concluding that it had to have been the work of white foreign invaders, such as the Phoenicians. They could recognize no predecessors for its construction, which would have required not just the mobilization of a huge amount of labor and resources, but great organization and skill. Eventually it was recognized by the whites as indigenous, the enclosures and buildings being typical of the local stone-building tradition. Both the pottery and the buildings are clearly part of a tradition that continues with the Shona people.

REMAINS OF AN EXTENSIVE EMPIRE
There are many disconnected walls and structures, which are probably the remains of former substantial enclosures or parts of wooden buildings. These remains are spread across a wide area. Daga (mud) huts were used to house the majority of people in the community.

MONT LASSOIS & VIX

LANGUEDOC-ROUSSILLON, FRANCE

—— GAZETTEER P. 146 – MAP REF. G6 ——

FOLLOWING THE SPREAD OF GREEK INFLUENCE into the western Mediterranean region in the 7th century BC, Greek city-states, such as Marseilles, were founded. The areas of Europe north and west of the Alps came into direct trading contact with these Greek colonies in the 6th century BC, and the corridor of the Rhône and Saône rivers through to the Seine valley became a major artery for trade. The Mediterranean world needed slaves, foodstuffs, and raw materials, while northwest Europe required luxuries, such as wine, and the vessels to serve it from. Mont Lassois was one of the centers of this trade, and the burial at Vix reflects the variety of luxury goods that was available at this time.

AERIAL VIEW OF MONT LASSOIS
This aerial view shows the wooded hilltop where the hillfort was situated, dominating the upper valley of the Seine River and the fertile land below. The famous Vix krater was found in a burial discovered in the village of Vix below the fortress site at Mont Lassois.

THE FORTRESS

The hill fort of Mont Lassois was an important fortress and residence of a powerful Celtic chieftain that developed from trade between the Rhône and Saône valleys. Although it has not been excavated very extensively, hundreds of brooches, much ornate pottery, Greek black-figure pottery, and fragments of Mediterranean wine amphorae have already been found, indicating both the high status of the occupants and the long-distance trade that they were engaged in. Mont Lassois was probably under the control of a wealthy, aristocratic chieftain in the second half of the 6th century BC.

THE BURIAL AT VIX

While Mont Lassois has not been fully excavated, a number of Celtic burials surrounding the settlement have been examined. The most spectacular finds were discovered in 1953 in the village of Vix, directly below the hill fort. Excavations revealed the tomb of a 30–35-year-old wealthy female, dubbed the "princess" of Vix, who had been buried under a large barrow or mound in the decades before 500 BC. Clearly, the local people at that time believed in some

EXCAVATING THE KRATER
The picture on the left shows the krater from the burial at Vix being excavated in 1953. One of the handles can be seen on the right-hand side.

form of afterlife because she had been buried with all the equipment she might need for her journey to another world. The body had been placed in a wooden burial chamber under a barrow, which was 138 ft (42 m) in diameter and 20 ft (6 m) high. The skeleton of the "princess" lay in the center of the burial chamber on the chassis of a wooden wagon, surrounded by bronze wagon fittings and a number of other grave goods. There were also bracelets, torques, a brooch, a necklace, and a gold diadem (a light, jeweled torc) weighing 17 oz (480 g), with winged horses on each end. A bronze basin from Tarquinia, dating from 520 BC, and two Greek cups, dating from 530–520 BC and 520–515 BC, helped to date the burial to the 20 or so years before 500 BC. The lavish nature of the burial indicates the high status and important role of women in the Celtic world during the Iron Age.

THE KRATER

In one corner of the chamber stood a great bronze krater (wine-mixing bowl), 66 in (164 cm) high and weighing 459 lb (208 kg), with the neck decorated with an embossed frieze of foot soldiers and chariots. All around were wine jugs, bowls, and cups. The krater is the most ostentatious example of imported goods from the Mediterranean world at this time and reflects the diplomatic activity between traders and local chieftains. This krater may have come from Tarentum or Sparta, probably arriving in sections to be put together by a Greek craftsman – there are Greek numbers under the frieze, which were hidden once the frieze had been soldered on.

VIX KRATER
The magnificent bronze krater is on display in a museum at Châtillon-sur-Seine in France. A frieze of warriors can be seen between the two handles, which have been sculpted into the shape of gorgons.

HILL OF TARA
COUNTY MEATH, REPUBLIC OF IRELAND
GAZETTEER P. 144 · MAP REF. D6

AERIAL VIEW OF TARA
This view shows the large oval ditch (Fortress of the Kings) around the hill, with the two joined ring forts, Cormac's House and Royal Seat, inside. The Stone of Destiny stands in the center of Cormac's House. Beyond the two forts is the small Mound of the Hostages, with the Rath (fort) of the Synods outside the ditch.

THE HILL OF TARA IS ONE OF the most important early sites in Irish history and one of the most complex. It is, in fact, a large collection of monuments and earthworks, centered on an imposing hilltop rising to about 507 ft (154 m), some of which have been investigated more fully than others. Excavations have taken place at Tara since the 19th century, including an attempt to find the Ark of the Covenant by a group of archaeologists in 1899. The wide-ranging age of the monuments, from the first Neolithic structure to the Iron Age and early Christian features, shows the long-term significance of the hill and demonstrates the importance of investigating a complex site in its entirety in order to understand its development, rather than concentrating on small periods of time.

SEAT OF THE HIGH KINGSHIP

Archaeological and historical evidence suggests that the Hill of Tara has been a crucial site in Irish culture since a very early date. Early Christian texts describe the site as the royal seat of the kings of Ireland and the focus of early dynasties, and it was clearly an important power base in the Iron Age and through to the post-Roman period. However, the presence of ritual and funerary monuments dating back to the Neolithic suggests that the hill also had strong ritual connections and had been in use for thousands of years. The earliest monument on the Hill of Tara is the Mound of the Hostages, a Neolithic passage tomb dating to *c.* 2130 BC. It is thought to still contain more than 100 individuals, buried with pottery and stone objects. Later, in the Bronze Age, other graves were dug on the mound, with the dead buried individually in pits. Pottery urns and bowls

with impressed geometric designs have been found, as well as, in one case, a whole necklace of jet, amber, and copper beads. Another early feature on the hill, known as the Banqueting Hall because of its long, sunken shape, was probably a ceremonial road leading to the hill. Probably Neolithic in origin, the hall served as an entrance way for many centuries. In the Iron Age, two ring forts with a series of ditches were built on the hill; the enclosure bank around it was probably constructed at this time too. In the center of one of the forts stands the Stone of Destiny, an enigmatic monument that clearly demonstrates the ritual significance of the hilltop. There are other forts surrounding the enclosure, including the Rath (fort) of the Synods, excavated in 1899.

MOUND OF THE HOSTAGES
The earliest monument at the Hill of Tara is the Mound of the Hostages, and this small passage grave may have been the starting point for all the later activity at the site. This photograph shows the entrance to the tomb, inside which a large number of burials were found.

THE STONE OF DESTINY
This phallic-shaped standing stone, in the center of one of the ring forts known as Cormac's House, was probably erected during the Iron Age. It is believed to have been the inauguration stone for the dynasty of the O'Neills, the kings of Tara. Its exact function, however, is unknown.

Stone-reinforced entrance

Low mound

THE CLASSICAL AGE

GREEK AND ROMAN SOCIETIES were city-based, commercialized, and multinational communities. Roman citizens living in Ephesus in Turkey would have recognized many familiar features in a small frontier town in Gaul, as well as many differences. When the Roman Empire finally collapsed in the 5th century AD, a huge administrative network of taxes and officials, supported by an infrastructure of roads, aqueducts, and trading routes, had been operating across a large area of the world for several centuries.

The materially oriented culture of the Classical period has left a large number of buildings, monuments, and works of art that are familiar to us today. Greek and Roman literature also remains an important part of our heritage. Despite this wealth of evidence, archaeological investigation of Classical sites is still necessary to provide a context for these great public works and to uncover the day-to-day economy and lives of Classical peoples.

① HADRIAN'S WALL
This fortified frontier was built in c. AD 122 by the emperor Hadrian to mark the limit of his empire in the north.

② TOCKENHAM
A high-quality Roman villa with mosaics, decorative ponds, and painted walls.

ROMAN ROADS
The Roman Empire flourished in Europe due to its excellent road system, which was built using high-quality materials and was cambered to allow drainage.

③ GRAND
A Gallo-Roman town with an advanced sewerage system and a renowned amphitheater.

④ PONT DU GARD
Built in the 1st century AD, this 60-mi- (40-km-) long stone aqueduct supplied water to the Roman city of Nîmes.

⑤ ROME
Traditionally believed to have been founded in 753 BC, Rome grew from a small village to become the capital of the Roman Empire.

⑥ HERCULANEUM
A town buried under volcanic mud after the eruption of Mount Vesuvius in AD 79.

⑦ POMPEII
Destroyed at the same time as Herculaneum, but buried under volcanic ash.

⑧ CARTHAGE
Using its maritime power, Carthage dominated the Mediterranean until it was captured by Rome in 146 BC.

⑨ SABRATHA
Originally a Carthaginian colony, this Roman city controlled the caravan trade across the Sahara Desert and was also a seaport. There are substantial remains of Roman civic and religious buildings, including this amphitheater.

ALEXANDER THE GREAT
Alexander the Great (356–323 BC) is famous for his unification of the Greek world. Using his stunning military expertise, he conquered the Persian Empire, bringing a wide area of the world – from South Asia and Egypt, to India and Pakistan – under Greek political control.

ATLANTIC OCEAN

NORTH SEA

BRITISH ISLES

Seine

GAUL

Loire

IBERIA

DALMATIA

AFRICA

Atlas Mountains

Sahara Desert

SCANDINAVIA

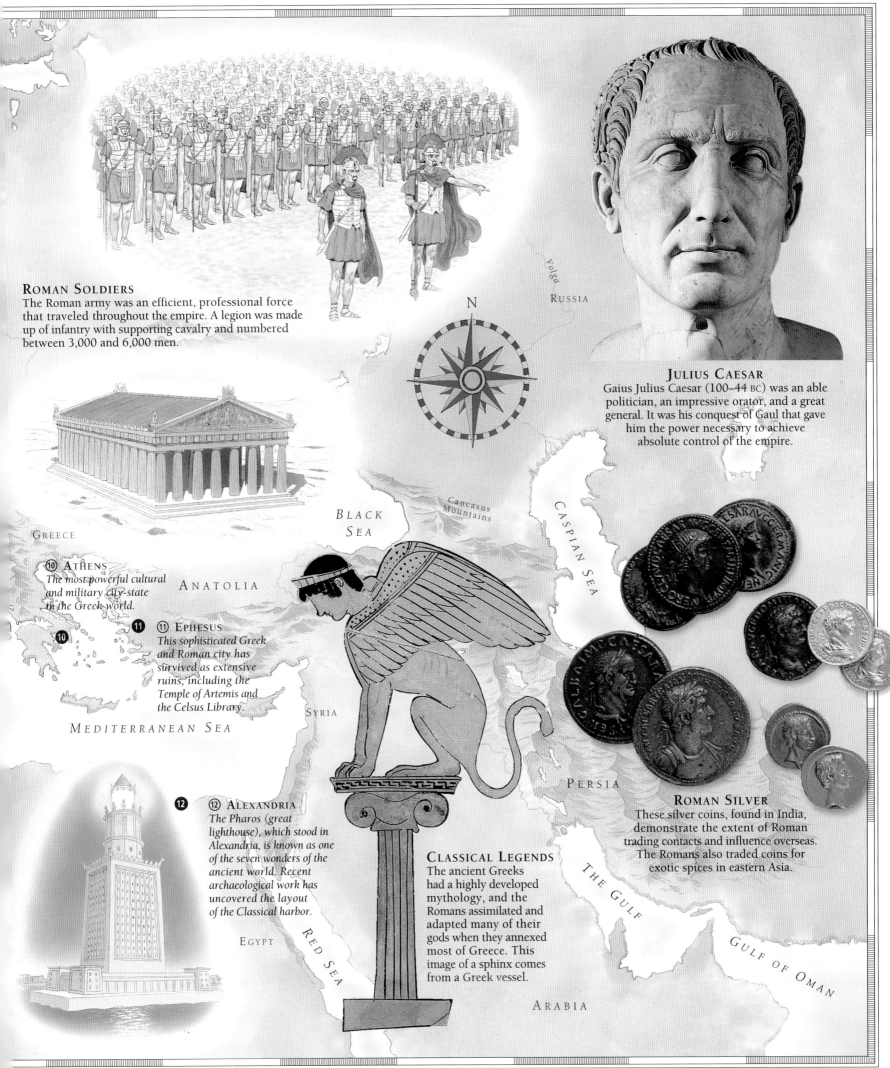

ROMAN SOLDIERS

The Roman army was an efficient, professional force that traveled throughout the empire. A legion was made up of infantry with supporting cavalry and numbered between 3,000 and 6,000 men.

JULIUS CAESAR

Gaius Julius Caesar (100–44 BC) was an able politician, an impressive orator, and a great general. It was his conquest of Gaul that gave him the power necessary to achieve absolute control of the empire.

GREECE

⑩ ATHENS
The most powerful cultural and military city-state in the Greek world.

ANATOLIA

⑪ EPHESUS
This sophisticated Greek and Roman city has survived as extensive ruins, including the Temple of Artemis and the Celsus Library.

MEDITERRANEAN SEA

BLACK SEA

Caucasus Mountains

Volga

RUSSIA

CASPIAN SEA

SYRIA

PERSIA

⑫ ALEXANDRIA
The Pharos (great lighthouse), which stood in Alexandria, is known as one of the seven wonders of the ancient world. Recent archaeological work has uncovered the layout of the Classical harbor.

EGYPT

RED SEA

CLASSICAL LEGENDS

The ancient Greeks had a highly developed mythology, and the Romans assimilated and adapted many of their gods when they annexed most of Greece. This image of a sphinx comes from a Greek vessel.

ROMAN SILVER

These silver coins, found in India, demonstrate the extent of Roman trading contacts and influence overseas. The Romans also traded coins for exotic spices in eastern Asia.

THE GULF

GULF OF OMAN

ARABIA

A ROMAN VILLA

TOCKENHAM, UNITED KINGDOM

GAZETTEER P. 144 – MAP REF. F8

SITE EVIDENCE

🔍 A large amount of Roman pottery, *tesserae*, and roof tiles have been found over many years in the fields at Tockenham.

🔍 Geophysical survey revealed the outlines of the walls of a Roman villa and other structures.

🔍 The combination of scattered finds and the basic outline of the building suggest a high-status villa with rich inhabitants.

DURING THE ROMAN PERIOD, large country estates, centered on spacious and luxurious residences called villas, divided up the landscape of Europe. These houses were stone-built and -roofed, were heated by complex systems of pipes, and had elaborate mosaic floors and painted plaster walls. Inside the villa, a wealth of pottery and furniture would have been used by its inhabitants, while outside, the workshops and outbuildings would have been busy with farming activities and the administration of the estate. Roman villas are a common feature in the ancient landscape of Britain and much of Europe, and Tockenham presents a classic example of how archaeologists find and investigate such sites. Complete floors and walls are rarely discovered; much more common are the outlines of building rubble and artifacts from the villa. Tockenham demonstrates how archaeologists can gather a wealth of information about Roman lifestyles by just piecing together a wide array of objects.

THE SITE

The local farmer had already uncovered a wide collection of Roman objects in the fields at Tockenham during several years of plowing, and wanted to investigate the site to find out what was beneath the soil to avoid damaging it further. To do this, surveying was carried out to discover the extent of the Roman finds and structures, and small trenches were dug to find out what survived below the surface. A plan was made of the earthworks on the surface, and geophysical surveys were carried out to locate buildings beneath.

Area of excavation

THREE-DIMENSIONAL PLAN

A survey using magnetometry quickly revealed where the buildings and walls were located. After this, resistivity was used to provide a more detailed picture of the features below the soil. From this information, a 3-D plot showing the villa and courtyard was created (see left).

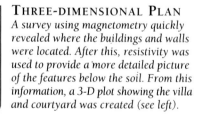

Mosaic squares come in regular sizes

Evidence of erosion

PLAN OF FINDS

The rectangular field shown above (adjacent to the villa) had been plowed recently, so field surveying could be carried out to retrieve archaeological material from the disturbed surface. Teams of people walked across the field collecting the finds, including pottery, stone fragments, and metalwork, which had been revealed by the plowing. The plan shows the higher concentration of Roman material, especially pottery, closer to the site of the villa, with a looser scattering beyond. Some prehistoric flints were also discovered.

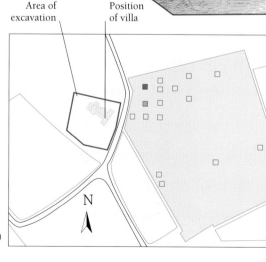

Area of excavation

Position of villa

▨ 1–5 pieces of pottery	6–10 pieces of pottery
■ 11+ pieces of pottery	├─────┤ 330 ft (100 m)

N

TESSERAE

Several small squares of different-colored stone and brick tiles used to make mosaics (known as tesserae) were found spread across the site.

Width: 1in (2.5cm)

Length:
1½in (4cm)

PIECE OF GLASS

A corner of a small square pane of glass, which would have been fixed into a frame, was found. Roman window glass was thick and green, unlike modern glass.

Intricate pattern

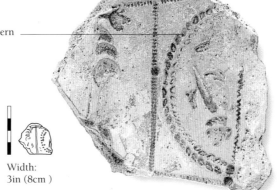

Width:
3in (8cm)

PIECE OF SAMIAN MOLD

Often found on villa sites, high-quality pottery called Samian ware was made in molds like the one above. This produced standard-shaped vessels with mass-produced relief designs, which were uniform throughout the Roman Empire.

Diameter:
2in (5cm)

COIN

Numerous Roman coins were found on the surface of the plowed soil. This heavily worn coin dates from the reign of the emperor Diocletian in the 3rd century AD.

Partially
eroded shell

Width:
2⅓in (7cm)

OYSTER SHELL

Large quantities of oysters were eaten in the Classical world. As a result, oyster shells are commonly found on Roman sites, surviving particularly well in chalky and limestone soils.

GEOPHYSICAL PLAN

The geophysical survey revealed the basic shape of the villa. In this plan, it is outlined in detail, with the walls shown as areas of high resistance (in black). By comparing it with other excavated villa sites, the archaeologists could predict that it was probably a "winged" type of villa, with a square dining room (*triclinium*) at the back, and an apsidal (round-ended) building to the northwest, which may have been the bathhouse.

├─────── 30 ft (9 m)

N

Courtyard
or garden

Apsidal (round-ended) building

Main villa
buildings

Probable
dining room

Combed pattern
on tile's surface

Large building
facing courtyard

Length:
12in (30cm)

ROOF TILE

Stone roof tiles would have been pegged, point down, to the rafters using iron nails. Here, the end of the tile containing the peg hole is missing. The orange residue is caused by being buried in clay.

Width:
8in (20cm)

HYPOCAUST TILE

This tile from a box flue pipe is part of a hypocaust heating system. The combed pattern helps the tile adhere to the concrete wall, and makes it instantly recognizable to archaeologists.

Length:
3ft 6in (1m)

Distinct fish-
head shape

FISH-HEAD SPOUT

This stone spout in the shape of a fish head was found near the site. It was thought to be modern until the villa was revealed, when it was indentified as part of a Roman pond decoration.

A ROMAN VILLA

INTERPRETING THE FINDS

ANALYSIS OF THE FINDS and of the survey results from Tockenham provides a detailed picture of the appearance of the villa and many glimpses of what life would have been like for its inhabitants. The materials that were used to build the villa can be studied, the techniques copied, and the interior decoration suggested, by comparing it with other similar sites. The various finds – the decorative spout, the *tesserae*, and the flue tiles – all reflect the Romans' inclination toward building comfortable and beautiful houses and gardens. The archaeological evidence also shows us how the builders and inhabitants of the villa used standard Roman building techniques, and even mass-produced materials, and how, although the community was on the fringes of the extensive Roman Empire, it was still influenced by Roman fashions and traditions.

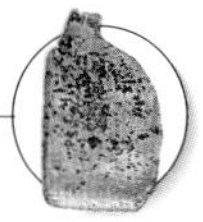

ARTIST'S IMPRESSION

From similar sites where buildings are well preserved, and from contemporary descriptions and pictures, it is possible to reconstruct what the villa may have looked like, even though only minimal foundations have survived. Set in a well-organized landscape of gardens, fields, orchards, and woods, the villa would have been the Roman equivalent of a manor house – the center of the farming estate and a luxurious home for a wealthy family. It would have had dining rooms, living rooms, and bedrooms – some of them heated and with mosaic floors and decorated walls, others painted. Outside, there may have been a park, with gardens and fishponds.

GLASS WINDOWS
We do not know how common glass windows were, but it is likely that only the bathhouse and the heated rooms were glazed. The other rooms probably had metal grilles and wooden shutters. These details add to the impression that this was a high-status villa.

Surrounding gardens

Window opening with shutters and no glass

Painted wall

Heated bathhouse

Slave carrying water

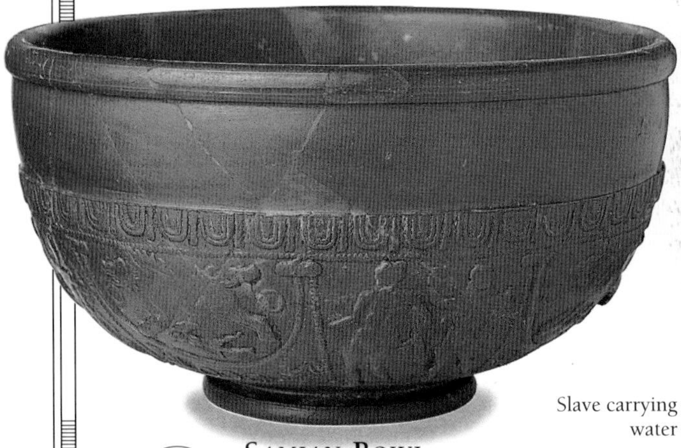

SAMIAN BOWL
High-quality bowls like this were imported over great distances from Gaul (modern-day France) and Germany to be used as tableware. The wealthy family of the villa may have used a bowl like this for serving fruit.

DECORATIVE POND
In the center of the front courtyard there was probably a shallow pond with fish-head spouts at either end. The pipe found here was unique in its high quality, with its closest comparisons found in Pompeii.

MONEY
The villa owner and his family would have been extremely wealthy. Most Roman coins were made of bronze and almost always depicted the head of the current emperor. Gold and silver coins were much rarer.

HYPOCAUST HEATING

This system of heating was based on under-floor hot-air circulation. A furnace situated outside the villa would have been fired with wood, heating the air that was fed into the flue pipes. The hot air would have circulated underneath the villa floor, which was supported on stone or ceramic pillars. It would then be fed through the walls in a series of pipes or flues made out of box tiles. The hot air and smoke would then escape through a series of chimneys or apertures. Only some rooms would have been heated, as the hypocaust would have required large quantities of wood and labor. This exposed hypocaust system was found at the Roman site of Chedworth in England.

Combed pattern to help fix the concrete

Air vent

Clay is used to make the box tile

COMPLETE BOX FLUE TILE

EXPOSED HYPOCAUST HEATING SYSTEM

Tower over the dining room

ROOF TILES
The overlapping tiles would have kept the roof watertight; different-colored tiles were sometimes used to create decorative patterns. They would have been mass-produced at a nearby quarry and chipped into standard shapes.

High ceiling

Painted paneled walls

Bed made of wood or bronze

FRESH OYSTERS
Oysters were a much more common food source in the Roman period than they are now and were extensively farmed. Many dishes were made using them, and several Roman oyster recipes are known.

CONCLUSIONS

🔎 The geophysical survey revealed the ground plan of a substantial, three-winged villa, set around a courtyard.

🔎 Although only a small area was excavated, the artifacts discovered tell us a lot about the construction of the villa.

🔎 The fish-head spout is an important find because the quality of its workmanship is unique in the archaeology of Roman Britain.

🔎 The evidence gives us a glimpse of the comfortable lifestyle enjoyed by the inhabitants.

MOSAIC FLOORS
Tesserae were put together to form mosaics, usually in geometric patterns or showing mythological scenes, like this portrait of "Spring" from Chedworth in England. At Tockenham, the mosaic floors may have been bought as a ready-made design or specially commissioned.

Design made up of *tesserae*

GLASS

MAKING GLASS IS A COMPLICATED PROCESS that has been adapted and developed over several thousand years. Glass has been in use since the first glazed objects were made in the eastern Mediterranean region (*c.* 3000 BC). Throughout its history, glass has been used for a wide range of objects, including cups, bottles, bowls, lamps, window panes, and jewelry. Because they could be decorated and were sometimes costly to produce, glass objects were often regarded as luxury items and works of art. As a result, there were places that excelled in high-quality glassmaking, such as medieval Venice and 18th-century London. Glassmaking reached its zenith during the Roman period, and few new techniques have developed since, apart from enameling and making lead crystal. The arrival of mechanization made glass a convenient and utilitarian material, and 19th- and 20th-century bottles and containers are common finds on archaeological sites.

GLASS INGREDIENTS

SILICA

POTASH

Glass is made by combining silica, obtained from grinding quartz, flint, or sand with oxide chemicals, such as potash (wood or plant ash), to strengthen it. Scrap glass can be added to help fuse the ingredients together. Other substances can also be added to provide color or consistency to the glass.

LOTUS CUP
—— 1250 BC ——

Before the invention of glass, the Egyptians were producing objects made of faience – a blue-glazed mixture of sand and soda, fired just below melting point. Faience-making began in Egypt in predynastic times (*c.* 3000 BC) when beads and pendants were made. Later, more sophisticated objects were produced. This Egyptian goblet, with its lotus-flower decoration, would have been a highly treasured object in that society.

1250 BC

Engraved petal decoration

Lotus-flower shape

—— KEY CHARACTERISTICS ——
Pattern is engraved onto the cup.
Vessel is shaped like a lotus flower.
Metal oxides create blue color.

PHOENICIAN GLASS
—— C. 950 BC ——

It was not until 1500 BC that glassworking was refined enough to make vessels, and by the 9th century BC the Phoenicians were trading glass in the eastern Mediterranean region. Glass vessels were made by dipping a sand-filled bag into a vat of molten glass, which coated the bag and took its shape. When the glass was set, the sand core would be emptied out. This vessel, called an *aryballos*, was made using this technique, but has additional decoration added by dripping pigment-dyed colored glass onto the vessel and then rolling it on a flat surface. These small containers were used for storing perfumes and unctions.

Bright-colored pattern

Small base

—— KEY CHARACTERISTICS ——
Made on a clay or sand core.
Small flasks and bottles common.
Bright colors used, especially blue and yellow.

ROMAN BLOWN VESSEL
—— 1ST CENTURY AD ——

Blown glass was probably developed in the eastern Mediterranean region in the 1st century BC. A huge Roman glass industry grew up, and glass objects became common household items. The Romans were extremely skilled glassmakers and perfected many complex techniques, such as producing colorless glass, painting glass, and engraving glass. When the Roman Empire collapsed, many of these techniques were lost, and glass became much simpler and plainer.

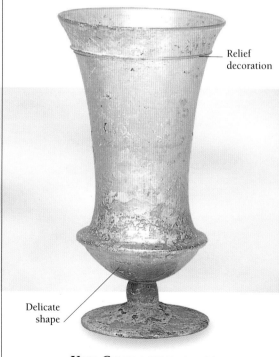

Relief decoration

Delicate shape

—— KEY CHARACTERISTICS ——
Glass was blown, not core-molded.
Much thinner, colorless glass was produced.
Complex shapes and wide variety of vessel designs.

GLASSBLOWING

The technique of glassblowing was perfected by the Romans, who were the first to make fine-stemmed vessels. Making glass objects by blowing was much quicker than with the core-built technique, and a much greater variety of shapes and designs could be made this way. Blowing a glass vessel requires powerful lungs and skill to ensure that the object is shaped and cooled at the correct rate. This is vital because any mistakes in the process will cause flaws or stresses that may affect the strength of the glass, perhaps causing it to break.

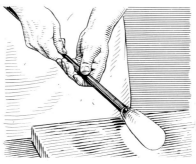

1 *A hollow metal rod, usually over 3 ft (1 m) long, is dipped into molten glass and manipulated until a lump of soft, hot glass has collected on the end. The lump is then rolled on a marver (a metal or wooden block) prior to blowing.*

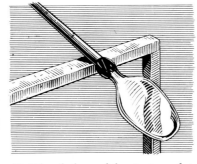

2 *When the lump of glass is at a perfect consistency, the glassmaker blows down the rod to form a bubble, which becomes the inside of the vessel. The glass can be shaped as it hardens, by rolling or by being blown into a mold.*

3 *Once the vessel has been formed, decoration can be applied while the glass is still hot and malleable. When finished, the vessel is cut away from the end of the rod using shears. It can also be knocked off the end with a sharp tap.*

ISLAMIC GLASS
14TH CENTURY AD

Glassmaking continued in eastern Europe after the fall of Rome and was combined with Islamic craftsmanship to produce new techniques for decorating vessels. Gilding and enameling, in which powdered pigments were painted onto the glass and fired to produce brightly colored designs, became popular in the 12th century.

Enameled glass pattern

KEY CHARACTERISTICS
Intricate patterns are used.
Gold leaf sometimes applied.
Design is painted onto the surface of the glass.

LEAD-CUT CRYSTAL
17TH CENTURY AD

Lead-cut crystal glass was invented in the 17th century. Adding lead oxide to the glass lowered the temperature to the required level to make it "crystal" clear and of a good enough quality to engrave. The engraving itself would have been done after cooling, often in a different workshop, and English vessels were sometimes sent abroad to engravers in Europe. Lead-cut crystal is heavier than other types of glass.

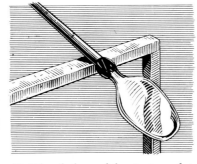

Design is engraved on glass

Thin delicate stem

KEY CHARACTERISTICS
Transparent color and even texture.
Engraved patterns and delicate decorations.
Simple and elegant designs and shapes.

COCA-COLA BOTTLE
AD 1957

Glassmaking techniques were revolutionized in the 19th century when mechanized processes were introduced, allowing the mass production of glass. With glassmaking becoming more scientific, clearer, better-quality vessels were also being made. By the early 20th century, glass bottles and containers had become commonplace, and some, such as this 1957 Coca-Cola bottle, became famous design classics.

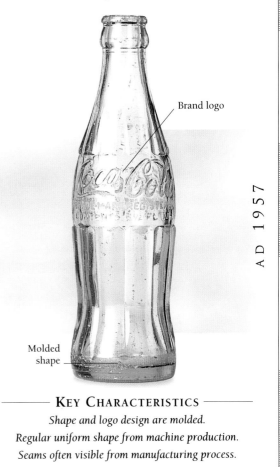

Brand logo

Molded shape

AD 1957

KEY CHARACTERISTICS
Shape and logo design are molded.
Regular uniform shape from machine production.
Seams often visible from manufacturing process.

COMPARATIVE SITES

EPHESUS

SELÇUK, TURKEY

GAZETTEER P. 138 – MAP REF. C2

THE GREAT CITIES OF THE CLASSICAL WORLD were sophisticated, highly developed urban centers, with complex and rigid bureaucratic systems, public administration networks, and a strong cultural and religious life. The abundance of inorganic artifacts and impressive structures that survives allows an unprecedented understanding of Classical cities, while the preservation of written sources gives a valuable insight into the culture. Ephesus is a good example of an important Greco-Roman city that grew from humble beginnings to become an economic center and a focus of religious devotion and leisure. Situated in what was then known as Asia Minor, Ephesus combined indigenous religious and cultural traditions with those of the Greek and Roman settlers to produce a rich and diverse society.

PUBLIC LATRINES
One of the most interesting aspects of the Classical period is the importance placed on public utilities. While many rich Roman households contained their own private facilities, public latrines were also built. Channels beneath the latrines would carry the waste away to a sewer. Roman settlements often had extensive water systems, which kept the cities supplied with freshwater.

THE STAR OF ASIA

Ephesus, once known as the "Star of Asia," was a flourishing and decadent Greco-Roman city, the prosperity of which relied on the trade networks of the Mediterranean Sea and on the popular cult of the goddess Artemis. The earliest inhabitants probably arrived during the Bronze Age, so it was already a settled site when the Greeks established a trading post here as part of their colonization of western Asia. Throughout its long history, the city retained its wide-ranging fame and importance because of its strong religious cults and economic prosperity. As new rulers took over, they added to the architectural, religious, and cultural diversity of the city rather than destroying it. The variety of buildings that have

MOTHER GODDESS
This statue, discovered in the town hall, is a combination of the Egyptian, Greek, and Hellenistic female goddesses Cyrene, Aphrodite, and Artemis. Her robes show real and mythical animals and signs of the zodiac. The four rows of ovoids probably represent breasts, which were considered to be symbols of fertility.

AERIAL VIEW OF PART OF EPHESUS
The remains we can see today are from the Roman remodeling of the city. The Arcadian processional way can be seen leading up to the theater from the harbor.

been found in Ephesus include public baths, gymnasia, a theater, a range of temples and shrines, fountains, monumental gates, a brothel, and public toilets.

In 296 BC, during the Hellenistic period, Lysimachus, Alexander the Great's successor, rebuilt the city as a trading port and expanded its area, enclosing it within a 5½-mi- (9-km-) long city wall. However, most of the buildings that have survived to this day date from the Roman period, when the emperors continued to build and remodel the city. The theater, for example, was originally built in the Hellenistic period, but was

enlarged by the Roman emperors Claudius, Nero, and Trajan. Many gods in the Classical world were adapted and amalgamated with those of other religions to ease political situations and integrate different societies. Ephesus was traditionally dedicated to the western Asian mother-goddess figure Cybele, and so, during the Greek colonization of the region, this devotion was transferred to the Greek goddess Artemis. Indeed, the huge Temple of Artemis was one of the seven wonders of the ancient world, and her cult attracted thousands of visitors to the city. This tradition was continued in later times by the transference of religious devotion to the Virgin Mary under the Christian church.

The strategic position and protected harbor made Ephesus ideally suited for commerce. The city was also at the start of the royal road to Nineveh, the ancient capital of the Assyrian empire to the east. However, the port began to silt up around 150 BC, partly due to engineering mistakes, and without this vital resource the city began to decline. The remains of the city are now found several miles inland.

THE LIBRARY OF CELSUS
Housing 12,000 books, stored as papyrus rolls in niches in the walls, the Library of Celsus was built by the Roman writer Aulus Cornelius Celsus in AD 110. It quickly became the intellectual center of the city.

CAST OF HUMAN BODY

In the 1860s, Guiseppe Fiorelli discovered that the long-since-disintegrated bodies buried by the volcano had left cavities in the hardened ash. By filling them with plaster, he could recreate the posture of the individual.

THE BURIED CITY

The town of Pompeii (dating as far back as the 6th century BC) had already been pillaged by treasure hunters before any real excavations began in the 1740s. It continued to be excavated throughout the 19th century, and, by 1860, scientific standards of recording had been introduced by the Italian archaeologist Guiseppe Fiorelli (1823–96), who also developed the technique of recovering human body casts (see above). The town covered 158 acres (64 ha) and was divided into blocks and numbered houses, so all finds could be traced to their exact locations. Only 75 percent of the town has been excavated, but it nevertheless shows the whole range of buildings that were usual in a Classical town. The central forum or marketplace was surrounded by a basilica and baths; the amphitheater, theater, and domestic buildings were all enclosed within the town wall, whose gates and towers survive today. Most of the enclosed area consists of houses that survive in remarkable detail, with furniture, wall paintings, colonnades, *atria* (the central courts of the houses), and gardens intact. Out in the paved streets, there were pedestrian walkways, water troughs for fountains, and graffiti and advertisements on the walls.

ARCH OF CALIGULA

Situated near the forum and macellum (market), this arch is just one of a number of triumphal arches in the city. It is built of brick, and would have been covered with plaster or cement to imitate marble.

POMPEII

BAY OF NAPLES, ITALY

GAZETTEER P. 147 – MAP REF. 17

ON AUGUST 24, AD 79, Mount Vesuvius, a volcano near the Bay of Naples, erupted and buried the Roman settlements of Pompeii, Herculaneum, Stabiae, and Oplontis in a deluge of ash and mud. This event was described by the Roman historian Pliny the Younger (AD 62–113), giving us a picture as vivid and graphic as that of any recent catastrophe. Just as with modern disasters, the survivors searched through the rubble to try to rescue people and possessions. Except for these contemporaneous disturbances, the sites remained largely untouched for 1,700 years. The buildings, their contents, and a wealth of vital evidence were preserved, along with thousands of bodies of people and animals. It is one of the richest groups of archaeological sites in the world, incomparable in its excellent preservation of material and presenting an exceptional view of everyday Roman life.

HERCULANEUM

BAY OF NAPLES, ITALY
GAZETTEER P. 147 – MAP REF. 18

HERCULANEUM WAS DISCOVERED in 1709 by well-diggers, and by 1735, it was possible to visit the amphitheater, making the town a celebrated tourist attraction. Systematic investigation began in 1738, aided by tunneling and blasting with gunpowder. The site has been a focus for renewed excavation recently, but there is still no complete plan – just a few streets and buildings have been mapped. It has been suggested that the town covered about 49 acres (20 ha) – a third of the size of Pompeii – but this remains an estimate until all the town walls are discovered. The site has provided archaeologists with information about the organic materials used by the Romans, which rarely survive. Wooden shutters, house partitions, pillars, furniture, and even a child's cot carbonized by the intense heat have been found. One of the most spectacular finds at Herculaneum is the House of the Papyri, which contained a huge library of 1,800 books.

SURVIVING STRUCTURES
This view over the buildings, streets, and courtyards shows how densely developed this area of the town was. Before the earthquake, the coastline was just beyond the foreground – the villas would have faced out over the sea.

MOSAIC OF NEPTUNE AND SALACIA
This mosaic is made of very small tessarae, so its highly detailed design is particularly clear. It was found in the courtyard of a wealthy wine merchant's house and shows Neptune and his wife, Salacia, in their summer dining room. Other maritime themed mosaics were found throughout Herculaneum, including a mosaic-lined grotto dedicated to nymphs.

GRAND

VOSGES, FRANCE
GAZETTEER P. 146 – MAP REF. H5

THIS PROVINCIAL GALLO-ROMAN town grew up around a successful healing center in an isolated clearing, rather than following the plan of a typical small Roman town. Many Roman settlements developed in this way, with town baths becoming the central focus. The sanctuary here was dedicated to Grannus, an important Apollo-like Celtic healing god, whose chief sanctuaries were based in the Vosges region of France. Grand is the modern name for what is thought to have been the Roman town of Grannum. Latin documents refer to cults of Mercury, Jupiter, and Grannus, and temples dedicated to Grannus, and probably also to Mars and Jupiter, have been excavated. Archaeological research has revealed a lot about the town layout, the road network, and various buildings. These included a large Greek-style basilica (with an impressive mosaic showing a comic mask and geometric motifs), various shops, a monumental fountain, and a large building near the cemetery that may have been a prison. There are a number of traces of the settlement, including mosaic floors and hypocaust heating systems, both inside and outside the town. Archaeologists discovered a sophisticated underground sewage network, up to 49 ft (15 m) deep, which served the whole town, with drains and deep wells. The renowned amphitheater was situated outside the town walls and was built up against a hill.

Intricate stone carving

STATUE OF MARSYAS
This fragment of a sculpture of the satyr Marsyas shows the cosmopolitan nature of the site at Grand. This satyr was linked with Athena, the goddess of wisdom, in the great Roman pantheon.

THE AMPHITHEATER
Built in the 1st century AD, the amphitheater was elliptical, with a maximum width of 451 ft (137.5 m) and an arena wall 6½ ft (2 m) high. It was one of the largest in Gaul.

THE DARK AGES

FOLLOWING THE COLLAPSE of the Roman Empire in the 5th century AD, there was a long period of disruption and change across Europe and the Mediterranean region. Out of this instability, early medieval states and kingdoms developed under leaders such as Charlemagne and Alfred the Great, while Muslim empires dominated the Mediterranean lands. The power of the Christian Church increased in Europe, but came under constant attack from Arabs in the south and Vikings in the north. The Vikings of Scandinavia, who were essentially seafaring traders, established new sea routes and became the first Europeans to land in America. Although this period has been known traditionally as the Dark Ages, new evidence has shed light on the varied cultures of this transitional era.

GREENLAND

BAFFIN BAY

DAVIS STRAIT

① ANTLER ARROWHEADS
Clear evidence of Viking long-distance seafaring is shown by these arrowheads, carved from reindeer antlers. Discovered in Greenland, they suggest that the Vikings hunted reindeer for food and materials.

LABRADOR SEA

CANADA

NEWFOUNDLAND

ATLANTIC OCEAN

② L'ANSE AUX MEADOWS
One of the few areas to be settled by the Vikings in America was L'Anse aux Meadows on the north coast of Newfoundland. The houses were built entirely out of sod, which was laid over wooden frames.

SOROR mea FLOREN
TINA accipe codicem
Quem tibi compo
sui feliciter
Amen

THE VISIGOTHS OF SPAIN
The Visigothic hold on Spain began in the 5th century AD and lasted until the Arab invasion in 711. During this time, intellectual life flourished, and the prelate and scholar Isidore of Seville (560–636) (shown left, giving a book to his sister) produced his *Etymologiae*.

N

WANDEL SEA

ARCTIC OCEAN

LOMBARDIC HORSEMAN

The Lombards took control of northern Italy from the 6th century AD onward. They were a disorganized people, with a fierce and savage reputation. Their numerous dukes ruled over small areas from Milan and Turin, and they seized Ravenna in 751. This horseman is engraved on a bronze plaque and comes from 7th-century Lombardy.

SVALBARD

VIKING LONGSHIP

Viking raids were carried out from ships that were wide and long, making them fast and easily maneuverable. Their shallow draft (depth of the hull) enabled them to sail close to the shore and up rivers. They were propelled by distinctive square sails and teams of rowers.

NORWEGIAN SEA

BARENTS SEA

NORWAY

Kölen Mountains

LAPLAND

Mezen

Dvina

③ **HOFSTADIR**
Viking colonists had reached Iceland by AD 860. The remains of this farmstead are typical of those found on the island.

❸
ICELAND

④ **IONA**
This island was granted to St. Columba in AD 563 for the foundation of one of the first Christian monasteries.

⑦ **OSEBERG**
This 9th-century AD ship was found under a barrow and contained the burial of a royal woman. It is the richest ship burial ever found.

⑤ **WINTERBOURNE GUNNER**
A number of Anglo-Saxon bronze objects, such as this shield boss, were found at this 5th-century AD burial site.

SWEDEN

⑧ **TRELLEBORG**
One of four specially built Viking forts found in Denmark, Trelleborg was occupied in the 10th century AD. Inside its circular ramparts were wooden barracks and workshops.

❹

NORTH SEA

DENMARK **❼** **❽**

BRITISH ISLES

❾ ⑨ **FEDDERSEN WIERDE**
An early settlement that consisted of houses built in a distinctive radial pattern on an artificial clay island. It had been abandoned by the mid-5th century AD.

RUSSIA

English channel

Rhine

❺
❻

FRANCE **❿**

⑩ **POUAN**
Some impressive swords and scabbards were excavated from this Merovingian graveyard.

GERMANY

Dnieper

⑫

⑫ **KIEV**
Founded in the 10th century AD, this fortified town soon became a center of eastern commerce.

Danube

⓫ ⑪ **RAVENNA**
Some of the most beautiful Christian and Byzantine art has been found in this city.

ITALY *GREECE*

⑥ **WEST STOW**
Farmsteads were situated here between AD 400 and 650. Wooden halls and sunken-floored cellars were found to contain a collection of everyday objects.

THE MEROVINGIANS
The Merovingian dynasty appeared in France after the collapse of the Roman Empire in the 5th century AD. Their cemeteries and tombs have revealed many artifacts, such as this fish-shaped fibula.

MEDITERRANEAN SEA

A SAXON BURIAL

WINTERBOURNE GUNNER, UNITED KINGDOM

—— GAZETTEER P. 144 – MAP REF. G8 ——

SITE EVIDENCE

○ Construction work has revealed Saxon burials over a wide area.

○ Previous investigations in the area had uncovered pieces of Saxon pottery.

○ Other local sites and finds provide a regional context for the cemetery.

○ Aerial photography has revealed a number of burial mounds with ring ditches around them in the immediate vicinity.

FROM THE END OF THE ROMAN PERIOD in the 5th century AD, southern and eastern England began to come under the influence of people from Germany, Denmark, and the Low Countries. At first, this consisted of raids, but later, incoming groups began to establish permanent settlements. Gradually, these newcomers worked their way into society and achieved an aristocratic takeover of the indigenous population. The Saxons, who came from what is now western Germany, are distinguished in the archaeology of this period by their wooden buildings and their burials, and almost all we know of their culture comes from the excavation of their graves. The discovery of a multiple-grave site at Winterbourne Gunner in southern England provided archaeologists with an excellent opportunity to study the people and their community. It was hoped that excavations would yield the nature and extent of the cemetery and the position of the Saxon settlement.

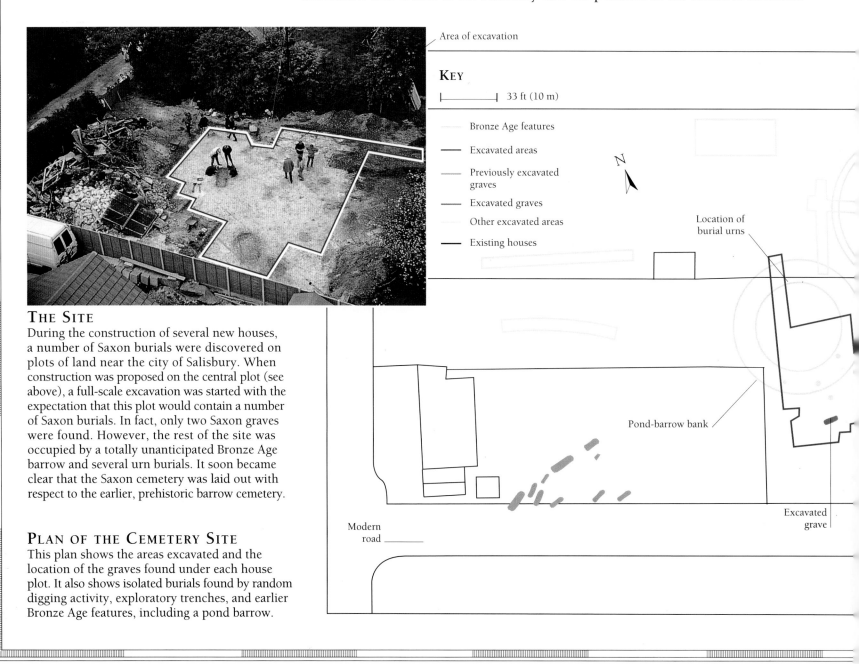

Area of excavation

KEY

|———————| 33 ft (10 m)

—— Bronze Age features

—— Excavated areas

—— Previously excavated graves

—— Excavated graves

—— Other excavated areas

—— Existing houses

N

Location of burial urns

Pond-barrow bank

Excavated grave

Modern road

THE SITE

During the construction of several new houses, a number of Saxon burials were discovered on plots of land near the city of Salisbury. When construction was proposed on the central plot (see above), a full-scale excavation was started with the expectation that this plot would contain a number of Saxon burials. In fact, only two Saxon graves were found. However, the rest of the site was occupied by a totally unanticipated Bronze Age barrow and several urn burials. It soon became clear that the Saxon cemetery was laid out with respect to the earlier, prehistoric barrow cemetery.

PLAN OF THE CEMETERY SITE

This plan shows the areas excavated and the location of the graves found under each house plot. It also shows isolated burials found by random digging activity, exploratory trenches, and earlier Bronze Age features, including a pond barrow.

PIN

An iron pin was found as a heavily rusted and corroded lump. During excavation, it was not possible to determine what this object was, and it was identified only after conservation.

Length: 1¼–2½in (3–6cm)

Decoration on shank

Roundel pinhead

SKELETON DETAIL

This close-up of the head and upper body of the skeleton shows the well-preserved skull of an adult female and the location of a brooch, a pin, and a bead. The woman was laid on her back with her hands in her lap, facing west, as was the custom. The second grave also contained a female skeleton in the same position.

EXCAVATED GRAVE

The skeletal remains were in good condition because the bodies had been buried in chalk. The bones were carefully excavated and cleaned, and the grave goods were left in position long enough to be recorded and drawn.

Corroded copper alloy

Circular pattern just visible

Brooch
Pin
Bead

Length: 1½in (4cm)

BROOCH

When this typically Saxon disk brooch was found, it was a green encrusted mass with no distinguishing features visible, suggesting only that it was a copper-alloy object. It was discovered in the last layers of earth in the excavated grave along with the other grave goods.

House built after excavation

House built after excavation

AMBER BEAD

While recognizable as a bead and only slightly corroded, it was not clear during excavation what material the bead was made of. After cleaning, it was identified as a perforated amber bead. Although no other beads were located in the grave, it can be presumed that this one was part of a necklace.

Length: 1¼in (3cm)

Corroded surface

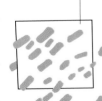

Previously excavated cemetery area

Coarse texture

Length: 2in (4.5cm)

POTTERY PIECES

A collection of Saxon potshards that were contemporary with the cemetery was discovered nearby. It is likely that these shards originated from the settlement that the cemetery served.

A SAXON BURIAL

INTERPRETING THE FINDS

FROM THE EXCAVATED EVIDENCE, it is possible to reconstruct a number of details about the buried individual, including her physical appearance. The body was clearly that of an adult female, who had been buried with some of her personal adornments. The location of the community settlement is suggested by the proximity of the pottery finds and provides an association of settlement and cemetery not usually discovered on archaeological sites. Only inorganic materials, such as the brooch and pin, survived in the soil, and any clothing that the woman might have been wearing had disintegrated long ago. The grave may have contained other goods that have since disappeared entirely, such as flowers, leather, baskets, and food. These basic domestic objects would have been intended to sustain the woman on her journey into the afterlife. The bead, brooch, and probably the pin, would have been prized objects, thought to be essential in the afterlife. Even with the range of objects found surrounding the body, archaeologists can only speculate about the funerary rites that would have taken place at her graveside.

Multicolored
glass bead

NECKLACE OF BEADS
It is common for Saxon burials to contain large quantities of beads. These are often highly colored and made of materials such as amber, crystal, and glass. The strings rarely survive, but the position of such beads in the graves suggest they were threaded and worn around the neck.

ARTIST'S IMPRESSION
From the graves and artifacts found at the site, it is possible to imagine the burial scene at Winterbourne Gunner. A fully clothed adult female has been laid on her back in the grave and surrounded with food, clothing, and useful objects for the afterlife. Mourners are in attendance, placing offerings into the grave. In the distance, the nearby settlement and the remains of the Bronze Age pond barrow can be seen. Other Saxon burials would have been made in the surrounding area, and they may have been marked in some way. The much larger mounds of Bronze Age barrows mark the higher ground above the settlement.

The settlement

Remains of
the prehistoric
pond barrow

Woman in typical
Saxon clothing

SAXON VESSEL
Saxon vessels were crudely made, poorly fired, and rarely decorated or glazed. This reconstruction of a pot is made up of many shards, and the irregular, loose shape and simple rim are typical of Saxon ware. The fabric included grits, sand, or crushed shell. These pots were used mainly for cooking and food preparation.

Head of
the family

Mourners at
the graveside

SAXON BROOCHES

A variety of types of Saxon brooch survive to provide clues about the styles of dress worn by various ethnic groups at different times. Generally, they were made of copper alloy, usually gilded, so that they would have appeared golden. But burial in the soil usually made them appear as corroded green shapes with fragments of gold leaf adhering. Most are decorated, with designs ranging from simple dot patterns to elaborate chip-carved, abstract shapes. Saucer brooches were frequently worn in pairs on the shoulders to fasten together simple tubular dresses, often with beads strung between them. They could also have been worn centrally on the chest. Cross-headed brooches could have been worn below the waist to hold a jacket together. By analyzing brooches and their positions on skeletons, archaeologists have sometimes been able to distinguish between Angle, Saxon, and Frisian groups.

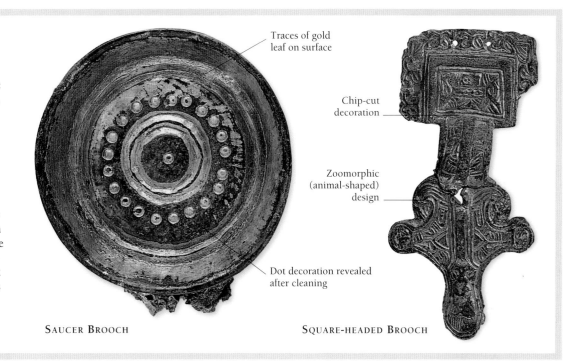

Traces of gold leaf on surface

Dot decoration revealed after cleaning

SAUCER BROOCH

Chip-cut decoration

Zoomorphic (animal-shaped) design

SQUARE-HEADED BROOCH

Farmland near the settlement

Neighboring groups coming to pay their respects to the family

The second grave

THE SKELETON

Examination of the bones, particularly the skull and pelvis, suggested that this was an adult female. The body had been laid in the supine position (on the back), with the head to the west, and the hands folded over the pelvis. The bones were well preserved in the chalk soil, but many of the smaller bones, such as those in the hands and feet, were missing. Using bone measurements, the archaeologists deduced that the woman had been of an average height.

CONCLUSIONS

🔍 The bones and the grave goods increase our knowledge of this Saxon cemetery, which has been revealed in a piecemeal fashion because of modern development.

🔍 The pottery gives us an unusual indication of the position of the settlement in relation to the cemetery.

🔍 The proximity of the village site to the prehistoric burial site shows that the Saxons respected the ancient burial ground and used the large Bronze Age barrows as the focus for their own cemetery.

Bowl of food for the afterlife

PIN FASTENERS

The pin was discovered to have a large, globular head. It would have been thrust through a cloak or other garments to hold them together. Pin fasteners are understood by archaeologists to be an alternative to the fibula (pin with a catch).

DECORATIVE BROOCH

The brooch was X-rayed to see beneath the corrosion. This revealed an intricate pattern in gold leaf, which implies that the individual had wealth and status in the community.

Grave dug into the chalk

COMPARATIVE SITES

OSEBERG

OSLO FJORD, NORWAY

GAZETTEER P. 145 – MAP REF. B7

IN 1904, A BEAUTIFUL SHIP from the early 9th century AD was discovered in a Viking burial mound on the west side of the Oslo Fjord in Norway and became known as the Oseberg burial. Because of the waterlogged conditions, the ship's contents, including fabrics and furniture, had been well preserved. The burial contained two bodies: one of a high-status woman, possibly Queen Asa – a powerful ruler known from historical Viking sagas – and the other probably her servant. The surviving timbers provided information about how the ship was constructed. It was a clinker-built ship, with each plank overlapping the one below, and with cleats, or wooden loops, lashed together and used instead of pegs or nails. The ship was 72 ft (22 m) long and 16 ft (5 m) wide and was probably a royal yacht, rather than a warship or merchant vessel. It was accompanied by a large collection of objects, including a wagon, four sleds, and several bedsteads. The ship has been reconstructed and is now on display in a museum near Oslo.

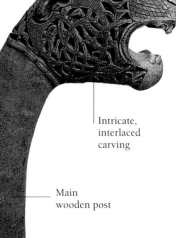

Intricate, interlaced carving

Main wooden post

SHIP'S HEAD
This carefully carved head of a beast is one of five from this ship. This one is rather conservative in style, although the carver has shown great skill. All the heads depict fierce-looking animals with snarling mouths. It is probable that the posts were used for some ritual purpose.

INSIDE THE BURIAL SHIP

The body, thought to be that of Queen Asa, was found surrounded by a wagon, four sleds, a saddle, several beds and tents, and a carved chest. The skeletons of ten horses and two oxen were also found in the burial. The red, black, and yellow tapestries found in the chest depict wagons and horses and provide further evidence of the Vikings' traveling tradition. All the wooden objects, including five posts (see above), were intricately carved with animal heads and geometric patterns. Their purpose is unknown, but they may have been intended to ward off evil spirits. Oseberg is the richest ship burial ever discovered, and the large quantity of high-quality wooden objects demonstrates the variety and skill of Viking craftsmanship.

SHIP UNDER EXCAVATION
This picture of the ship under excavation in 1904 shows its excellent state of preservation. The decking and the beautifully carved prow of the ship are clearly visible.

BURIAL CHEST
An oak chest made from a single log and decorated with silver rivets and iron bands was found on the ship. It had an elaborate locking system and contained a set of tools.

Silver rivets

Animal-shaped iron rod

L'ANSE AUX MEADOWS
NEWFOUNDLAND, CANADA
GAZETTEER P. 137 – MAP REF. P5

THIS SITE ON THE northernmost tip of Newfoundland is the only archaeological evidence we have that the Vikings reached the American continent. The historical sagas of the Norse Vikings describe the travels of Leif the Lucky, son of Eric the Red, in about AD 1001. It is thought from these descriptions that he set off from Greenland, sailed to Baffin Island and Labrador, and finally reached Newfoundland, which he called "Vinland." There was no archaeological evidence to support these historical sources until the discovery of the site at L' Anse aux Meadows. Helge and Anne Ingstad discovered the settlement in the 1950s using local information and the old sagas. They excavated several turf-built structures of unmistakably Norse character, overlooking Epaves Bay. Radiocarbon dating of the artifacts (of which there were only around 150), and the presence of metals that were not used by indigenous groups, suggest a Viking settlement. The buildings had turf-built walls and roofs, and the door frames were made of wood. Inside the houses, there were hearths and benches. Stray Viking finds, such as coins, have been discovered in eastern North America, but these suggest only trading activities or temporary settlements. Only with further fieldwork to locate other sites or finds of imported objects from Scandinavia will it be possible to assess the degree of contact between the Vikings and Native American groups.

THE LANDING SITE
When the Vikings reached what they called "Vinland," now identified as Newfoundland, they described an island with a projecting cape and a well-wooded coastline. This picture shows the landing site, near the turf houses, at the northernmost point of the cape.

RECONSTRUCTED VIKING HOUSES
Some of the eight structures at L'Anse aux Meadows have been reconstructed by archaeologists, using excavated evidence as a guide. The houses were grouped in pairs, and were built from squares of stacked turf, with wooden lintels and doors that were probably made of flotsam.

POUAN
AUBE, FRANCE
GAZETTEER P. 146 – MAP REF. G5

AFTER THE COLLAPSE of the Roman Empire in continental Europe in about AD 476, several groups emerged as powerful rulers in its former provinces. These groups had existed under Roman government and had settled within and around the fringes of the empire. No longer under imperial control, they became kingdoms in their own right. Groups like the Franks and Burgundians in Gaul (modern-day France), the Vandals in North Africa, and the Visigoths in Spain filled the political vacuum left by the fall of Rome and fought one another for control of large territories.

Pouan is a Merovingian cemetery near the town of Troyes and dates from between AD 450 and 500. The Merovingians were a warlike group of Franks, whose distinctive culture spread across France in the 5th century. Pouan is a site that dates to earlier Merovingian history, from the time of the first kings, Childeric (d. 481) and his son, Clovis (465–511), who are known about from historical documents.

The graves at Pouan contained decorated swords and scabbards, along with many jewelry items, such as brooches and gold pins. Such artifacts are very important to archaeologists because they demonstrate how fashions and culture changed from the Roman traditions and ideas to new, medieval ones. For example, jewelry, such as the fibula brooch (right), are Roman in design, but Frankish in decoration and style. They also tell us about the wealth and trading contacts of their owners.

In about AD 500, 11 years before his death, Clovis and his people converted to Christianity. By this time, the Merovingian kingdom was one of the strongest powers in Europe, rivaling Theodoric's Ostrogothic kingdom at Ravenna (see pp. 88–9). This blend of Christian beliefs, mixed with elements of Roman and post-Roman cultures, laid the foundations of early medieval Europe, with its emerging nation-states.

MEROVINGIAN SWORD

Colored enamel inlay

JEWELRY
The cemetery contained many richly decorated items of jewelry, including this fibula brooch with its brightly colored enamel inlays. It would have been worn as a dress fastening, probably on the shoulder, to keep folds of material together. The Merovingian culture is famous for the high quality of its craftsmanship and the richness of its jewelry and weapons.

FIBULA BROOCH

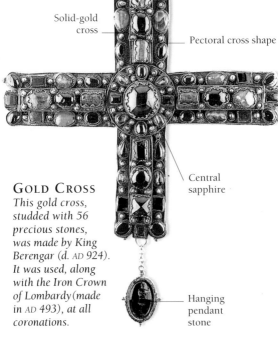

GOLD CROSS
This gold cross, studded with 56 precious stones, was made by King Berengar (d. AD 924). It was used, along with the Iron Crown of Lombardy (made in AD 493), at all coronations.

Solid-gold cross

Pectoral cross shape

Central sapphire

Hanging pendant stone

RAVENNA
EMILIA-ROMAGNA, ITALY
GAZETTEER P. 147 – MAP REF. 16

AN ANCIENT CITY IN NORTHERN ITALY, Ravenna illustrates the complex political and cultural struggles that took place across Europe in the wake of the collapse of the Roman Empire. Ravenna had been an important Roman port and the headquarters of the imperial fleet. After the invasion of Italy by the Visigoths in AD 402, Ravenna became the capital of the western empire. Soon the focus of the Roman Empire shifted to Constantinople, and it was during this period that the Ostrogothic king Theodoric (AD 493–526) became a key figure. He was a general under the empire, but a rival for supreme power in Italy, and was regarded as a king by his army. Ravenna fell under the Byzantine rule of Justinian (AD 482–565) in AD 540, and many impressive buildings were constructed.

CULTURAL CENTER

While recent excavations have revealed remains of an ancient settlement and foundations from its time as a Roman port, it is from its most turbulent period that the most lasting monuments remain. Theodoric, who stationed his army in Ravenna, built and modeled much of the architecture on Roman styles, while remaining Germanic in his cultural outlook. For example, his mausoleum outside Ravenna is built in a Roman style, even though the dome, built out of a single block of limestone weighing 330 tons (300 tonnes), covers the ruler in a Germanic fashion. Theodoric built St. Apollinare Nuovo in AD 490. It was named after the first Bishop of Ravenna, Apollinaris, and its interior is famous for its two rows of mosaics showing gifts being given to Christ and the Virgin Mary. The Neonian Baptistry, built near some Roman bathhouse remains, is the city's oldest monument and has an impressive mosaic of the baptism of Jesus.

In AD 540, Ravenna was recaptured by the armies of the Byzantine emperor Justinian, and it became an outpost of the imperial court and a center of political, cultural, and intellectual life. There are many churches with finely made and highly colored mosaics in Ravenna. The most impressive Byzantine structure, however, is the church of St. Vitale (AD 547), which has a number of impressive mosaics of Justinian, his queen, Theodora, and their household guards. One famous scene shows Theodora and her attendants bestowing gifts on the church. The early Christian churches of Ravenna show that despite having been shaped by many cultures, the architecture and ideas of Roman society still exerted a very strong influence on the people who followed.

MOSAIC OF AN EMPEROR
This mosaic of the emperor Justinian is found in St. Apollinare Nuovo, which was built in AD 490. Justinian ruled the Eastern Roman Empire in Constantinople from AD 527 to 565.

ST. APOLLINARE, CLASSE
Built on the site of a previous temple to the Roman god Apollo, construction of this basilica was started by the Ostrogoths in AD 535 and was completed under the rule of Emperor Justinian.

Window columns

Campanile (bell tower)

Portico across the west front of the church

Main church

THE MEDIEVAL AGE

EMERGING AFTER THE STATES and chiefdoms of the Dark Ages, the medieval period laid the foundations for many modern European states. At the same time, the Mediterranean region and western Asia were dominated by the Byzantine and Islamic Empires, which have left distinct remains. In western Europe, many nations were based on the rigid hierarchy of the feudal system, from the king and his lords down to the peasants, while a separate Christian hierarchy existed, with the pope at its head. Thus medieval society was divided into three distinct groups – those who fought (the upper classes), those who prayed (the clergy), and those who worked (the peasants). These three types are reflected in the castles, churches, monasteries, and villages still remaining. While trade and crusades brought many developments, by the 16th century, changes in religious and scientific thought and a growth in world trade signaled the end of this way of life.

MEDIEVAL PEASANTS
Many people in the medieval world were peasant farmers, who lived in the countryside and worked the fields to supply food for themselves and the rest of society. Medieval agriculture was very labor intensive, with many jobs to be done throughout the year.

NORWAY

SCOTLAND

IRELAND

BRITISH ISLES

① HYLTON
A typical example of a late-medieval castle, decorated with elaborate heraldic designs.

② OXFORD
In the 12th century, Oxford University became one of the world's most prestigious centers of learning and attracted many European students.

ENGLAND

Thames

ENGLISH CHANNEL

Seine

Rhine

③ CANTERBURY
This town became a pilgrimage center after the death of Archbishop Becket in 1170. This reliquary held his bones.

④ PARIS
A medieval city and cultural center, dominated by its monasteries, the university, and the royal palace.

Loire

FRANCE

⑩ CLUNY
The wealthiest and most influential monastery of the medieval world, founded in AD 909.

⑤ SAUMUR
Towering high above its town, the Château de Saumur was built in the 14th century by King Louis I of Anjou.

Alps

⑨ CARCASSONNE
The citadel of Carcassonne was built as a strategic fortress by the Trencavels and flourished from the 13th century until the mid-17th century.

⑥ SANTIAGO DE COMPOSTELA
The bones of St. James the Great (who died in AD 44) were found here in AD 813, and it was a thriving pilgrimage center by the 11th century.

Pyrenees

PORTUGAL

SPAIN

⑦ ÁVILA
A well-preserved medieval city with a complete town wall, a cathedral, churches, and houses, built between the 11th and 13th centuries.

Tagus

N

Guadalquivir

⑧ GRANADA
Built on the site of earlier fortifications, the Alhambra was the 14th-century palace of the Islamic king Mohammed I. The beautiful buildings contain a series of gardens and courtyards.

⑫ NOVGOROD

This 9th-century Russian trading center was found in an excellent state of preservation because it had been built largely of wood, which had been preserved by waterlogged conditions.

North European Plain

RUSSIA

POLAND

⑪ ROTHENBURG

This town was rebuilt after an earthquake in 1356, and, as a result, it survives today as a perfect example of a walled medieval city.

GERMANY

⑬ WARSAW

By the end of the 14th century, Warsaw was a fast-growing town with a castle, a hospital, and a city council.

Carpathian Mountains

⑭ BUCHAREST

Known as the Paris of the east, Bucharest expanded in the 15th century, when Vlad the Impaler built his many palaces there.

Danube

⑮ VENICE

Venice was a medieval trading center based on hundreds of tiny islands. It grew up while supplying the Fourth Crusade, and became famous for the production of high-quality glass, such as this early 14th-century beaker.

ITALY

MAPPA MUNDI

The above map was drawn in 1290 by Richard of Haldingham and measures 5½ x 4½ ft (1.65 x 1.35 m) with north to the left. The map is a symbolic representation of the medieval world and is centered on Jerusalem, the religious focus of the Christian kingdoms. Many mythical beasts are shown inhabiting the outer edges of the map, where the world was believed to end.

BLACK
SEA

⑯ CONSTANTINOPLE

By 1453, the Byzantine Empire collapsed and Constantinople became a Muslim city, with the Blue Mosque and Topkapi Palace built among the Roman remains.

TURKEY

Taurus Mountains

SYRIA

GUTENBERG PRESS

The first printing press was made by Johannes Gutenberg in Germany in the 1450s, combining several earlier inventions. Initially Bibles were printed, followed by Classical and scientific literature. Printing brought literature to a wide audience and contributed to the Renaissance – the cultural revolution marking the end of the medieval period.

⑰ KRAK DES CHEVALIERS

A "state-of-the-art" crusader castle begun in 1144 by the military Knights Hospitaller, it was positioned to defend the road from Antioch to Jerusalem, so that pilgrims could travel more safely. It was eventually abandoned in 1271.

MEDITERRANEAN SEA

A MEDIEVAL CASTLE

HYLTON, UNITED KINGDOM

GAZETTEER P. 144 – MAP REF. G5

SITE EVIDENCE

🔎 Geophysical survey and evaluation trenches had located the additional missing buildings of the castle.

🔎 An earthwork survey of the area surrounding the castle had revealed the lost gardens, ponds, and park laid out during the Tudor period.

🔎 The heraldic evidence visible on the extant building provided additional information on the owners of the castle.

ONE OF THE MOST CHARACTERISTIC FEATURES of medieval life was the fortified residences of landowners, known as castles, which dominated the landscape of much of Europe. Many castles began as timber and turf-built structures, gradually being replaced by stone. Further modifications were required as warfare became more complex, with attackers using stone-throwing machines and cannons. However, castles were residences as well as fortresses, and, by the end of the Middle Ages, many had become ostentatious homes rather than true military bases. By this stage many had palatial wings and formal gardens, with only token defenses. Hylton Castle was the ancestral home of the Hyltons, and was built by Sir William Hylton between 1390 and 1430. All that survives of the castle today is the elaborate gatehouse that contained the Hylton family residential complex. Archaeologists, therefore, wanted to establish the position of the rest of the castle and its grounds.

THE SITE

The surviving building was examined by analyzing the stonework and by studying old prints and drawings. It appeared that the gatehouse was built as a large, complex, self-contained residential tower in the form of a gatehouse and later evolved into a Tudor house. A geophysical survey and evaluation trenches were used to try to locate the surrounding buildings. A full earthwork survey was carried out to discover whether the features were associated with the castle.

Areas of excavation

Elaborate battlements

GEOPHYSICAL DATA

The area to the west of the gatehouse was surveyed first, because it was believed to be the location of the castle gardens. However, the geophysical data showed a graveled area, probably the castle courtyard, rather than any gardens. The survey was expanded to the east, and this showed a number of wall lines, probable cobbled areas, and more extensive evidence of gardens.

Possible drive leading up to castle gates

High resistance indicating cobbled area or gravel.

N

Vaulted foundation (cellars of 18th-century extension)

THE CASTLE

This view shows the surviving tower, formerly the gatehouse, from the northwest. The elaborate battlements with complex crenallations and the remains of stone statues can be seen.

KEY

▫ High resistance: rubble or natural features

▪ High resistance: walls or gardens

▫ Extant buildings

|————————| 33 ft (10 m)

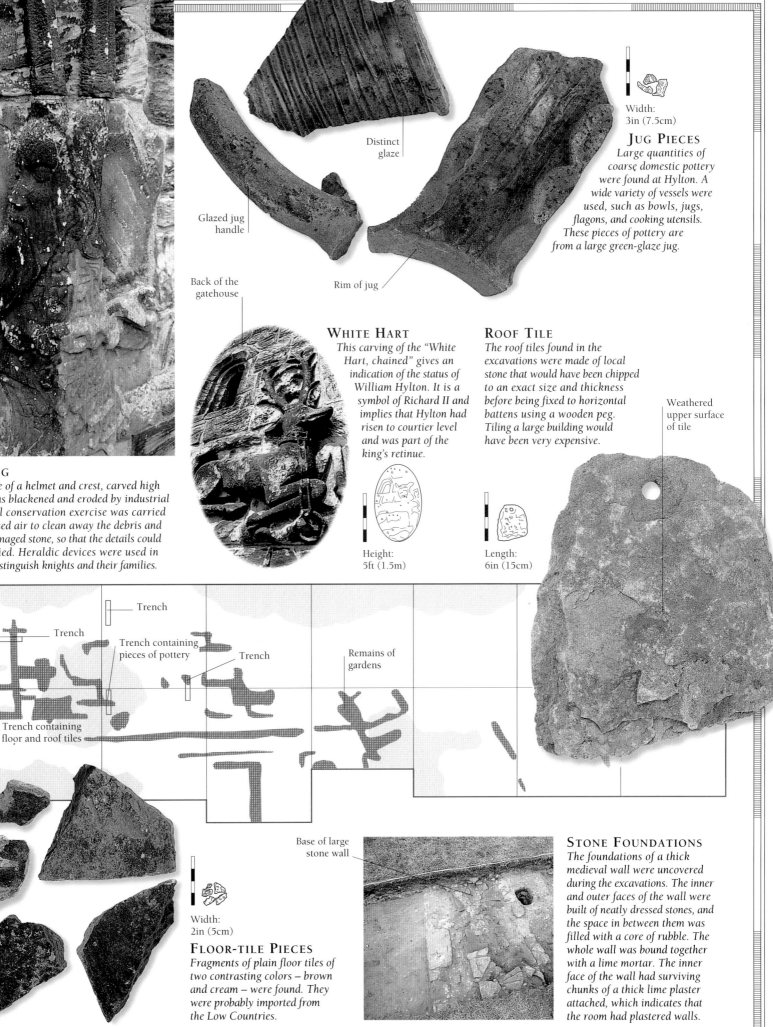

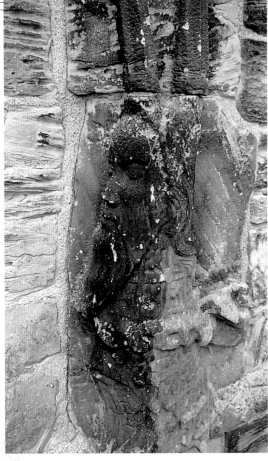

Width:
3in (7.5cm)

JUG PIECES
Large quantities of coarse domestic pottery were found at Hylton. A wide variety of vessels were used, such as bowls, jugs, flagons, and cooking utensils. These pieces of pottery are from a large green-glaze jug.

Distinct glaze

Glazed jug handle

Back of the gatehouse

Rim of jug

HELM CARVING
This heraldic device of a helmet and crest, carved high up on the tower, was blackened and eroded by industrial pollution. A careful conservation exercise was carried out, using compressed air to clean away the debris and outside layers of damaged stone, so that the details could be drawn and studied. Heraldic devices were used in medieval times to distinguish knights and their families.

WHITE HART
This carving of the "White Hart, chained" gives an indication of the status of William Hylton. It is a symbol of Richard II and implies that Hylton had risen to courtier level and was part of the king's retinue.

Height:
5ft (1.5m)

ROOF TILE
The roof tiles found in the excavations were made of local stone that would have been chipped to an exact size and thickness before being fixed to horizontal battens using a wooden peg. Tiling a large building would have been very expensive.

Length:
6in (15cm)

Weathered upper surface of tile

Trench

Trench

Trench containing pieces of pottery

Trench

Remains of gardens

Trench containing floor and roof tiles

Tile shards

Width:
2in (5cm)

FLOOR-TILE PIECES
Fragments of plain floor tiles of two contrasting colors – brown and cream – were found. They were probably imported from the Low Countries.

Base of large stone wall

STONE FOUNDATIONS
The foundations of a thick medieval wall were uncovered during the excavations. The inner and outer faces of the wall were built of neatly dressed stones, and the space in between them was filled with a core of rubble. The whole wall was bound together with a lime mortar. The inner face of the wall had surviving chunks of a thick lime plaster attached, which indicates that the room had plastered walls.

A MEDIEVAL CASTLE

INTERPRETING THE FINDS

FROM THE VARIOUS SURVEYS, DOCUMENTS, and excavations undertaken, and from the existing structural evidence at Hylton, we can begin to reconstruct the appearance of the castle in the later Middle Ages. The grounds are not as easy to interpret because the medieval gardens may not have existed or, at most, may have formed only a basis for the later Tudor landscaping. Although the investigations at Hylton were not extensive, many parallels can be drawn with other similar castle sites, and as a result archaeologists are able to suggest a great deal about the physical appearance of the site and the lifestyle of the lord and his household. Hylton is a typical small castle site, but it is important archaeologically because very few others have had such extensive geophysical and earthwork surveys done to complement the work on the buildings themselves. Combined with the archaeologists' existing knowledge of the medieval period, the finds that were made provide enough information to suggest the day-to-day lifestyle of the inhabitants of the castle. Hylton is unique in the quality and elaboration of its architectural and heraldic detail, and following examination and cleaning of the stonework, the ostentatiousness of its owners can be fully appreciated.

JUG
From the fragments of pottery found, we can reconstruct the type and size of the vessel, along with its decoration and height. This type of jug would have been used to serve wine, ale, and water during mealtimes.

Thatched or shingled roofs

ARTIST'S IMPRESSION
From the limited excavation and geophysical survey, and by comparison with similar sites, we can suggest the layout of the castle at Hylton. Beyond the gatehouse would have been a courtyard surrounded by domestic buildings. The gatehouse itself would have been a private residence for the lord and his family, while opposite this, across the courtyard, would have been a large hall used for communal feasting and entertainment. The rest of the courtyard buildings would have included private rooms, kitchens, and storerooms.

Gatehouse tower with the lord's private accommodation

Hylton coat of arms

Portcullis

THE HYLTON HELMET
After cleaning, the details of the stone carving were revealed and could be studied. The carving appeared to show a helmet decorated with the head of Moses wearing horns. This unusual image stems from a mistranslation of the Bible. The Book of Exodus describes Moses as coming down from Mount Sinai in a beam of light, and the Hebrew word for beam of light is very similar to the Latin word for horns. The carving would originally have been painted in bright colors, and would have been highly visible.

Long, approaching driveway

Stables and barns for visitors' horses

ROOF TILES

Stone roof tiles were used on the main hall. They would have been coursed and overlapped, with the largest at the eaves and the smallest at the ridge, to produce a weather-tight roof. The weight of the tiles suggests that the roof was heavily reinforced with timber.

Possible range of private rooms

Top table

Hanging tapestries

Main dining hall

FLOOR TILES

In England, tiles were used mainly in churches, castles, and palaces. These tiles, imported from the Low Countries, were much larger and thicker than English ones, with no pattern and a single-colored glaze.

Screened passage

Kitchen

Service end of the hall

Porch entrance to screens passage

Window with stone surround

Possible range of service buildings

STONE FOUNDATIONS

The width of the wall and the size of stones in the foundations give some idea of the dimensions and height of the buildings, and indicate that it was a solidly built, large-scale medieval hall.

ROYAL LIVERY

William Hylton probably achieved his courtier status during the Hundred Years' War with France (1284–1415). During this time he would have gained a retinue of associates, who would have been bound to him in return for certain privileges and would have worn a livery (uniform) to indicate their allegiance. Hylton himself was part of Richard II's retinue (as indicated by the White Hart carving) and would have worn the king's livery. This detail from the *Wilton Diptych* shows Richard II kneeling before the Virgin Mary and wearing a badge and a cloak embroidered with the White Hart pattern. John the Baptist stands above holding a lamb.

John the Baptist

King Richard II

The White Hart

THE WILTON DIPTYCH (DETAIL)

CONCLUSIONS

🔎 Geophysical survey and evaluation trenches revealed evidence of substantial buildings near the surviving gatehouse tower.

🔎 The excavated finds indicate that some rooms had tiled floors and plastered walls, under solid, stone-tiled roofs.

🔎 The heraldry provides a striking, pictorial evidence of the ancestry and powerful connections of the lords of Hylton castle.

🔎 The earthwork survey also revealed later gardens of Tudor date.

COMPARATIVE SITES

CLUNY

SAONE-ET-LOIRE, FRANCE

GAZETTEER P. 146 – MAP REF. G6

CLUNY WAS FOUNDED by William of Aquitaine in AD 909, in the Burgundy region of France. Although under the direct control of the pope, Cluny was in effect a self-governing institution and spawned about 2,000 daughter monasteries all over Europe. It was the focus for the donation of property and privileges by a wide variety of people, and, as a result, it acquired large amounts of wealth and vast estates. These estates required efficient management and administration to ensure that the large numbers of monks and servants could be supported from the supplies they produced. It soon became one of the largest, wealthiest, and most powerful monasteries in medieval Europe, a fact that is reflected in the remaining buildings.

Elaborate cross

Transept tower

Tower with Romanesque detail

Latin inscription

Figure playing lyre

CARVED CAPITAL
This capital of a pillar from the church at Cluny dates from the 12th century. The architecture of the abbey was famed for its extravagance, and the decoration often depicted detailed scenes from the Bible and everyday life. The capital would have been painted in bright colors.

THE MEDIEVAL ABBEY

Archaeological study of monasteries such as Cluny can provide a great deal of information about the lifestyle of their inhabitants. It can also tell us about the domestic arrangements of these great economic institutions and the activities and landscapes that were necessary to support them. However, very little survives of the vast complex of medieval buildings that formerly existed at Cluny. Parts of the west tower and the great west doorway remain, but the most impressive portion, and the one that gives the best impression of what the abbey was like during the Middle Ages, is the south transept. Professor Kenneth Conant has painstakingly reconstructed, on paper and in models, the original arrangements within the abbey, using surviving architectural fragments, documentary sources, and selective excavation. He has, for example, been able to show that the second monastic church

at Cluny, which stood between 1010 and 1050, was retained in part within the later cloister when the new abbey church was built to the north after 1088.

MONASTIC LIFE

The architecture of the monastery reflects its wealth and the rich lifestyle of its inhabitants. By the 11th century, the church had been rebuilt on a massive scale for the third time. In addition to the religious structures, there were domestic buildings, including storage rooms, workshops, stables, yards, kitchens, and guesthouses, so that Cluny would have looked more like a small, bustling town than a place of worship. Its extravagance was well known throughout Europe and not always approved of. Indeed, some subsequent monasteries deliberately avoided reproducing its decorative architecture and extravagant lifestyle. Despite the success of these austere orders, such as the Cistercians, Cluny remained influential throughout the medieval period, and, at its height, its liturgy, architecture, and wealthy estates were recognized throughout the continent. The vast complex of buildings survived until the French Revolution (1789), when it was almost completely destroyed.

THE ABBEY RUINS
Little of the original abbey remains after most of it was destroyed in the French Revolution. However, the impressive Romanesque south transept tower still exists, and soars above the later buildings.

KRAK DES CHEVALIERS
QALAT AL-HISN, SYRIA
GAZETTEER P. 148 · MAP REF. E3

FROM THE END OF THE 11TH CENTURY, knights and soldiers from western Europe traveled to the Holy Land, under the blessing of the papacy, to try to take the area from the Muslims and establish a Christian kingdom. A number of military orders of warrior-monks were, therefore, set up to fight the Muslims and protect and support pilgrims traveling to Jerusalem. The Knights Hospitaller, or Knights of the Order of St. John, initially cared for pilgrims in Jerusalem, but soon became involved in fighting – building or taking over existing castles, such as Krak des Chevaliers in Syria. This was originally a Muslim stronghold, but was granted to the order by Raymond II of Tripoli in 1144. The fort was strengthened by the incoming knights and withheld many sieges until it was regained by the Muslims in 1271.

THE CRUSADERS IN BATTLE
This manuscript shows heavily armed crusaders wearing distinctive heraldic designs and fighting the Muslims. Warfare was considered perfectly compatible with Christianity in the Middle Ages and was even encouraged by religious leaders. There were eight crusades in all, spanning three centuries.

THE CASTLE

The massive walls and concentric ring of towers at Krak des Chevaliers were mostly rebuilt and strengthened in the 13th century, possibly after an earthquake. This was one of the greatest castles in the world and a masterpiece of medieval military design. Inside the defenses it contained nine water cisterns, so it could survive sieges, as well as extensive accommodation for its large garrison, which would have consisted of the knights, monks, their servants, and staff. Because it was owned by a religious order, the castle also contained chapels and a cloister, like other monasteries. The castle survived attacks by Saladin (1137–93), the most famous Muslim military leader, but it fell in 1271, after its defenses had been undermined by a month of intensive tunneling, and probably also because its crop supplies had been burned by its attackers. The Christian forces finally withdrew from the Holy Land in 1291.

CARVED CROSSES
Most medieval castles contained chapels, which were an important focus for a religious stronghold like Krak des Chevaliers. These crosses are carved into the wall of the stairs leading to St. Helena's Chapel in the castle.

High, flanking tower

Main part of the castle

Outer, curtain wall

Battlement tower

VIEW OF KRAK DES CHEVALIERS
The castle stands on top of an isolated hill and has a large complex of buildings. It can be approached on the level only from one direction. The huge inner walls, positioned close to the outer wall, subjected attackers to two lines of counterattack.

CARCASSONNE

LANGUEDOC-ROUSSILLON, FRANCE

GAZETTEER P. 146 – MAP REF. G7

AT FIRST GLANCE, THE TOWN of Carcassonne appears to be one of the best-preserved medieval cities in Europe, with its castle, cathedral, numerous buildings, and magnificent fortifications. In fact, much of the city was restored in the mid-19th century by the architectural historian Eugène Viollet-le-Duc (1814–79) and, as such, retains the plan, form, and atmosphere of a thriving medieval city. It was originally a Gallo-Roman site, and some of the fortifications of this period survive as the basis of the later architecture, most of which dates to the 13th century. The town was important in medieval times, since it controlled the main trade route between the Mediterranean and the Atlantic and dominated the economy of the local region.

THE WALLED CITY
The controversial restoration of Carcassonne in the 19th century has left the town looking like new, with many of the ruins having been reconstructed. This view shows the citadel looming magnificently above the surrounding landscape, as it would have in medieval times.

PORTE NARBONNAISE
Built in 1280, these huge sandstone defensive towers mark the main entrance to the citadel. The defenses include portcullises, iron doors, a moat, and a drawbridge.

Windows along battlements

A MEDIEVAL FORTRESS

The roughly oval site of Carcassonne, which had first been occupied by the Iberians as early as the 5th century BC, was chosen by the Romans in the 1st century AD, and was occupied by the Visigoths in the 5th century. After a brief period of Muslim occupation in the 8th century, it was controlled by the Franks, and was under the rule of the counts of Toulouse for 400 years. It suffered two sieges, in 1209 and in 1240; the first one lasted for two weeks, and the city seems to have had only one line of defense at that time. Although the walls were damaged in the second siege, the city was not taken. As a result of the attacks, the city was repaired and reinforced in the mid-13th century and made impregnable.

Stone-built towers

Tile-covered roof

THE SIEGE STONE
The Basilique St.-Nazaire contains a Siege Stone, which depicts the siege of Carcassonne by crusaders in 1209. The crusaders were sent by the pope, who wanted to take the town from the Cathars, whom he deemed heretics.

THE FORTIFICATIONS

The fortifications that protected the town until it ceased to act as a frontier fortress in 1659 have enabled many medieval structures to survive. The turreted outer ramparts were built by King Louis IX and his son Philip the Bold, who added the heavily fortified Porte Narbonnaise to the inner ramparts. Within the city walls, the Château Comtal, originally a palace, was turned into a fortified castle, with five towers and defensive wooden galleries, and was cut off from the rest of the city by a moat and a barbican. Not all buildings had a defensive purpose; the cathedral, the Basilique St.-Nazaire, started in the 11th century by Pope Urban II (1035–99), was mostly rebuilt in the Gothic style at the turn of the 14th century, although the Romanesque nave remains. Remains from the medieval period have been housed in a museum and include Gothic windows, stone missiles, and Romanesque murals.

Scallop shell

PILGRIM BADGE

As a symbol of the completion of a holy mission, medieval people would wear distinctive pilgrim badges. These badges have been found all across Europe, and those from Santiago de Compostela portray St. James and a scallop shell. He is usually depicted as an elderly, bearded man dressed as a pilgrim with a hat and staff.

SANTIAGO DE COMPOSTELA

GALICIA, SPAIN

GAZETTEER P. 146 – MAP REF. E7

ONE OF THE TRADEMARKS of the Middle Ages was the importance placed on the act of pilgrimage. Santiago de Compostela was the third most popular Christian pilgrimage center after Rome and Jerusalem, and pilgrims undertook the long and perilous journey here from all over Europe. A shrine-church was built at the end of the 9th century to mark the site of the tomb of St. James – the apostle traditionally believed to have traveled to Spain. It was not until the start of the 10th century that the cult became established, when William Firebrace (AD 979–1029), Count of Aquitaine, made a famous pilgrimage here, and a monastery and shrine were developed by the Cluniacs. The popularity of the cult spread beyond Spain, and a new church was started in 1076. It was finished by 1122, but was developed over the following centuries, taking in the architectural styles and ideas that accompanied the pilgrims.

A SHRINE OF CHRISTENDOM

Carved halo

By the 12th century, large numbers of pilgrims were making their way to the shrine at Compostela. There were four main routes across France and into Spain: one from beyond St. Gilles; another from Le Puy; Vezelay; or Tours; one from the Low Countries and Normandy; and the last by the sea from the north. All the overland routes converged at the Pyrénées and followed the road through Galicia to Compostela. As the number of pilgrims increased, the route was marked by more and more elaborate Romanesque churches and monasteries, many of them under the influence of Cluny (see opposite page).

The cathedral at Compostela is one of the most impressive and complete monuments of medieval Christendom. The Romanesque church that can be seen today mainly dates from the 11th and 12th centuries, although the richly sculptured Baroque west façade was added in the 18th century. Its twin towers reach a height of 243 ft (74 m) and dominate the Pórtico de la Gloria below, where the pilgrims would have entered. The crypt, which is part of the foundations of the original basilica, is said to contain the relics of St. James. Other important buildings at Compostela include the Hotel de los Reyes Católicos, built as an inn and hospital for sick pilgrims, and the Convento de San Paio d'Antealtares, erected to house the shrine of St. James, now in the cathedral.

St. James's staff

THE SPIRITUAL CENTER

The huge cathedral dominates the medieval town and is still in the center of the town today. The rich and extensive carvings on the 18th-century west front reflect the popularity and investment in the cult.

THE PÓRTICO DE LA GLORIA

The stunning Pórtico de la Gloria (Doorway of Glory), with its statues of apostles and prophets, greeted the pilgrims on entering the cathedral. This sculpture of St. James stands below a carving of Christ.

NEW SETTLEMENTS

FOLLOWING THE DISCOVERY of new land in the Americas, Africa, and Asia by the Europeans in the 15th and 16th centuries, great opportunities for trade appeared, and most European countries pushed for an extension of their economic interests. The search for gold, spices, slaves, and other commodities was led by the Spanish and Portuguese, and France, Britain, and the Netherlands soon joined the fray.

It quickly became apparent that there was more than just trade potential in these new areas. Pressure brought about by population growth across Europe led to a hunt for more land and resources, and new colonies abroad seemed to provide a solution. The initial problem faced by these fledgling colonies was survival. Many people died of disease or hunger within their first year in the colony, and the only work that was available was physically very hard. In addition, the settlers had to rely on the indigenous peoples for advice on local plant and animal resources and for access to the land. This advice was exchanged for goods wanted by the locals – beads, metal tools, and eventually firearms to replace arrows and spears. In effect, often highly technological European societies met head-on with local, essentially Stone Age cultures, and the rights of the indigenous people were generally ignored.

① QUEBEC
One of a number of French forts along the St. Lawrence River, this was a combined settlement, store, and fortress.

② NEW YORK
A Dutch trading post that was taken over by the English in 1664.

③ ST. MARY'S CITY
English Catholic settlers established this city in 1634. The staple crop was tobacco.

④ ST. AUGUSTINE
A colony founded by the Spanish in 1565 after they had slaughtered the established French Protestant settlers.

⑤ ST. KITTS
The first Caribbean island to be settled by the English, in 1623. Sugar harvested by slaves was sent to Europe.

⑥ POTOSÍ
In 1545, the Spanish discovered silver in a Bolivian mountain and established a successful mining settlement there.

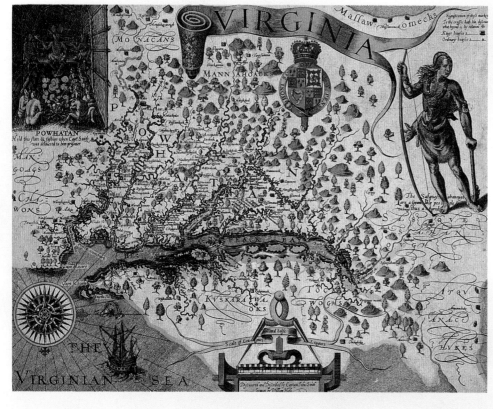

MAP OF VIRGINIA

In 1607, Jamestown in Virginia became the first permanent English settlement in North America. At first, the settlers were plagued by disease and tensions with the native peoples. This map by colonist John Smith (1580–1631) shows Virginia in 1624, by which time some stability had been achieved.

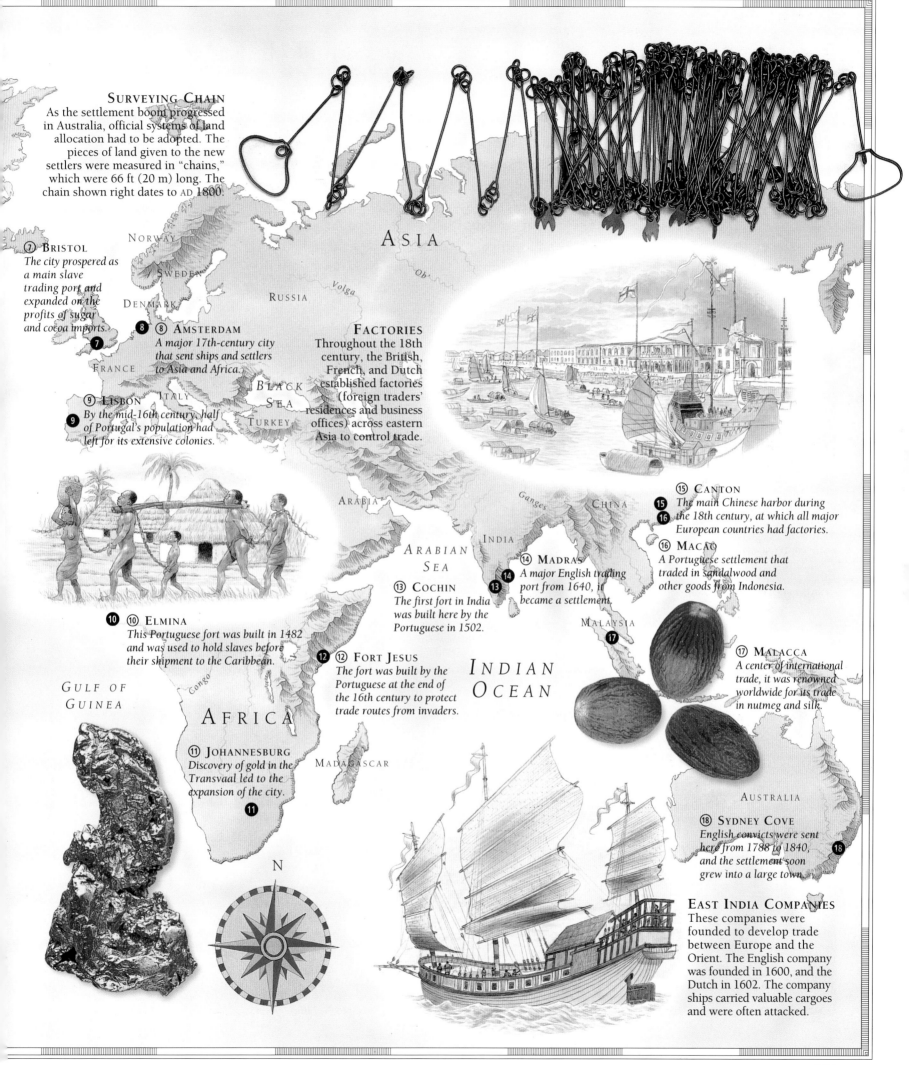

SURVEYING CHAIN
As the settlement boom progressed in Australia, official systems of land allocation had to be adopted. The pieces of land given to the new settlers were measured in "chains," which were 66 ft (20 m) long. The chain shown right dates to AD 1800.

ASIA

⑦ BRISTOL
The city prospered as a main slave trading port and expanded on the profits of sugar and cocoa imports.

⑧ AMSTERDAM
A major 17th-century city that sent ships and settlers to Asia and Africa.

⑨ LISBON
By the mid-16th century, half of Portugal's population had left for its extensive colonies.

FACTORIES
Throughout the 18th century, the British, French, and Dutch established factories (foreign traders' residences and business offices) across eastern Asia to control trade.

⑮ CANTON
The main Chinese harbor during the 18th century, at which all major European countries had factories.

⑯ MACAO
A Portuguese settlement that traded in sandalwood and other goods from Indonesia.

⑭ MADRAS
A major English trading port from 1640, it became a settlement.

⑬ COCHIN
The first fort in India was built here by the Portuguese in 1502.

⑩ ELMINA
This Portuguese fort was built in 1482 and was used to hold slaves before their shipment to the Caribbean.

⑫ FORT JESUS
The fort was built by the Portuguese at the end of the 16th century to protect trade routes from invaders.

⑰ MALACCA
A center of international trade, it was renowned worldwide for its trade in nutmeg and silk.

GULF OF GUINEA

AFRICA

⑪ JOHANNESBURG
Discovery of gold in the Transvaal led to the expansion of the city.

INDIAN OCEAN

AUSTRALIA

⑱ SYDNEY COVE
English convicts were sent here from 1788 to 1840, and the settlement soon grew into a large town.

EAST INDIA COMPANIES
These companies were founded to develop trade between Europe and the Orient. The English company was founded in 1600, and the Dutch in 1602. The company ships carried valuable cargoes and were often attacked.

N

A COLONIAL SETTLEMENT

ST. MARY'S CITY, UNITED STATES

GAZETTEER P. 137 – MAP REF. N7

SITE EVIDENCE

○ Documents refer to St. Mary's City as the first city of Maryland and indicate where the settlement was located.

○ Evidence of a large brick building had been obtained from old documents and from an earlier incomplete excavation by Henry Chandlee Forman in 1938.

○ Field surveying around the area of the Merchant's House had produced surface finds of oyster shells, bricks, and pottery.

ON MARCH 27, 1634, after four months at sea, a group of 140 English Catholic colonists entered Chesapeake Bay on America's east coast. After an initial exploration of the Potomac River, a bargain was struck with the local Native Americans, the Yaocomico, and the settlers began to build a storehouse and fortified guardhouse. This new settlement was named St. Mary's, and, over the following 61 years, the town developed into the capital of Maryland, with the colonists building houses, meeting halls, and a chapel with a cemetery. The original settlement area has been preserved by the opening of a colonial museum, and archaeologists have already uncovered artifacts, burials, and various building foundations. By following up an earlier excavation in the area known as St. Peter's, archaeologists hoped to discover more about the governor's mansion and to be able to place it in context with the other sites, such as the cemetery and the fort.

THE SITE

With relatively little development having occurred in the area, large parts of the site had been left undisturbed. Many of the fields were open to geophysical survey, and previous field surveys had produced several objects. Also, by using early maps and surveys, it was possible to narrow down the targets for excavation. The relative rarity of surface finds, most of which appeared in the top layer of the soil, meant that all the material from the trenches would have to be screened.

DIGITAL MAP OF THE AREA

With such a wide-ranging and multifaceted site, it was important to get an overall view of the different areas in relation to one another. There did not appear to be any obvious center to the settlement, but by using GPS (see p. 123) the archaeologists hoped to be able to predict the layout of the town. This map, with contours and modern roads marked, helped the archaeologists choose possible areas for investigation.

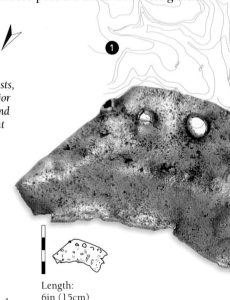

① ST. JOHN'S
Previously excavated by archaeologists, this site revealed one of the first major 17th-century buildings in the area and gave the archaeologists some insight into the architecture of the period. It also demonstrated how little European architectural styles had been adapted for frontier conditions. The house had been constructed in 1638 by John Lewgar, the first Secretary of State for the colony, and was later occupied by Charles Calvert, the 3rd Earl of Baltimore.

THE ARRIVAL OF THE COLONISTS
Paintings and engravings of the period helped provide further information on the settlers, including details of dress and, in particular, the appearance of the indigenous people. This painting shows the landing of the Maryland colonists with Leonard Calvert, the son of Lord Baltimore, who had initiated and organized the emigration.

Length:
6in (15cm)

COPPER FRAGMENT
Copper was rare in North America, so the copper objects brought by the settlers became valuable as sources of material for the Native Americans. They were usually recycled by the local peoples into new objects, such as this scraper found at the St. John's site.

SKELETON

Excavation in the cemetery adjoining the chapel area revealed a number of skeletons and over 500 graves. This skull had evidence of a pipe facet (a groove worn into a tooth), created by sucking on a pipe stem for a long period.

Pipe facet

Fused lead

MELTED SHOT

The settlers stored guns, gunpowder, and lead shot in the cellar of St. Peter's, where there was an explosion in 1694. This piece of melted lead shot was found during the excavations.

Length: 5in (15cm)

Exterior wall

GEOPHYSICAL IMPRESSION OF ST. PETER'S

The geophysical survey revealed a clear square pattern, and strong signals implied an explosion that had wiped out the random magnetic field and created a uniform one.

② ST. PETER'S SITE ❷

Excavations at this site revealed substantial brick foundations, which suggested an elaborate mansion. The way that the bricks were positioned gave an indication of the direction the building was facing.

③ CEMETERY SITE

The cemetery, which had already been partially excavated, was investigated further using geophysics. The results suggested one of the largest colonial cemeteries in America. ❸

④ LEONARD CALVERT'S HOUSE

Occupied by the first chancellor and governor of the colony, this house was the first state building and the core of the developing settlement. ❹

⑤ MERCHANT'S HOUSE

Located by geophysics while searching for evidence of a fort, this house had a brick chimney base and produced finds including English pottery, a Dutch sculpture, and a Native American pipe. ❺

Face decoration

Quartz stone

Length: 2½in (6cm)

PIPE BOWL

Several Native American pipe bowls were found across the site, including this one with a face decoration, found at the Merchant's House. It may have been traded with the settlers for other goods such as hoes or axes.

PROJECTILE POINTS

Made from local quartz, these projectile points, found at St. Peter's, reveal the level of technological advancement of the Native Americans.

Length: 2in (5cm)

Metal bell

Riveted edge

BELL

Found during the excavtion of the St. John's site, this bell was probably intended for use in falconry, where it would have been attached to the leg of a bird of prey. However, these bells proved to be attractive objects to the Native Americans and were then imported in large numbers for trading purposes.

Width: 1in (3cm)

Details modeled in plaster

RELIGIOUS SCULPTURE

This statue of a Madonna's head was found in the Merchant's House and is likely to have belonged to a Dutch man named Van Sweringen. Similar examples of religious sculptures have been found in Amsterdam in the Netherlands.

Height: 2in (5cm)

A COLONIAL SETTLEMENT

INTERPRETING THE FINDS

BY ANALYZING THE VARIOUS ARTIFACTS and foundations found at the different sites, archaeologists were able to learn a huge amount about the living conditions and lifestyles of the colonists. Locating the brick walls of what may well have been one of the first and finest brick buildings in the New World was the most exciting achievement for the archaeologists. The details of its structure, generated by analysis of the foundations and from archival research, created a picture of an elaborate building made of local bricks with terra-cotta roof tiles. The construction of St. Peter's (probably from 1670 to 1680) proves that the colonists were comfortably settled in their town and had begun to establish a form of local government. This confidence was due, in part, to their good relationship with the Native Americans, who had provided guidance on the growing of tobacco and the wherabouts of local food sources. Tobacco played a vital part in the colony's economy and it soon became the foremost commodity, dictating the economic conditions of the community. This overdependence on a single crop, and the movement of the Maryland capital to Annapolis in 1695, led to the eventual decline of the city.

Individual
tobacco leaves

STAPLE CROP
By 1637, tobacco had become the staple crop of the colony and had become valued as a currency in itself. Colonists traded with it and priced goods according to its value.

ARTIST'S IMPRESSION
Lord Baltimore's dream had been to transplant an English manor and its lord to the New World and to establish a community with an intrinsic social order. The ultimate symbol of this dream is the mansion house of St. Peter's. Research into similar buildings of the same period in England provided some of the necessary architectural detail, and the aerial photographs of crop marks suggested the presence of a walled garden. Details of the brickwork and the discovery of English and Native objects all point to an elaborate, large house in its own grounds.

CHRISTIANITY
The key investors in the settlement were Catholics, who were escaping religious persecution in England. Both the Catholics and the Jesuits brought religious icons with them, and the chapel was one of the first buildings to be built. By the end of the 17th century, the settlers had converted some of the local people.

Field of
tobacco plants

BRICKWORK
Excavations on the site of the mansion revealed substantial brick foundations. The settlers had built the mansion in a distinctive style, using the more attractive Flemish bond on the front of the building and the more common English bond for the rear. This enabled the archaeologists to confirm whether the front or back of the structure faced the rest of the town.

Flight made of
duck, goose, or
eagle feathers

ENGLISH BOND
The bricks are laid with alternate layers of headers and stretchers (end or side on).

FLEMISH BOND
The bricks are laid with alternate headers and stretchers to create a pleasing pattern.

Quiver made
of deerskin

ARROWHEADS
Stone and wooden tools were the main implements of the local people, and these quartz arrowheads would have been attached to shafts with feather flights. The indigenous peoples relied on these weapons for all their hunting requirements.

ARMORY

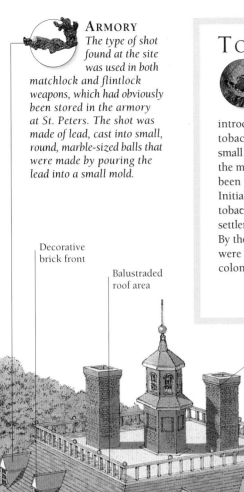

The type of shot found at the site was used in both matchlock and flintlock weapons, which had obviously been stored in the armory at St. Peters. The shot was made of lead, cast into small, round, marble-sized balls that were made by pouring the lead into a small mold.

TOBACCO SMOKING

Used in ceremonies and for medicinal purposes, tobacco was an important element of Native American culture and was soon introduced to the English settlers. The local tobacco was bitter and was smoked only in small amounts, so the settlers chose to cultivate the milder Oranoco variety, which had already been introduced to the area by the Spanish. Initially planted just between tree stumps, tobacco soon became the main crop of the settlement and was planted across large areas. By the 1670s, over 300,000 lb (600,000 kg) were being exported each year, and the colony was thriving economically.

NATIVE AMERICAN SMOKING TOBACCO

Decorative
brick front

Balustraded
roof area

English-style
roof details

Woodland full of
game, including
wild turkey

Corn was the staple
food of the locals and
the settlers

Wooden "worm"
fences were used to
control and protect
the animals

Cattle and other
animals were
raised by
the settlers

Native Americans trading
with the new settlers

TRADING TOOLS

Traders exchanged a wide range of metal objects with the Native Americans, including hoes, axes, and copper kettles. The kettles were sometimes broken up and used for jewelry or scrapers.

TRADING TRINKETS

Along with obvious trading items, such as copper (which was available only in the far north of America), the Native Americans also favored unusual items, including bells and brass objects. These were usually traded for knives and cloth.

CONCLUSIONS

◉ A substantial and sophisticated mansion had been built here at the height of the colony's economic and social stability.

◉ The mansion had been destroyed in an explosion; the subsequent fire was hot enough to fuse the shot and to affect the geophysical signal generated by the bricks.

◉ The site had yielded many objects that implied a trading link between the two coexisting cultures. The Native Americans were using the imported materials and adapting them for different uses.

COMPARATIVE SITES

NEW YORK

NEW YORK STATE, UNITED STATES

GAZETTEER P. 137 - MAP REF. N7

FOUNDED IN 1624 BY DUTCH COLONISTS, New York is one of the oldest cities in the United States and has a long and varied history of manufacturing and trading activities. Its ideal location for settlement and shipping has enabled it to continue to grow until the present day, so most archaeological evidence about its earliest inhabitants has been lost beneath the modern city. Archaeologists must turn to maps and contemporary records to understand the city's early development. The street plan at the southern end of Manhattan Island does preserve the lines of the earliest fort, its narrow, irregular streets in contrast to the later grid plan laid out in 1811. The oldest buildings in the city date to the mid-18th century.

WALL STREET
The initial good relations with the Native Americans did not last long, and the settlers were forced to build a defensive wall around the edge of the settlement. Wall Street, in the heart of New York's financial district in lower Manhattan, is named after the defensive palisade that was built in the area in 1653.

THE FIRST INHABITANTS

This area of the eastern seaboard of the United States was first inhabited by Algonquin-speaking Native Americans, who lived in caves north of the city. The English navigator Henry Hudson (d.1611) was looking for a northern route to the Spice Islands and China when, in 1609, he came across a navigable river (later named the Hudson) and an excellent port.

However, Dutch colonists were the first Europeans to settle here in 1621, which makes New York, then called New Amsterdam, one of the oldest cities in North America. They bought Manhattan Island (then Manahattan) from the Native Americans in exchange for trinkets, and built a settlement at its southern end. New Amsterdam was based on the area that is now lower Manhattan. The twisting streets and the outline of the fort (later called Fort George by the English) are surviving remnants of this period. The settlers used the Native American trails on the island; the main north–south route, lying on higher ground, is now the lower half of Broadway and cuts across the grid system. A Dutch engineer later laid out a grid-line streetplan, dividing up the island and some of Brooklyn. The settlers' economy was based on trading with the Native Americans for furs, which were then taken to Europe by the Dutch West India Company.

ENGLISH CONTROL

The Dutch had settled for trade rather than for colonial expansion, but the combination of the English threat, bad relations with the Native Americans, and internal problems ended their supremacy in the region. In 1664, the English took control and renamed the settlement New York, and it remained in their hands until the Revolutionary War (1776–83), when the federated states broke away from English rule. Shipping remained a major focus for trade and contributed greatly to the rise of New York as a successful modern city.

Despite the problems of excavation in a modern city, relics of earlier times can still be found. Excavations of a federal courthouse yielded foundation walls, cellar floors, courtyards, and almost one million artifacts.

DUTCH DECK GUN
The Dutch ship The Tiger was destroyed in 1613, and its discovery by workmen in 1916 has provided archaeologists with some of the earliest archaeological remains in New York. This breech-loading, swivel deck gun from the ship was one of several found at the site.

FORT GEORGE, 1750
Fort Amsterdam was renamed Fort George by the English in 1664, and the fort continued to control naval access to both rivers, while also defending the settlers against Native American attacks. The haphazard street plan of the present-day financial district reflects the early settlement clustered around the fort for security.

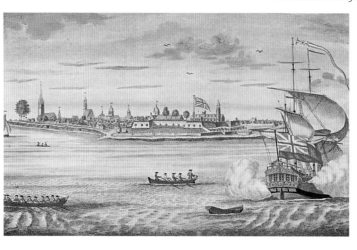

EARLY MAP OF NEW YORK
This map, from 1761, shows the southern half of Manhattan Island and Brooklyn. It indicates the defensive and trading advantages of the island, with New Amsterdam on the southern tip.

FORT JESUS

MOMBASA, KENYA

GAZETTEER P. 142 – MAP REF. G5

IN THE 16TH CENTURY, as European nations expanded into new areas of trade, the east coast of Africa became dominated by Portuguese colonies. Fort Jesus was not primarily a trading port for the exchange of goods, but a military post to control the coastline and protect the shipping route from southern Africa to India, which was often threatened by the Islamic caliphates of the Arabian peninsula. Protection was also needed against the Dutch, who had rival trading routes and wanted to destroy the Portuguese presence in the area. The fort was built of huge rubble and earth ramparts, over 13 ft (4 m) thick, on a coral ridge at the southern end of Mombasa Harbor. It survives in its 16th-century layout with few alterations, and is square in plan, with large bastions at each corner to house gun emplacements. The fort, now a museum, displays artifacts found there during excavations in the early 1960s. It also contains the substantial finds salvaged from a 17th-century Portuguese wreck, which was discovered in the harbor at Mombasa.

A COLONIAL FORT

Fort Jesus was an imposing-looking fortress with a large outer wall that had a sheer drop down to the sea. The walls and a corner bastion of the fort can be seen in the background of this picture. Now a museum, the fort also contains artifacts salvaged from the frigate the St. Antonio de Tanna. These finds include ceramic vessels, barrels, and weapons. The ship itself would have been armed with several swivel guns.

DEFENDING THE FORT

Fort Jesus suffered many attacks and raids during its life because of its important strategic position and close proximity to the town of Mombasa. Muslims, who had settled in Mombasa in the 11th century, led many attempts to destroy Portuguese interests in the area, and the fort was occupied and reclaimed several times, with much fighting.

In 1696, the Sultan of Oman besieged the fort for almost three years. Despite aid from local coastal peoples, disease brought by the reinforcements killed all the Portuguese and left the fort defended by just a handful of local men and women. It eventually fell because most of the garrison had been decimated by disease.

The St. Antonio de Tanna was a Portuguese frigate that had been sent to Fort Jesus in 1697 to aid the besieged inhabitants. However, it ran aground just below the walls of the fort and sank. When the ship was found in the 1960s, archaeologists used its artifacts, as well as historical references, to identify it as the St. Antonio de Tanna. Several seasons of excavation have produced thousands of artifacts, including musket balls and other weaponry, shoes, wooden barrels, and Chinese and Indian ceramics and metalwork. The timbers have been carefully recorded so that the ship's construction and design can also be studied. All this has enabled archaeologists to build a picture of life aboard a 17th-century trading vessel.

EXCAVATING THE WRECK

Timbers and a swivel gun of the St. Antonio de Tanna can be seen sticking out of the seafloor. The archaeologist floats above the wreck and removes mud and sand using a suction pump.

REMOVING THE ARTIFACTS

A diver removes a figurehead from the wreck after it has been recorded in its exact location. This type of find gives the archaeologists an idea of the wealth and status of the ship and its owner.

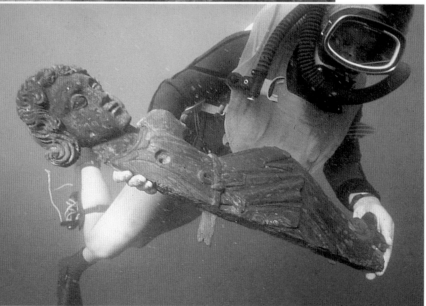

SYDNEY COVE

NEW SOUTH WALES, AUSTRALIA

GAZETTEER P. 156 – MAP REF. E7

WHEN CAPTAIN COOK DISCOVERED BOTANY BAY in Australia in 1770, he suggested to the British government that it might be a suitable location for a penal colony, in order to relieve overcrowding problems in British prisons. However, when the first shiploads of convicts arrived in 1788, Captain Arthur Phillip (1738–1814) decided that Botany Bay was not ideal and instead established the colony at a site farther north. This site had been named Port Jackson by Cook, but was renamed Sydney Cove, after the British Secretary of State at that time. Phillip described Sydney Cove as "the finest harbor in the world," and so more convicts were transported there in 1793. The first free settlers arrived five years later. Prisoners built roads and buildings from whatever materials were available and, out of these, a town developed. Farms were soon established in the surrounding area, and the arrival of free settlers with practical skills and trades led to the growth of a self-reliant, thriving city, no longer just associated with the penal colony. The last convicts were sent to Sydney Cove in 1840, but by then, the city had become established as a successful maritime trading port, with impressive public buildings and a growing population.

Gold sovereign

THE GOLD RUSH

Gold was discovered in Australia in 1851, sparking off a gold rush similar to the one that had taken place in the American West. Large numbers of people soon arrived at mining camps hoping to make their fortune. These two early gold-rush coins date from 1852.

BEGINNINGS OF A MODERN CITY

The natural advantages of the cove for shipping, combined with the discovery of attractive resources such as gold and unclaimed land, hastened Sydney's expansion from its unpromising beginnings. For the archaeologist, continued development presents problems when trying to understand the early history of such settlements, which can be largely lost beneath the modern city. Indeed, the original harbor is now surrounded by dense building. Much of the early settlement's layout and history must be reconstructed from contemporary documents and accounts, and from early maps and plans like the one pictured left. The oldest district of Sydney – the Rocks, where Captain Phillip landed – does survive in part and has been heavily restored as a tourist attraction, containing warehouses and cottages that belonged to the early, developing town.

PROSPECT OF SYDNEY IN 1888

This plan of the harbor shows the extent of development at the end of the 19th century. Sydney had become a thriving city, as shown by the number of steamships sailing in and out of the harbor.

CADMAN'S COTTAGE

This sandstone house is Sydney's oldest surviving building. Built in 1816 as barracks for the crew of the governor's boat, it was named after an ex-convict turned coxswain.

THE INDUSTRIAL AGE

THE INVENTION OF STEAM POWER lay at the heart of the Industrial Revolution, and the ensuing construction dramatically altered the landscape of many countries. The world's first industrial centers grew up close to large coalfields – the Ruhr in Germany, Birmingham in England, and both eastern and western Pennsylvania in the United States all had an accessible and seemingly inexhaustable supply of coal to support their burgeoning industries. Steam power, combined with the new inventions of the age, increased output and led to the new concept of mass production. The precision of the machinery improved product quality, and the interchangeability of mass-produced parts underlined the accuracy and consistency of the machining process. Objects produced by such processes no longer bear the stamp of an individual craftsman. This makes identification difficult, since all the objects manufactured from one type of machine will look the same. Because of this uniformity, and because of the sheer number of objects from this period, archaeologists have relied more on documents and plans to trace the origins and distribution of their finds.

THE FORD FACTORY
Entrepreneur Henry Ford (1863–1947) combined the interchangeability of parts with mass production of components to revolutionize the car industry. Based in Detroit, Michigan, he created a fast, economic process that became a model of industrialization that would spread worldwide.

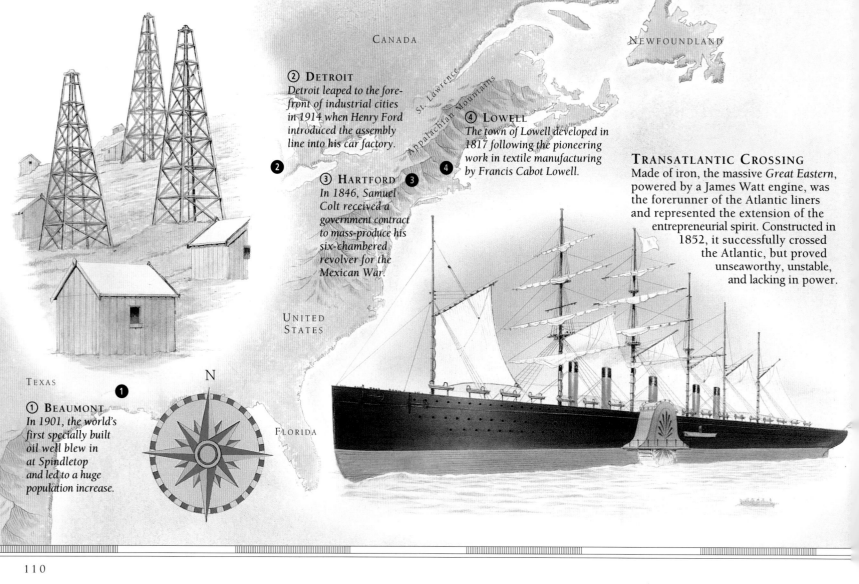

CANADA

NEWFOUNDLAND

St. Lawrence

Appalachian Mountains

② DETROIT
Detroit leaped to the fore-front of industrial cities in 1914 when Henry Ford introduced the assembly line into his car factory.

④ LOWELL
The town of Lowell developed in 1817 following the pioneering work in textile manufacturing by Francis Cabot Lowell.

③ HARTFORD
In 1846, Samuel Colt received a government contract to mass-produce his six-chambered revolver for the Mexican War.

TRANSATLANTIC CROSSING
Made of iron, the massive *Great Eastern*, powered by a James Watt engine, was the forerunner of the Atlantic liners and represented the extension of the entrepreneurial spirit. Constructed in 1852, it successfully crossed the Atlantic, but proved unseaworthy, unstable, and lacking in power.

UNITED STATES

TEXAS

N

① BEAUMONT
In 1901, the world's first specially built oil well blew in at Spindletop and led to a huge population increase.

FLORIDA

MATTHEW BOULTON

Born in Birmingham, Matthew Boulton (1728–1809) was one of the greatest inventors and entrepreneurs of his time. Built on barren heathland, his "manufactory" was the first to use production-line methods. His work on steam engines with James Watt made him a key figure of the Industrial Age. He once told King George III, "I sell, sir, what all the world desires – power."

⑤ STEPHENSON'S "ROCKET"
Built in 1829 by Newcastle-born inventor George Stephenson and his son Robert, the "Rocket" had copper tubes inside its boiler for efficiency, and was capable of 30 mi (50 km) per hour, an astonishing speed for the time.

⑥ LANCASHIRE COTTON MILLS
The mass production of textiles in the mills around Manchester required high-precision, reliable machines. This led to improvements in many industrial processes.

⑦ BIRMINGHAM
The work of Boulton, Watt, and William Murdock – the inventor of gas lighting – made Birmingham a major center of the Industrial Age.

⑩ ESSEN
Rich coal and iron supplies, and, later, a railroad network, made Essen into one of the world's largest industrial centers of its day.

⑧ COALBROOKDALE
Site of the world's first iron bridge, it was here in 1709 that iron was first produced from a coke-fueled furnace.

⑨ DARTMOUTH
The world's first atmospheric steam engine was built by inventor Thomas Newcomen and was first used to extract water from a coalmine in 1712 .

⑪ LE GRAND HORNU
An industrial village built by Henri de Gorge-Legrand that included houses, stores, and offices, using plentiful supplies of local coal.

ATLANTIC OCEAN

THE KRUPPS STEEL WORKS

By 1870, Germany had become the dominant industrial power in Europe by concentrating on steel-making. Much of the later steel production centered on making weapons.

⑫ LE CREUSOT
The site of large-scale pig-iron production in coke-fired furnaces from 1780 onward.

A FACTORY

BIRMINGHAM, UNITED KINGDOM

GAZETTEER P. 144 – MAP REF. G7

IN 1761, MATTHEW BOULTON (engineer, 1728–1809) established one of the world's first factories, the Soho Manufactory, on a site in Birmingham in central England. Initially, the factory made buckles, buttons, and coins using water power. But in 1774 Boulton was joined by James Watt (engineer, 1736–1819), and they developed an advanced form of the steam engine, which improved the production capabilities of the factory. Although no buildings remain, the approximate location of the factory had been established by local archaeologists, using plans drawn up by Boulton and others. The archaeologists needed to locate enough factory remains to place the whole complex on the map. They also had to discover whether any remains had been left intact under the subsequent housing to enable the identification of individual buildings. With these points in mind, the archaeologists chose to concentrate on the mint and the principal building.

SITE EVIDENCE

🔍 From existing documents, the archaeologists knew the approximate location of the factory.

🔍 Plans drawn up by the Board of Health in 1850–5 gave the layout of the factory buildings across the site.

🔍 Surface finds, including bits of shell, had been found in the gardens, and a trial excavation had located a small section of brickwork.

THE SITE
This site was unusual because a lot of existing documentary evidence was associated with it – plans, drawings, and historical writings – so the real challenge facing the archaeologists was to locate the evidence in the ground. This was difficult because of its position in a suburban area, with many of the key buildings being located directly under houses and in people's backyards. Once the local residents had agreed, the archaeologists cleared the gardens and assessed the site with ground radar.

Area of excavation

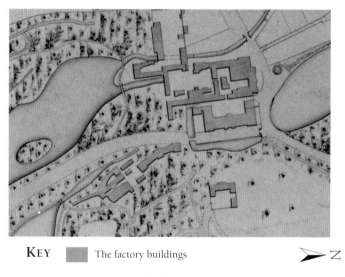

ORIGINAL MAP
This old map shows the position of all the former buildings, as well as natural features that are still visible. Archaeologists can use a dated document to help locate specific buildings, but such a map may be speculative and could be misleading.

KEY ▨ The factory buildings

⊳—Z

 400 ft (120 m)

DOCUMENTARY EVIDENCE
As with many industrial sites, the Soho Manufactory had created a wealth of documentary evidence in the form of plans, engravings (see right), and sales particulars. One key document revealed that the mint's steam engine had been sold in 1850, along with 203 ft (62 m) of pipe that had powered the mint's new cutting-out room. This evidence enabled the archaeologists to locate precisely where the steam-engine room and mint were in relation to each other and to dig a trench in the correct place.

The principal building

Copper patina

The cutting-out room

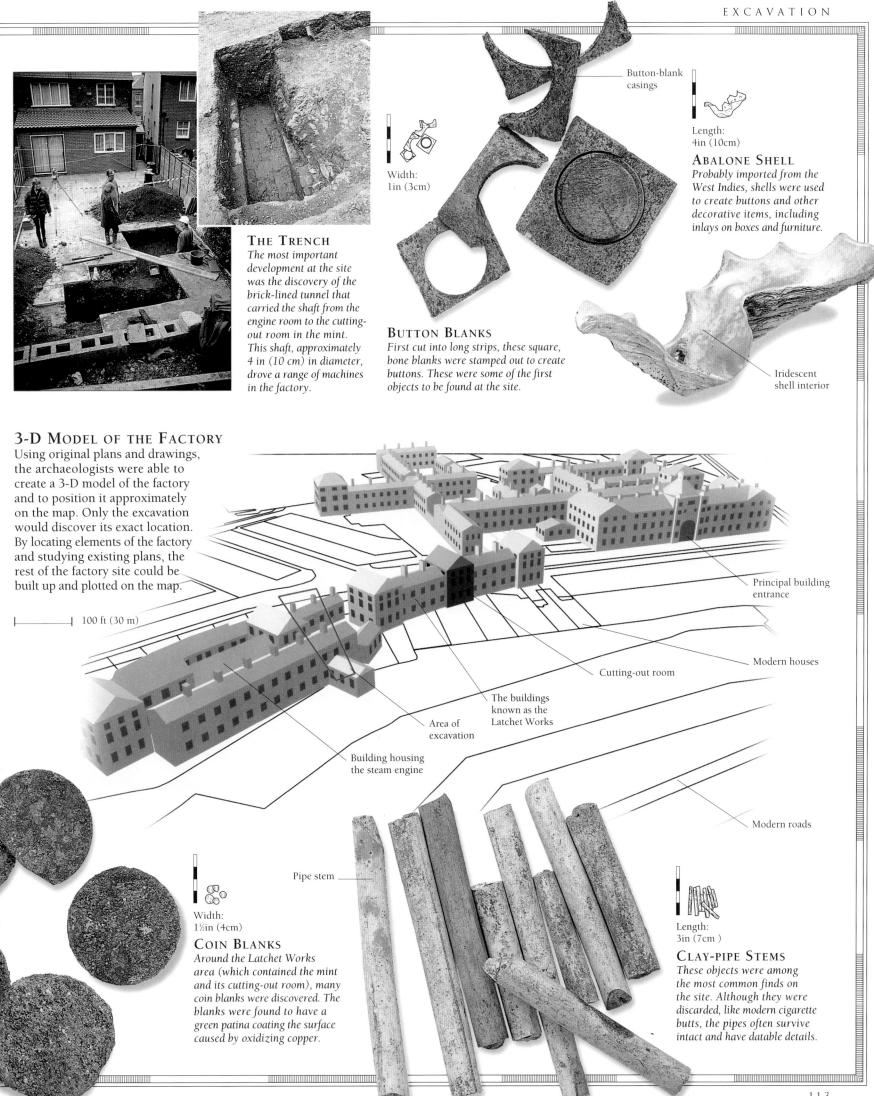

THE TRENCH
The most important development at the site was the discovery of the brick-lined tunnel that carried the shaft from the engine room to the cutting-out room in the mint. This shaft, approximately 4 in (10 cm) in diameter, drove a range of machines in the factory.

Button-blank casings

Length:
4in (10cm)

ABALONE SHELL
Probably imported from the West Indies, shells were used to create buttons and other decorative items, including inlays on boxes and furniture.

Iridescent shell interior

Width:
1in (3cm)

BUTTON BLANKS
First cut into long strips, these square, bone blanks were stamped out to create buttons. These were some of the first objects to be found at the site.

3-D MODEL OF THE FACTORY
Using original plans and drawings, the archaeologists were able to create a 3-D model of the factory and to position it approximately on the map. Only the excavation would discover its exact location. By locating elements of the factory and studying existing plans, the rest of the factory site could be built up and plotted on the map.

├─────┤ 100 ft (30 m)

Principal building entrance

Modern houses

Cutting-out room

The buildings known as the Latchet Works

Area of excavation

Modern roads

Building housing the steam engine

Pipe stem

Width:
1½in (4cm)

COIN BLANKS
Around the Latchet Works area (which contained the mint and its cutting-out room), many coin blanks were discovered. The blanks were found to have a green patina coating the surface caused by oxidizing copper.

Length:
3in (7cm)

CLAY-PIPE STEMS
These objects were among the most common finds on the site. Although they were discarded, like modern cigarette butts, the pipes often survive intact and have datable details.

A FACTORY

INTERPRETING THE FINDS

BY FOLLOWING THE SHAFT that had been located in the trench (along the length described in the sale document of 1850), the archaeologists came to the center of the old latchet (transferable shoe buckle) works. It was already known from existing documents that the cutting-out room had been housed in a building originally erected in 1794 as a latchet works and that the center room of this building was three stories high. The additional finds of coin blanks and button casings in the trench also served to reinforce the archaeologists' belief that they had found the cutting-out room. This central room was known to have had a Neoclassical interior, with the ceiling supported by Classical columns in the Doric style. Beneath the floor lay the main gear wheels, which converted the spinning shaft into a drive for the presses. This was the heart of a steam-driven process that made objects at a rate of one per second. At the other end of the shaft, the archaeologists discovered the sump that provided the water for the engine.

SHELL BUTTONS
The stamping process could be adapted to a number of products, and it is believed that buttons were manufactured here using a variety of materials, including abalone shell and ivory. Once the basic round shape of the button had been cut out, it was decorated with carving and inlaid materials.

Carved and inlaid center

Viewing balcony

ARTIST'S IMPRESSION
By combining the engravings and plans and incorporating the new archaeological discoveries, it is possible to reconstruct the cutting-out room. The engravings gave extra details of features that do not appear in the plans, and the size of the room could be calculated from descriptions in Boulton's archives. Drawings for the Bombay and Calcutta mints, which were larger replicas of the Soho original, showed the layout of the room and its internal decorations. Although the room was beautifully designed, the cutting-out machines would have been noisy and dirty to operate.

Rod of iron linking the piston to the wheel

Levers linking the piston to a rod of iron that goes through the wall to the wheel above

Hermetically sealed chamber to create a vacuum for pumping the piston

Trolley containing copper strips ready for cutting out

THE STEAM ENGINE

Boulton and Watt steam engines were first developed at the Soho Manufactory and soon powered all the machinery in the factory. Watt's invention of a separate condenser made them even more efficient, and they soon became mass produced. Until 1805, Boulton exported his engine designs and therafter sold complete engines in kit form to power machines and mine pumps as far away as India and the West Indies. Watt's inventive genius and Boulton's entrepreneurial skills helped sell the engines to a number of different industries that later used mass production, often utilizing new systems of labor similar to those pioneered in Birmingham.

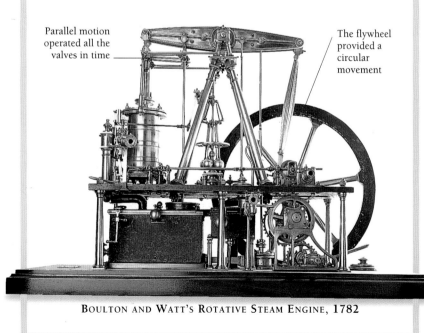

Parallel motion operated all the valves in time

The flywheel provided a circular movement

BOULTON AND WATT'S ROTATIVE STEAM ENGINE, 1782

DRIVE SHAFT
The tunnel containing the drive shaft from the steam engine entered the cutting-out room 3 ft (1 m) below the ground. The shaft was attached to a pinion, which powered the vertical shaft up out of the floor.

Horizontally
spinning wheel
with cams to
produce upward
movement

Distinctive
semicircular
window

The room is
designed to look
like a Neoclassical
rotunda

Doric columns

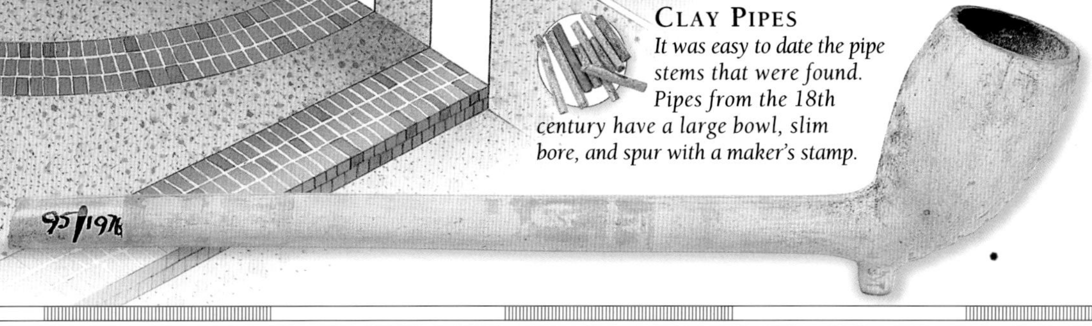

COIN STAMPING

The final process in coin
production was the stamping.
The coins went from the
cutting-out room to be milled
(given an edge) and then to be heated ready
for stamping. The coin presses were simple
enough to be operated by children and
could produce 40,000 coins per hour.

**A COIN
STAMP AT
THE SOHO
WORKS**

Elephant
symbolizing
the British
dependency
of Ceylon

COPPER PENNY, 1802

COINING

The number of coin
blanks confirmed the
presence of cutting-out
machines. Strips of metal would
be fed through the machine, and
standard-size holes would be
cut to produce the blanks. Then,
the coins would be milled,
bleached, washed, and dried.
After a final annealing, they
would be stamped with a design.

OLD DEBRIS

At a lower level,
beneath the floor
of the mint, the
remains of the earlier button-
making industry were found.

Brick floor

CLAY PIPES

It was easy to date the pipe
stems that were found.
Pipes from the 18th
century have a large bowl, slim
bore, and spur with a maker's stamp.

CONCLUSIONS

🔍 The exact position of the mint building
was located, enabling the rest of the factory
to be mapped accurately.

🔍 The position of the steam engine and
the route of the tunnels that carried the drive
shaft to the rest of the mint were located.

🔍 By combining intensive documentary
research with good excavation, archaeologists
demonstrated that it is possible to get excellent
results, even from an area that was as intensively
redeveloped as the Soho area of Birmingham.

COINAGE

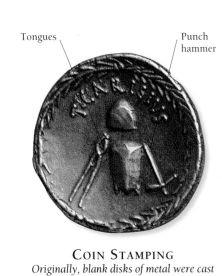

COIN STAMPING

Originally, blank disks of metal were cast in molds and designs were stamped on each individual disk using a punch, to make a coin. From the 18th century, the process became mechanized, so large numbers of standard coins could be produced quickly.

Tongues

Punch hammer

AS SOCIETIES BECAME MORE COMPLEX and sophisticated economies emerged, coins developed to provide a standard of value recognized by everyone, to replace earlier systems of bartering and exchange. Coinage was also used to make powerful political statements and to indicate national or local identity to a wide audience. Initially, the metal content of each coin reflected its actual value, but gradually coins came to represent a specific value without containing a proportional metal element. Coins have constantly evolved and changed over the centuries, undergoing periods of minting, reissuing, and alterations in design that reflect changing political and economic circumstances. Because of this, they can provide important evidence for archaeologists, as they can often be dated very precisely. By applying stratigraphy techniques (see p. 125), coins can be used to date other artifacts and whole sites associated with them.

LYDIAN COINS
— 700 BC —

The earliest-known coins were made in Lydia, Turkey, in the 7th century BC. They were made from lumps of electrum, a mixture of gold and silver, to an exact weight, and stamped with the mark of the Lydian kings to guarantee their value. Three weights were produced: a 1 stater of just over ½ oz (14 g), a ⅙ stater, and a 1/24 stater. Later, separate gold and silver coins were made, with the silver ones being half the weight of the gold staters. These weights became the forerunners of the standards used on Persian coins.

700 BC

1 STATER COIN

The king's mark

⅙ STATER COIN

Irregular shape

1/24 STATER COIN

Lion's-head design

— KEY CHARACTERISTICS —
Finished coin had an irregular shape.
Each coin had a standard weight.
Stamp indicated guaranteed value.

ROMAN COINS
— 300 BC —

Roman coins were originally copied from Greek ones, but the emperors soon began to produce gold, silver, and bronze coins with their own portraits, to be distributed throughout the empire. Coins were also issued to celebrate important events, such as famous victories and major festivals, and can be found as far afield as India.

BRONZE SESTERTIUS OF EMPEROR NERO

Name and title of the emperor

ROMAN SILVER COIN

Romulus and Remus

Marks made by stamping

— KEY CHARACTERISTICS —
Different metals used for different values.
Portrait head of Roman rulers.
Wording states titles, names, and dates.

DARK AGE COINS
— AD 650 ONWARD —

The circulation of coins became very restricted in the Dark Ages because, in many regions, the centralized governments that had controlled coin production collapsed. Instead, local money-makers and rulers produced coins for use in their local area. This means that coins from this period are not commonly found. Many hoards of later western European coins have been found in Scandinavia because they had been extracted as a form of tax by the invading Danes. Most Dark Age coins carried the name and location of the mint, the moneymaker, and the portrait of the local ruler.

SAXON SILVER PENNY

Alfred the Great

VIKING COIN

Legend reads "Eric the King"

— KEY CHARACTERISTICS —
Most Dark Age coins were made of silver.
A wide variety of coins were minted.
Only small numbers of each coin type survive.

MINTING A COIN

From an early date, coin-making became concentrated in the hands of specialized money-makers who produced their coins in workshops called mints. Because of the value of the metals used in coin production, the mints were closely regulated, and the coins they produced were closely monitored and controlled by local rulers and governments. In the late 18th century, before mechanization, coins would have been cut out and stamped individually by hand, making coin production a very slow and labor-intensive process.

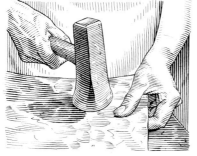

1 *The blank coin disks are struck out of a sheet of metal, which has been flattened to the required thickness using a hammer and anvil. The sheet of metal will have to be frequently reheated to maintain malleability.*

2 *The sheet is cut into strips and then small squares of metal. The corners of these squares are trimmed off using large shears to create circular blanks. Any excess metal would have been melted down again and reused.*

3 *A design is stamped onto the disk using a hammer and two dies. A patterned die is recessed into a holder, and the disk is placed on top. Another patterned die is then hammered down on top, leaving a motif on both sides.*

GROAT AND NOBLE
— 14TH CENTURY AD —

As trade increased during the medieval period, more valuable coins were introduced alongside the standard silver English penny. The gold noble was worth 80 silver pennies, while the silver groat was worth only four silver pennies.

NOBLE

Royal coat of arms

Ships indicate trade

Portrait of King Edward III

GROAT

— KEY CHARACTERISTICS —

Greater weight of metal to increase value.

Complex designs stamped on the coins.

Many more coins in circulation.

SILVER THALER
— 16TH CENTURY AD —

Silver mines in Bohemia were used to produce large coins called thalers in the 16th century. Each coin carried a portrait of the local duke or ruler and was stamped clearly with the minting date. "Thaler" is the origin of the modern word "dollar."

Count Stephen of Slick

Duke of Saxony

— KEY CHARACTERISTICS —

All large-sized silver coins.

Depict a wide range of local rulers.

Easily dateable from inscription.

CARTWHEEL COIN
— LATE 18TH CENTURY AD —

The first official coin in Britain was minted in 1672, but by the 18th century, there was a shortage of coins because of the increasing demands of industry. Many people began to produce forgeries and private-issue coins, as well as restamping imported silver coins. As a result, in 1797 Matthew Boulton (see pp. 112–15) was granted a license to manufacture official 1 and 2 pence copper coins. They were mass-produced to a standard design using steam power and were very large, containing the actual value of copper. In 1811, he began producing lighter coins containing less copper, and a mechanized mint was built on Tower Hill in London.

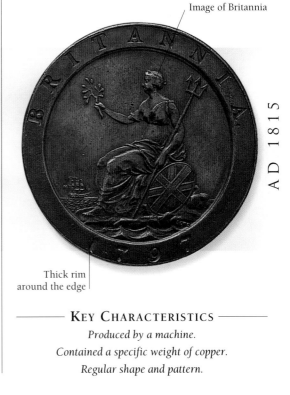

Image of Britannia

Thick rim around the edge

AD 1815

— KEY CHARACTERISTICS —

Produced by a machine.

Contained a specific weight of copper.

Regular shape and pattern.

COMPARATIVE SITES

A FEAT OF ENGINEERING

The bridge at Coalbrookdale was designed in a semi-circular arch and demonstrated not only the strength and versatility of coke-smelted iron, but also the design and engineering skills that would revolutionize the world.

COALBROOKDALE

SHROPSHIRE, UNITED KINGDOM

—— GAZETTEER P .144 – MAP REF. F7 ——

IN THE EARLY 18TH CENTURY, at Coalbrookdale in England, the combination of abundant natural resources and the inventive genius of ironmaster Abraham Darby (1678–1717) produced one of the most important developments of the Industrial Revolution by the successful use of coke to smelt iron. Coalbrookdale, as a whole, provides a complex problem of interpretation for archaeologists, particularly with regard to the dating of the various phases of the site. They have had to interpret not only the more obvious features, such as the standing buildings, but also secondary features, such as the complex water systems of the furnace pools and the remains of coal and iron pits. Industrial entrepreneurs, like the later generations of the Darby family, were constantly improving and developing their sites, and, in the case of the original furnace site, for example, there are many layers of evidence dating from 1638 (before Darby was born) through to 1777. Coalbrookdale is a multifaceted site containing the remains of several industries, including coal mining, iron-making, and pottery production, as well as the ultimate testament to the Darbys' skill – the iron bridge across the Severn River.

INDUSTRIAL INNOVATION

In order to produce usable iron, the ore has to be smelted with a material that is rich in carbon. Originally this was done using charcoal. This process produced a material that could be hammered into shape and was known as wrought iron. By increasing the temperature and blowing air into the furnace, the iron ore would melt and run off as a liquid that could be poured into molds. This was called cast iron, and it could be used to create a wide range of shapes.

At the beginning of the 18th century, Abraham Darby made a crucial technical breakthrough by developing the use of coke – the solid substance left after the gases have been extracted from the coal – to produce iron successfully. Previous foundaries had relied upon an increasingly limited supply of charcoal, which was also too soft to be used in larger furnaces. Local forests were being exhausted at an alarming rate, and the provision of this essential raw material could not keep up with the increasing output of new objects. Coal was too impure to create high-quality iron, but by partially burning off its impurities – mainly sulfur – to create coke, and by adding limestone to

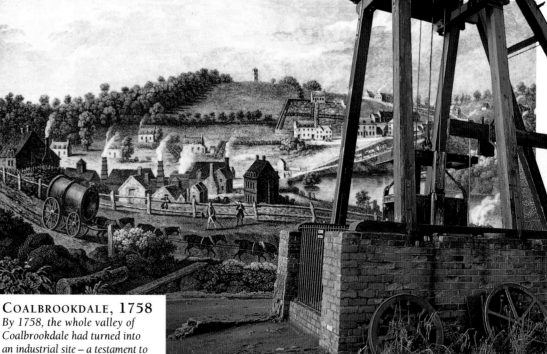

COALBROOKDALE, 1758
By 1758, the whole valley of Coalbrookdale had turned into an industrial site – a testament to how the industrial age changed the landscape. Furnaces producing coke and smelting iron lit up the whole valley at night.

THE WORLD'S FIRST IRON BRIDGE

The bridge across the Severn River at Ironbridge (as the town was subsequently named), was built by the third Abraham Darby in 1779. It was the first iron bridge in the world and was much copied. Cast iron was eventually replaced by steel, which was stronger and cheaper.

Translucent, lustrous finish

COALPORT CHINA

By the late 18th century, Coalbrookdale was thriving and other industries were drawn to the area. The Coalport pottery was established by Rose, Blakeway and Co. in 1796, to manufacture fine-quality tablewares.

carry away those impurities, Darby found a method that avoided the use of charcoal. In addition, he discovered that, by reheating the iron in a second furnace, its quality was improved and it was less brittle and easier to cast.

These new innovations vastly increased the range of objects that could be produced. Darby's first patent of 1707 was to produce cooking pots cast in sand, and, in the early years of the factory, iron cooking pots were the main product. However, the new cast iron could create more complex structures, and the factory developed new techniques of molding in sand that could create more sophisticated shapes. In 1709, Darby refined the use of coke to smelt iron

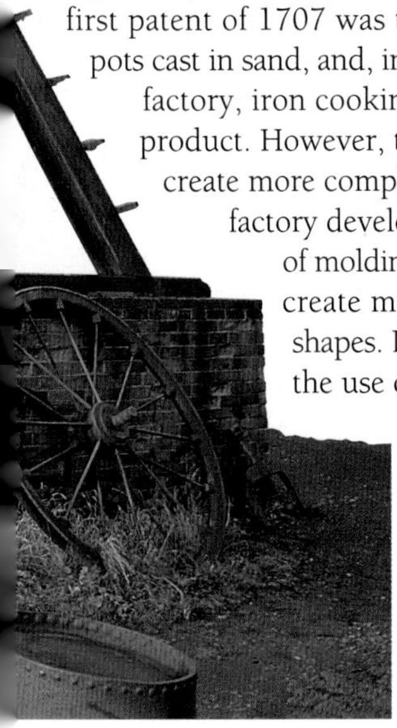

COAL MINING

The local coal supplies were exploited to the full, providing the basic raw material for the coke which, when combined with the ore, created iron.

and, by 1713, over 11 tons (10 tonnes) of iron were being produced a week. Abraham Darby died in 1717, and his son, also named Abraham, inherited the factory and continued in his father's manner of innovation, introducing steam engines to circulate the water. Cast iron could now be used to make the cylinders for the new steam engines, while pig iron (iron that contained impurities) was being produced for use in forges. All this made it possible for the second Abraham Darby to succeed in making iron that was pure enough to be split into rods, to supply the growing market for nails. In the mid-18th century, Darby's foundary, now under the management of the third Abraham Darby, produced the first iron rails – forerunners of the tracks that would carry the first steam engine. The mining engineer Richard Trevithick (1771–1833) would later come to Coalbrookdale to purchase the iron boilers that he used to create the first steam locomotives.

This new cast-iron material could be used for a huge range of products, but is, perhaps, most impressively illustrated by Darby's iron bridge. Erected in 1779 and crossing the Severn

River, it was the first completely iron bridge in the world. It was designed so that the component pieces formed a semicircular arch in two 70-ft (21-m) halves. It took three months to build and used a prefabrication technique that involved slotting the cast sections together. The fact that it remained in use until 1950, carrying large volumes of traffic across the Severn River, is a testament to its strength.

Three generations of the Darby family all worked on this site, and it was their skills that formed the technological basis for the Industrial Revolution. Indeed, it was the iron from the Coalbrookdale furnaces, combined with Boulton and Watt's steam power (see pp. 114–17), that powered the Industrial Revolution.

ESSEN

RUHR VALLEY, GERMANY

GAZETTEER P. 146 – MAP REF. H4

THE RUHR REGION OF GERMANY was the cradle of industrialization in continental Europe. From the late 18th century onward, abundant supplies of coal and coke led to the establishment of huge mines and factories to smelt iron and steel. Although, technologically, England was still the world leader in iron and steel production, Germany was its main rival because of the huge quantities of materials that it produced. This advantage was heightened in the early 19th century, when England's supply links with continental Europe were weakened due to the Napoleonic War. The Krupp factory, situated near Essen in the Ruhr valley, became one of the biggest steel-making factories in Germany. It was established in 1812, but did not achieve real success until the 1860s, when the pioneering industrialist Alfred Krupp introduced steam power to his factory and began to mass-produce cast-steel objects. Its success continued, particularly when the munitions factory was created.

IMPRESSIVE ARTIFACTS
The Krupps factory was destroyed during the Allied bombing raids on Essen at the end of World War II. This photograph shows one of the huge cannons that were being manufactured for the German army. Such objects make dramatic finds for archaeologists.

RAPID EXPANSION

From 1587 to 1968, the many generations of the Krupp dynasty dominated the industries of the German city of Essen. Arndt Krupp (d.1624) arrived before a plague epidemic and bought large areas of land from the fleeing locals. It was not until Friedrich Krupp (1787–1826) founded the cast-steel factory in 1811 that the family became wealthy. The factory initially produced steel rollers and dies for making coins and household items like utensils. By the time his son Alfred (1812–87) was in charge, the technique of producing steel had been perfected, and he manufactured rails and seamless-steel railroad wheels. Alfred also sold his machines for making cast-steel objects to companies all over Europe. In response to the Franco-Prussian War (1870–1), Krupp began producing cannons, gun barrels, and munitions, and by the end of the 19th century the factory was the biggest weapons producer in Germany. By the time of his death, Alfred Krupp had received many honors and had armed 46 nations – laying the groundwork for the holocaust that would begin in 1914.

Although Alfred Krupp spent most of his life making weapons of destruction, he was, ironically, considered to be a model employer, offering sick pay, pensions, and reasonable housing to his employees.

In fact, the success of the enterprise can be judged by the number of employees in the factory. The Krupp factory began with just seven employees in 1812, but the number had risen to 16,000 by 1873, and to a staggering 68,300 by 1912. It was Friedrich Krupp (1854–1902) who engineered the great expansion of the company – by inventing shells that pierced the steel that he had sold to armies and navies. He thus created a cycle of offensive and defensive products that made one another obsolete. After Friedrich's death, his daughter Bertha (1886–1950) married Gustav von Bohlen und Halbach (1870–1950), who took over the business and named the "Big Bertha" cannon after his wife. Gustav was famous for his Nazi sympathies, and his son Alfred (1907–67) was involved in seizing Jewish property and employing slave labor from concentration camps. He was convicted of war crimes in 1948 and was jailed for six years. There are many Krupps sites across Germany, and archaeologists have access to huge amounts of documentation. Despite its less than savory past, the Krupps factory at Essen is invaluable for understanding the industrial archaeology of Germany.

THE FACTORY IN 1865
By this date, Alfred Krupp had just introduced steam power to the factory, and the company was expanding rapidly. The Essen area was fortunate to have excellent water supplies from the Ruhr – water was essential for the cooling processes in steel production.

MUNITIONS FACTORY
The picture on the left shows the factory floor during World War I, with hundreds of mass-produced gun barrels lined up. Krupps supplied munitions to fighting powers across Europe, not just to the German army.

LE GRAND HORNU
BORINAGE, BELGIUM
GAZETTEER P. 146 – MAP REF. G5

LE GRAND HORNU LIES IN THE HEART of the Borinage region of Belgium and is a spectacular monument to the Industrial Revolution in Europe. Even today, much survives of the steam factory that was begun in the early 19th century, and the surrounding landscape bears the scars of its industrial past.

The region is well known for its coal mining, and by the beginning of the 19th century the area already had coal pits drained by steam engines. A French industrial pioneer, Henri de Gorge-Legrand, bought the operation in 1810 and built new factories near the pits and houses for the workers. By 1831, de Gorge had dramatically expanded the enterprise and had added an engineering works to manufacture steam engines and locomotives to be exported across Europe.

THE WORKERS' DWELLINGS
The workers' houses were built as two-story terraces along cobbled streets immediately adjacent to the coalfield and de Gorge's grand residence. They have all been altered and extended since the closure of the business, as can be seen in the foreground of this picture.

AN INDUSTRIAL VILLAGE

Belgium was the first country in continental Europe to have an Industrial Revolution. Like Britain, it had natural reserves of coal and iron ore; in addition it was ideally positioned between France and Germany. The rapid pace of industrialization led to the construction of entirely new types of buildings at developing centers like Le Grand Hornu, including huge factories, engine rooms, chimneys for steam houses and large numbers of workers' dwellings. Le Grand Hornu is an oustanding surviving example of this type of architecture. De Gorge had the foundry, stores, offices, and assembly shops (designed by the architect Bruno Renard) built around an elliptical courtyard, 460 ft (140 m) long by 262 ft (80 m) wide, linked with continuous arcading. In 1831 a large house for de Gorges's family was begun across one end of the courtyard, but it was not completed before his death and was not used until 1843 when it became an administrative building. Many of the buildings survive, converted into offices or as roofless shells. The large numbers of workers required by the new factories led de Gorge, like Krupp at Essen and many other contemporary industrialists, to provide houses for his labor force. By 1830, the industrial complex was huge and employed 1,500 workers, and de Gorge built more than 400 terraced houses, each containing six rooms, some of which survive.

The business was extremely successful and the mines produced over 165,000 tons of coal a year throughout the 19th century. The factories themselves used vast quantities of the coal to make a range of metal objects.

RESTORATION

The business at Le Grand Hornu went into decline in the 1950s, and the buildings fell into disrepair. In 1971 the whole site was bought by an architect, Henri Guchez, who restored many of the buildings. There is still much work to be done at the site, and archaeologists may yet discover further buildings, mine shafts, or artifacts in the vicinity.

THE FACTORY BUILDINGS
This photograph shows one of the factory buildings with a steam chimney dominating one end of the central oval courtyard.

Chimney for steam to escape

Neoclassical facades

REMAINING STRUCTURES
Coal mining ceased at Le Grand Hornu in the 1950s, but archaeologists have plenty of standing structures available for analysis. The number of artifacts discovered is limited but whole tramway locomotive engines have survived in reasonable condition.

GEOPHYSICAL SURVEYING

OVER THE PAST 50 years, scientists working with archaeologists have spent a great deal of time developing methods to discover what lies beneath the soil before excavation has begun. The appearance of microchips and small computers has made technical equipment more accessible to archaeologists, and several different methods have been developed.

Although anyone can carry out the collection of data, it is the job of the geophysicist to interpret the results, understand problems in data presentation, and produce archaeological information that is useful in the understanding of a site. This involves recognizing modern features and geologic and natural aspects, and separating them from the archaeology. The most important development in geophysics has been the invention of small computers that can be fixed to the survey equipment to collect the data.

Portable computers allow this complex scientific data to be processed and presented quickly and efficiently at the site. There are three methods of geophysical survey – magnetometry (gradiometry), magnetic susceptibility, and resistivity. The first two of these techniques are the most widely used.

MAGNETOMETRY

Magnetometry measures disturbances in the Earth's magnetic field caused by local features in the ground. The magnetometer has two sensors, one above the other, and the machine is set using the points of the compass. The machine measures the difference between the two sensors, the top giving the magnetic field and the lower any effects from features in the ground. The difference in these readings will indicate any local disturbance. The operators must be sure that there is no metal in their clothing or footwear, which would affect the magnetometer readings.

Usually, sites are laid out on a grid basis for surveying. Readings are taken over a number of 66-ft (20-m) square grids, from which the data are collected. Following downloading and processing of the information, printouts are obtained with details of the survey depicted in dot-density, grayscale, or trace-line formats. The technique is used to get a general overview of a wide area and to indicate whether there is anything of archaeological interest present before a more detailed survey is undertaken. Magnetometry is particularly good at revealing walls, pits, and ditches.

RESISTIVITY

Resistivity measures the resistance of features under the soil to an electric current passed through the ground between two probes. Readings are taken at regular intervals along a grid. The machine takes a reading when there is contact between the two probes. Features such as buried walls or stony surfaces provide high resistance to the current, whereas pits and ditches, with high water and humic content, allow the current to pass through easily. The machine measures the difference in the readings and compares them to a

RESISTIVITY METER
The picture shows a resistivity meter in use on a site. The site is measured into grids and probes are inserted at intervals, usually 20 in (50 cm) or 3 ft (1 m) apart. The white box contains the computer that logs the information.

general background reading, which is provided by two probes located some distance away, but in the same type of subsoil. The quality of information can vary considerably, depending on the environmental conditions, such as the dampness of the soil. Resistance surveys are best carried out in spring or fall when the seasons are changing and the ground is neither extremely dry nor extremely wet.

OTHER METHODS

Magnetic susceptibility, or "mag. sus." as it is more commonly known, is primarily used to locate areas of former burning, or of burned clay, which might indicate occupation in the form of hearths or burned structures. The instrument is used to analyze the mineral content and magnetizability of soil particles. By using this technique, details of disturbance to the ground, or heating, which affects the magnetic behavior of particles, can be identified. Mag. sus. is a very sensitive and flexible technique.

GEOPHYSICAL PRINTOUT
Above, a grayscale printout of the data from Navan (see pp. 60–3) and right, an interpretation of it. The key shows how the located features have been interpreted. This is an early Christian banked and ditched enclosure with internal structures. The archaeological features are surrounded by signals from geology and the background "noise" of the soil.

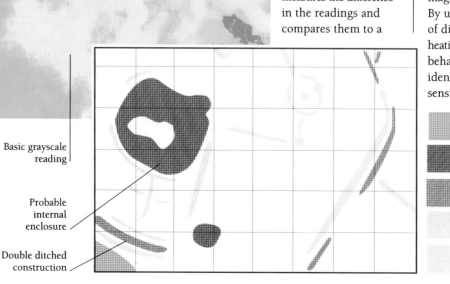

Basic grayscale reading

Probable internal enclosure

Double ditched construction

▨	HIGH RESISTANCE – NATURAL
▨	HIGH RESISTANCE – NATURAL/ARCHAEOLOGY
▨	HIGH RESISTANCE – BANK
▨	LOW RESISTANCE – DITCH
▨	HIGH RESISTANCE – ARCHAEOLOGY

GLOBAL POSITIONING BY SATELLITE (GPS)

A GPS SURVEY
The "rover" receiver is carried on a pole in a backpack. The operator receives information from a number of satelittes (at least four), and records the data from a specific position.

GLOBAL POSITIONING by satellite (GPS) is an accurate way of finding the position and height of points on the Earth's surface, and, as such, it enables archaeologists and others to carry out detailed mapping to a very high standard on any sites that they deal with. GPS is widely used by all types of surveyors and military bodies, as well as by archaeologists, and relies on the use of American military satellites that are put into space to assist global navigation and missile control.

MAPPING A SITE

GPS uses two receivers on the ground, which pick up the signals transmitted by the satellites. Since the height, position, and courses of the satellites are known, provided the receivers can pick up the signals from enough satellites (usually four or more), a position on the ground can be calculated by resection.

To carry out an accurate survey by GPS, the data from the satellites must be collected by both receivers simultaneously. One receiver is fixed at a known position, and the second, the "rover," is moved around the site to the points being surveyed. Readings from the two receivers are stored in computers and can be compared so a contour map of the area covered by the "rover" can be constructed in great detail. Printouts are made later.

GPS data are recorded from the latitude and longitude, but can be converted to fit into the US Geological Survey, and therefore can be related to published maps. GPS has proved to be a very accurate and time-saving tool for the archaeologist.

Area of high ground

Flat area or ditch

SURVEY PLOT
Here, the data recorded have been used to produce a three-dimensional model of the topography of the site.

GROUND RADAR

GROUND RADAR SURVEY
Here, at the site of the Soho Works in Birmingham (see pp. 112–15), a survey is being carried out with ground radar in an attempt to record brick walls and a brick-lined well.

SINCE THE 1930s, when radar (RAdio Direction And Ranging) was developed for detecting aircraft, it has been adapted for use in many different areas, including space exploration, satellite imaging, and civil engineering. Since 1989, attempts have been made to use it in archaeology, and by 1992, the first radar tests were finally carried out on archaeological sites.

SURVEYING THE SITE

The ground radar emits waves of electromagnetic radiation. As these strike the boundary between two different materials, or between a material and a void, they are bounced back to a receiver. The returning signals are recorded, and a printout of a section through the site can be made.

So far, while the method seems to work well, there have been some problems with interpretation of the information. The data recorded are a reflection of the time taken for the signal to return, so the cross section that is produced is a reflection of different return times, and not necessarily of the disposition of the layers in the ground. Nevertheless, radar does seem to be a useful method for finding, for example, buried river channels or voids, such as crypts or tunnels. Although the detail may not be completely clear, complex stratification on an archaeological site can be definitely indicated.

Radar is most successfully used on urban sites because high levels of metals, concrete, electricity, and other services can interfere with the results of resistivity and magnetometry. The deep, complicated stratigraphy on urban sites is more suited to radar, because it often appears mixed and unclear using the other methods.

RADAR DISPLAY
The results from the survey in Birmingham are being printed out. The lines indicate the different time reflections in the ground, rather than individual archaeological layers.

Radar is more commonly used in the US than in other countries, where magnetic and resistance methods are more popular.

Provided that problems with physics, the instruments, and interpretation of the results can be overcome, it is likely that ground radar will become a very revealing tool for archaeologists in the future.

FIELD SURVEYING

ONE OF THE simplest and most effective ways of locating areas of archaeological activity, or previously unknown sites, is field surveying. This is best employed after fields have been plowed, harrowed, and preferably rolled, because these activities bring finds to the surface. It does not matter if crops have been planted, but it helps if the field has been rained on, since the finds are thereby washed and are more visible. Depending on the aim of the project, almost anything that is found can be important. Pottery is of particular interest, since, if it is found in quantity, it can indicate settlement sites. Flint and exotic stone can often indicate prehistoric activity, whether in permanent settlements or hunting areas. Concentrations of modern ceramics, metalwork, brick, and tile are more likely to indicate modern activity, and are less likely to be of interest to the archaeologist. Soil colors and textures might also be noted, as dark humic areas may indicate areas of past occupation.

METHODS

The method of surveying used is dependent on the aim of the project. To recover artifacts, the collection area could be divided into a regular series of squares of varying sizes. It is more likely that some sort of sampling strategy will be used, so that only a small percentage of the area will be examined. In this case, the area will be divided into squares and only certain ones walked – a random sample – or lines across the area will be walked at regular intervals of 33 ft (10 m) or 66 ft (20 m). Again, finds of pottery, flint, or other artifacts are recorded.

Surveying is an effective and efficient way to locate previously unsuspected archaeological sites, or to put known sites into their landscape and chronological context. Provided that the analysis of finds is carefully carried out, and the implications of results are appreciated, it can be a simple and cost-effective

FIELD SURVEYING
People equipped with bags and labels are systematically searching a plowed field. The more people used, the more effective field surveying is.

Distinct
Roman design

POTTERY SHARD
Different types and quantities of pottery are very important in field surveying. This shard of Samian ware would usually indicate a Roman site.

way of recovering archaeological data over a wide area. This technique does depend on the materials surviving in the ground, and these are usually inorganic objects that reflect earlier human activities at the site. Consequently, surveying is less effective as a technique at archaeological sites where there are few or no inorganic objects.

AERIAL PHOTOGRAPHY

AERIAL PHOTOGRAPHY has revolutionized archaeology in two ways – it can be used to locate new sites without disturbance and can also put existing sites into their landscape context. An aerial view can reveal perspectives not visible on the ground – for example, one mound can be seen as a part of a larger landscape, and an isolated bank or ditch as part of a large-scale field system. Aerial photography became widespread as a result of military reconnaissance during World War II and has become more accessible and inexpensive as light aircraft and better cameras have been developed.

There are three main techniques that utilize aerial photography. The first technique, earthworks surveying, reveals irregularities on the ground that are not clear when viewed from surface level. These details can be seen very easily from the air as low sunlight hits the banks and throws the ditches into shadow. The second technique, photographing a field, can reveal differences in coloring that relate to buried features and structures. Finally, crop marks can occur where different levels of moisture and depths of soil created by archaeological features produce different growth and ripening rates of crops. Buried pits and ditches will produce taller, later-ripening crops, while stony areas produce more withered, earlier-ripening crops.

Dark lines of rich soil

Chalk bedrock

SITE OVERVIEWS
In this cereal field (above left) dark lines appear where the crop is growing over deeper, moisture-rich soil in the ditches along a trackway. In the plowed field (below left), the action of the plow has revealed the chalk bedrock (light areas) and the shape of a prehistoric or Roman field system (dark areas).

EARTHWORKS
The rectangular crofts and main road through this deserted village can be seen in low sunlight. The village is bounded by a ditch and surrounded by a ridge and furrow. These earthworks are typical of deserted settlement sites found all over the British Isles.

EXCAVATION TECHNIQUES

EXCAVATION IS STILL the main method used by archaeologists to investigate sites. The aim of an archaeological excavation is to retrieve information from the ground in a systematic way, so that the order in which the site was established, developed, and collapsed can be understood. Since the archaeologist has to destroy the site in order to do this, it is vital that the excavation is

recorded and the results made available for other archaeologists. Only a small area of the site may be excavated, in order to preserve the site for future archaeologists. All the features are accounted for in their individual contexts, and all finds from them are carefully recorded.

METHODS

Two main methods are used. The first of these is to dig trenches. Sections are cut across the site to see whether there are layers of occupation beneath the soil and to see what they contain. Although

little will be gleaned about the layout of any structures, the overall stratigraphy and chronology of the site are usually ascertained. Finds from the various layers will help to date them. Excavation stops at the sterile soil

The second method is to open a large area, remove the topsoil, and expose the archaeology below. Since, in the past, so many structures were built of wood, all that will be left of them are postholes and wooden slots. It is only by open-air excavation that changes in soil color and texture, and thus the plans of such structures, can be recorded. Again, all features and their related finds are recorded in their context.

Other methods include the use of heavy machinery to strip turf or topsoil from sites. Likewise, trenches can be cut across sites mechanically. Elsewhere, everything is done by hand, using small trowels, brushes,

CLEARING THE SURFACE
Here, excavators using trowels are carefully clearing the surface to reveal changes in color and texture, which may reveal traces of wooden buildings.

spades, shovels, hoes, and sieves. The excavation depends on the aim of the project, what is to be recorded, and whether the site is likely to be destroyed by modern development. It has become common to dig test pits, as well as carry out evaluations with small trenches, to determine if there is anything of archaeological interest on the site.

Layers of deposits

A SECTION
A cut through the ground such as this reveals various layers of deposits and can provide a cross-sectional history of a site.

POSTHOLE ANALYSIS

SINCE MANY STRUCTURES from earlier times were built of wood, one of the most common features of archaeological sites is the posthole, the socket that once held up the wooden post. Except in unusual circumstances – for example, when a structure is waterlogged, or if a timber has been burned and, therefore, carbonized – the timbers will rot away, and the archaeologist will be able to see only the "ghosts" of their former existence.

CONSTRUCTION

In ancient times, the usual method of construction for wooden buildings was to dig a series of pits, placing posts in them, and packing soil and stones around them for support – rather like putting up a wooden fence today. The differences between the ground material and the post filling survive for many centuries. If the post has rotted *in situ*, or was cut off at ground level, remains of it

can be recognized, either as a stain in the soil or as a more fibrous filling material. If the post has been removed, a void or loosely packed material may remain.

Careful excavation and cleaning of the ground surface into which the postholes were dug can reveal a difference in texture and color between the natural geology, the packing of the posthole (which may contain large stones), and the site of the post itself. The pits are usually examined by sectioning them to record a cross section of the posthole, post site, and fill. By recording the situation and distribution of the postholes across a site, the plans of various buildings at the site can

be determined. Understanding the patterns of postholes and identifying different buildings from the depths and fills of the postholes is a difficult

task, especially when there are very large numbers of them; however, it can provide vital information about the plan of an archaeological site.

ANGLO-SAXON HOUSE   Posthole
This house is defined by a trench that revealed the vertical wall timbers. Inside the house the postholes represent the partitions between rooms, while outside they supported buttress timbers.

Void of square post

DENDROCHRONOLOGY

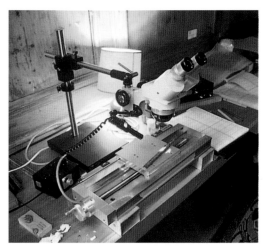

AS TREES AGE each year, they add one or two growth rings to their girth, which vary in width according to the tree's response to its environment. Any branch of a tree will show these variations. By looking at the ring pattern of living trees and comparing the sequence of

Varying widths of ring seen in section

rings with timbers from older trees of the same species, which are in turn compared to yet older specimens, a long chronology can be constructed from overlapping patterns. A core is taken from a living tree, and its ring pattern is compared with the rings from the beam of a surviving old building. The structure itself can be compared with earlier roof timbers and then with timbers from archaeological excavations. In this way, a reliable chronology can be built up by working backward, eventually dating back several thousand years. Before this stage is reached,

TRUNK SECTION
This section of a tree trunk shows the pattern of rings of the heartwood; the bark and sapwood are not present. The varying widths of the rings can be clearly seen.

there may be several "floating" sequences, where the rings overlap, but do not relate to the master chronology, so that further research on other samples is required. When any timber is found on an archaeological site, its ring pattern can be compared with the known sequence.

PRECISE DATING
If the sapwood and bark are present in the wood, then a precise felling date can usually be obtained for the tree. Since much wood was worked "green" in the past and not seasoned, the felling date could even be the date of construction. So far this method has been applied to only certain types of tree – Douglas fir and pinyon pine in the United States, pine in the eastern Mediterranean region, and oak in

MEASURING RINGS
The rings and their widths are counted automatically and analyzed by computer, which compares their pattern to known sequences for each species and region.

northern Europe. Research on other species is continuing across the world. The principal use of dendrochronology has been to correct or verify radiocarbon dates, and the two techniques are used in tandem to provide as accurate a dating as possible.

RADIOCARBON DATING

THE FIRST ABSOLUTE, scientific dating method used in archaeology was radiocarbon dating, and it remains the most widely practiced. It is based on the fact that all living things contain carbon, of which there are both stable and radioactive types. Carbon-14 is the radioactive form and once an organism dies, the level of carbon-14 in it decays, shedding atomic particles at a constant rate. This rate can be measured using scientific equipment, and from it a date for the object can be obtained.

RATE OF DECAY
This rate of decay was first worked out by Willard Frank Libby (1908–80) in the US, who discovered that after 5,568 years half the amount of carbon-14 in an object will have decayed. Many corrections have been made to his method since it was developed, including the revision of the half-life to 5,730 years. The

discovery that the atmospheric production rate of carbon-14 has not been constant through time has necessitated the application of a correction factor (calibration). Radiocarbon ages are reported initially in uncalibrated radiocarbon dates that assume no atmospheric flux. The dates are adjusted to calendar years by taking this flux into account. Therefore, a radiocarbon date of 2000 BC actually represents a calendar date of *c.* 2350 BC.

INCREASING ACCURACY
Earlier dating methods measured just the radioactive emissions from objects, but with "acceleration dating," the radioactive component of the whole sample is measured, thus producing more accurate dates from smaller amounts of material. Radiocarbon dating is best known for the dating of single objects, but it is mainly used for samples from excavations. A range

of samples, rather than a single specimen, is taken from well-defined archaeological sites, and it is important that the samples are not contaminated by modern materials. The method can be used on anything that was once alive – wood, charcoal, bone, or shell – and once the amount of radioactive carbon has been measured, then the year of the object's death, which is when the radiocarbon proportion began to decay, can be ascertained. Although there are some problems with the technique, it is still a useful archaeological tool, and, since the late 1970s, it has revolutionized our knowledge of the absolute or calendar dates of many sites and objects around the world.

RADIOCARBON LABORATORY
The dating process is carried out by physicists and takes places in specially built laboratories. New technology (accelerator mass spectrometry) can count the radiocarbon atoms directly, so smaller samples can be used.

ENVIRONMENTAL ANALYSIS

ANALYZING SAMPLES
Microscopic samples are examined by specialists using powerful microscopes. This technique can be used to examine insect remains or to count pollen grains and seeds, amongst others.

MATERIALS USUALLY studied by biologists can also provide important information for archaeologists about former landscapes, economies, and environments. Wet environments, such as peat bogs, lake sediments, and waterlogged deposits, can preserve macrofossils – leaves, twigs, seeds, and nuts – and microfossils, such as pollen grains. Pollen can give us a clear view of the vegetation – woodland, scrubland, or open country – and, if we have a sequence of samples, we can study changes through time and see the impact of people on their natural environment.

MACROFOSSILS

The macrofossils from a site can reveal something of the diet of the site's inhabitants, aspects of their technology, and their management of the landscape, while insects and fungi can give a detailed picture of the local environment. Elsewhere,

in chalk and limestone areas, microscopic land snails can be used in the same way as pollen grains. This is because certain species of mollusks prefer particular environments, some inhabiting damp woodland, while others enjoy dry grassland. So the changes in their population can be very good indicators of the local environment. Using a flotation tank, archaeologists in the field can obtain macrofossils from archaeological deposits, which can then be examined by specialists.

Other samples are taken by environmental archaeologists when they remove a stratified core of soil or peat. The analysis of such samples is carried out in a laboratory by specialists. Some environments, such as the very dry conditions found in the deserts of Egypt and South America, also preserve materials that can be analyzed in this way. Frozen conditions preserve other types of

evidence. The work of archaeologists often relies on the preservation of biological materials that would usually rot away, found in special environments. It also relies on the presumption that past species required the same conditions as similar ones do today. Analysis of these microscopic findings can provide essential information for understanding aspects of our ancestors' health and daily life.

FLOTATION TANK
The sample of soil is placed on a mesh on the top of a barrel. Water is pumped through and floats off the lighter, vegetative material, which is collected in two fine-meshed sieves.

MARINE ARCHAEOLOGY

FROM THE EARLIEST times, people have used seas, lakes, and rivers as an important resource for food, transportation, and power. Marine archaeology is therefore an important partner of land archaeology. Investigation under water was made possible by the invention of the Aqua-Lung and has become

mainstream as underwater technology improved during the 20th century.

The principles of underwater excavation are the same as for that on land. Sites are surveyed, recorded, and excavated; the same approaches are used, although adapted for wet conditions. Some practicalities make excavation more difficult, such as the limitations placed on divers and poor visibility. Some, however, are beneficial, such as the ability to float over the site and the easy removal of the spoil. The preservation of materials under marine conditions is very different from that on dry land, with organic materials, such as wood, often surviving very well, while metal objects become sealed in mineral deposits. Objects from water sites need careful conservation if they are to survive once excavated.

Marine archaeology is used to investigate two key areas of archaeology. First, in many parts of the world, coastlines and water levels have altered considerably since earlier times, and submerged ground surfaces and structures can be excavated along with harbors and sea defenses. Second, marine archaeology can facilitate the recovery of shipwrecks and their contents. They can tell us much about the technological achievement of a society, as well as its trading and economic contacts. Shipwrecks are important to archaeologists because they are in "closed context" (artifacts relate to one specific moment). One of the biggest problems faced by marine archaeologists is the difficulty

in protecting underwater sites from treasure-hunters. If they have not already been looted by the time the archaeologists arrive, they are particularly difficult to protect.

USING A LIFT
The diver is carefully using a lift to suck away overlying deposits of silt and sand, in order to expose the site and its associated objects.

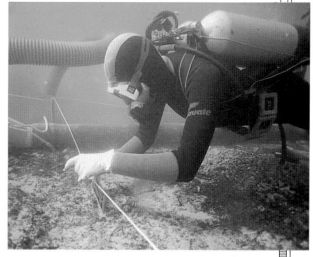

WORKING IN THE DEPTHS
The diver is working within a previously positioned grid of lines and pegs, so that the area being excavated can be related to the overall plan.

POT RECONSTRUCTION

UNTIL THE DEVELOPMENT of modern materials such as plastic, pottery was the most widely used material for storage, cooking, and tableware. However, pottery is easily breakable, and is useless once broken – it cannot be recycled and must be discarded. Once clay is fired to make pottery, it becomes very durable and survives well in most soil conditions. Its wide distribution and durability make pottery one of the most common finds on sites dating from the Neolithic period onward.

Pottery is very useful to archaeologists for dating and identifying sites. Its common occurrence, the developments in its technology (types of temper, firing

temperatures, and quality of the clay), and changes in decoration and style enable archaeologists to produce ceramic sequences for each site. In any archaeological field or excavation project where pottery is found, a type series is developed. This is a representative sample of the different fabrics and surface applications – glazes, slips, and decorations – encountered. The type series then acts as a reference collection against which any shard discovered can be compared and identified. The type series can then be compared with other local collections of pottery and those from farther afield. Usually, there will be some background

knowledge of kilns, vessel types, shape, and decorative features, so that the shards can be identified, provenanced, and dated. Inevitably, a small number of shards will be unidentifiable, but by studying their

Vessel spout

PIECING TOGETHER
The shards are sorted into different types, and the rims, bases, and body shards are separated. Each shard is washed using water and a soft brush.

Water-soluble glue

GLUING THE PIECES
Shards are fitted together using glue and tape to support the drying joints. A water-soluble inert glue is used, so that the pot is not chemically damaged.

SKULL RECONSTRUCTION

RECONSTRUCTING FACES from skulls is a rapidly developing technique. It is often used in forensic science for victim identification, but it can also be useful for archaeological cases. Modern experiments have shown that there is a direct link between the shape and form of the skull and the appearance of the face, and that the face can be reconstructed using scientific methods. The skull shown here is that of Horemkenesi, an ancient Egyptian court official, whose mummified body was found in a tomb near the pyramids of Giza, and whose life was detailed in inscriptions. Specialists at Bristol City Museum, in England, have completed a long-term project to unwrap and reconstruct Horemkenesi's remains.

Initially, the skull must be cleaned and studied – reconstruction can be carried out only if enough of the skull survives. Because skulls are often fragile, and may need to be

Nose cavity

THE ORIGINAL SKULL
Horemkenesi's skull, complete and fairly well preserved, determines the shape of the final facial reconstruction.

studied again or reburied, a plaster cast is taken for the reconstruction to be built upon. This is usually done by wrapping the skull in aluminum foil for protection, and making a mold, which is filled with plaster to create a solid skull cast.

For a reconstruction to be as accurate as possible, background information about the individual is gathered from investigation of the skeleton and the context of the burial. Ideally, the age, sex, and ethnic group of the person should be known. Statistical tables have been built up in recent decades, using dissected bodies and ultrasonic probing, which records the average depth of flesh at certain key points over the skull. These vary according to sex and ethnic origin, so it is essential that these facts are ascertained. Pegs are inserted into about 30 places on the cast to indicate the depth of flesh in each area. Unfortunately, these measurements cannot take into account the weight of the individual, as it is not possible to discover this from the skull.

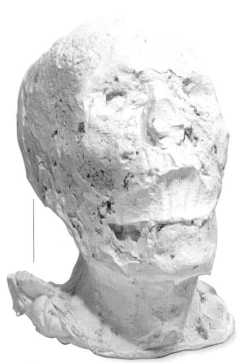

COPYING THE SKULL
A cast is made of the skull to use as the basis for the reconstruction, and then layers of muscle and tissue are built up over the cast of the skull, using clay. A neck has been added to make the head look more lifelike and in proportion.

typological context, archaeologists can gain some information while research from other sites is compiled. Most pottery finds are small, broken fragments called shards. Complete or near-complete vessels are rarely found, but the importance and abundance of ceramics means a great deal of attention is paid to the process of reconstructing shards into vessels. Pottery shards fall into four types – rims, bases, body shards, and special features, such as handles, spouts, or legs. Some shards provide more information than others; body shards provide little information about the original vessel, but just one rim or base shard can reveal the diameter of the vessel. In some cases, where diagnostic shards (fragments that provide information about the shape and size of the vessel) are present, it is possible to reconstruct the whole pot. As little as ten percent of the pot is enough to do this if a representative profile of the vessel survives – the rest can be built with modern materials.

Recently developed scientific techniques have increased what archaeologists can learn from pottery shards. Thin sectioning provides archaeologists with a cross section of the fabric of the pot for microscopic analysis, so that archaeologists can identify geologic inclusions and other materials in the fabric and identify where they came from. In some rare cases, lipid analysis can be carried out to identify minute traces of substances once held in the pot that have seeped into the fabric, such as fats.

Adding plaster

SEALING THE CRACKS
Once the pot is dry, gaps between the shards are filled using plaster. It is later smoothed to make the fill unobtrusive. It is then ready to be painted.

Vessel spout

A RECONSTRUCTED JAR
By using a type sequence, this vessel has been revealed as a "hole-mouth" jar. It was used for storing olive oil, which would be poured off through the spout.

Several other key features on the skull are added at this stage – the edges of the eyes and mouth are marked, and the nose is scaffolded. The nose is difficult to reconstruct because, although the skull gives some indication of the size and shape, the lack of bone in the nose means that it is partly an estimation.

Clay is used to build up the face and follows the anatomical structure of the muscles and tissues. Marks on the skull, indicating the position of the muscles, provide information about their strength and position – it is this that largely determines the character of the face. More clay strips are added and smoothed to provide skin, using the pegs as guidelines for depth. Superficial features, such as eyelids and eyebrows, are then sculpted. The mouth can be difficult, as there is no indication on the skull for its shape or fullness. Ears are also largely guesswork. Finally, details are added to the reconstruction based on archaeological evidence. The face can be aged according to indications from the skeleton, and other cultural details can be deduced from evidence such as hairstyles and jewelry, which tell us about the status of an individual.

FACES OF THE PAST

In archaeology, unlike forensic reconstruction, there is little possibility of achieving a perfect likeness. However, this technique can be useful at least in giving some idea of what actual individuals in the past may have looked like. This is especially important for known historical characters, whose casts can be compared with their portrayals in contemporary sculpture and art. It can also be used to give some idea of what some of our earliest human ancestors may have looked like.

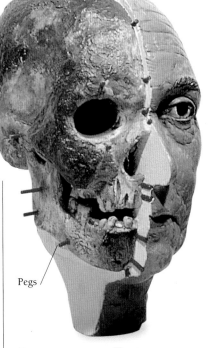

Pegs

BUILDING THE FACE
This cut-away reconstruction shows the pegs fixed to the replica skull at key points to indicate the depth of flesh that is to be built up. The other side shows the finished clay version. Horemkenesi's decayed teeth can be seen on the unfinished side.

Bronze finish

BRONZE CAST
A bronze cast of the finished reconstruction is ready for display in the museum. This is what Horemkenesi may well have looked like. It is probably a close enough likeness that those who knew him well – friends and relatives – would have recognized him.

EXCAVATING HUMAN REMAINS

DURING ARCHAEOLOGICAL excavations, evidence of human remains need more detailed and specialized treatment than any other finds or features. This is because they are more complicated and can

A SKELETON
This body was carefully placed in the grave after funerary rituals had been carried out. Individuals were often buried close to their dead relatives.

provide so much more information about individuals in the past. For human skeletons there are often many religious and social considerations that must be respected during investigation because certain groups, such as the Inuits and Orthodox Jews, question the validity and morality of the excavation of their ancestors.

Skeletons are found in many different states of preservation, depending on the materials that they were buried in, how old the body is, and, most important, what the ground conditions are. Bones do not survive well in acidic soils, and at some sites, like the ship burial at Sutton Hoo in England, they have decayed completely, leaving only stains in the soil. In waterlogged conditions, where skin and soft

tissue can survive, bones often lose their structure and are reduced to collagen. However, at many other sites skeletons are well preserved, and the teeth (which are made of the toughest material in the body) are usually the best preserved of all.

CAREFUL REMOVAL
When graves are discovered, the archaeologist excavates them according to the rules of stratigraphy. The uppermost layer of soil is the most recent, so the grave is emptied in the opposite sequence to the way it was originally dug and filled. The way the grave is positioned in relation to other features, finds, and burials can be best established using this method. Sometimes, evidence for the manner in which the body

was buried is found in the grave, such as finds of fragments of cloth, animal skins, wood, nails, or even small shroud pins.

Once the skeleton has been encountered in the excavation, small tools, often made of wood so as not to damage the bone, are used to clean it and reveal all the bones without moving them. It is then recorded *in situ* using photographs, scale drawings, and specialized recording sheets, so that the exact position and condition of all the bones and artifacts are precisely recorded. Without this stage, important information about the method and type of burial, the condition of the bones, and unusual characteristics of the skeletons could be lost.

After the recording has been completed, the skeleton can be removed and taken to a laboratory. The bones can often be very fragile, so they are carefully packed and supported during transit. If scientific testing is to be carried out, samples of the bone are extracted from the skeleton and sealed separately. These samples are then used for DNA

ESTABLISHING SEX
The shape of the skull, especially around the eyes (orbits), the chin, and the teeth, differs between males and females. This, along with any artifacts, usually makes it possible to determine the sex and social status of the skeleton.

CULTURAL CLUES
Occasionally, hair, shoes, brooches, and even metal stains on the skeleton can give evidence about burial preparations. It can reveal how hair was worn or how coins were placed in the deceased's mouth to pay the way to the afterlife.

Six-strand braid of hair

Diseased foot

EVIDENCE OF DISEASE
A surprising number of skeletons from one ancient cemetery showed evidence of psariatic arthritis. A predisposition to develop this condition can be inherited, but environment can also play a part, such as high levels of lead in the diet.

testing, and sometimes for electron spin resonance (ESR). From analysis of the stable isotopes, the importance of fish, animals, and plants in the diet of the individual can be revealed.

LABORATORY WORK
In the laboratory, the bones are fully cleaned and studied by an osteologist. If enough of the skeleton remains and is in good enough condition, characteristics of the dead person can be revealed. In addition to the sex of the individual, age and height can be estimated using comparisons with modern populations. Signs of disease, ill health, and injury can also be found. If there are stomach contents, they can reveal information about the local diet. The most interesting conclusions can be drawn when a large number of bodies, often in cemeteries, are excavated – instead of reconstructing the appearance of one individual, the health and lifestyle of whole populations can be revealed.

CONSERVATION OF FINDS

CONSERVING STONE

Many of the greatest archaeological monuments around the world, from the temples of the Maya in Central America to the cathedrals of Europe, are built of stone. Despite the material's durability, ancient stonework can need intensive conservation. Modern industrial and traffic pollution, particularly in the form of acid rain, has accelerated the decay of many monuments at a staggering rate. Sites in or near big

CLEANING THE STONE
For standing buildings, the stonework must be cleaned and consolidated in situ. Here, a high-pressure microspray is being used to get rid of the dirt that has accumulated over time.

cities are often those thought to be the most threatened. The situation of the Acropolis in Athens or the pyramids in Egypt shows how the spread of cities can affect the monuments of the past.

Cleaning the stonework of buildings involves first removing the debris, such as soot and pollution, from the surface, which is often done using a spray of fine sand or water. Any buildings that retain traces of ancient paint or plaster will require special treatment. Once the moldings are clean, any damage to them can be assessed, and the stonework consolidated using chemicals. In some cases, where the building has decayed to a great extent, new stone must be applied to the exterior to replace damaged sections and, after weathering, it will blend in with the original stone.

PRESERVING FABRIC

Fabric rarely survives beneath the soil, and is usually found preserved in very wet, very dry, or extremely cold conditions, where bacteria and microorganisms cannot survive and start the process of decay. It is, therefore, not surprising that the permafrost of the Arctic, deserts, and marshlands have provided an impressive number of textile remains. When fabrics are found, they are generally very fragile and are usually lifted from the excavation with soil still around them. They are then cleaned in the laboratory.

The biggest problem for wet fabric is the effect of bacteria and mold and of drying out, which can cause cracking and disintegration. When conserving dry fabrics, their brittleness and any insects must be taken into account. Often, refrigeration or freezing is the best way to stop decay. After cleaning, the fabrics are stored at a stable humidity, under the correct lighting and temperature conditions, and are monitored carefully for bacteria and insects. The continuing advances in science have helped conservators in their work, so that there is little that cannot be conserved for future generations, providing the costs can be met.

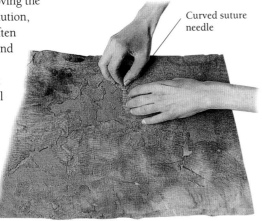

Curved suture needle

CONSERVING A LINEN TUNIC
Artifacts made of fabric can often be extremely brittle. Extreme care is taken when repairing the damaged areas, using a similar silk thread.

X-RAYING AND CONSERVING METALWORK

Metal objects are very sensitive to environmental conditions, particularly moisture, when they are buried, so they are often very decayed and corroded when they are discovered by archaeologists. The objects require careful conservation in order to prevent further corrosion once they come into contact with the air, and to reveal the original appearance of the object.

Metal objects are often x-rayed before conservation to establish what is underneath the outermost corroded layer. Sometimes the corrosion is removed by careful abrasion to reveal the features inside. Iron rusts, so it must be kept as dry as possible in an airtight container. Copper-alloy objects often need treatment to prevent "bronze disease" – the familiar verdigris color that can destroy the objects. Conservation cannot restore the original gold color of artifacts.

A layer of corroded material often builds up around the outside of the metals that have been buried, and this can become so large that the original object in the center of the mass is completely obscured. At underwater sites, the reaction between irons and salts produces a large concreted mass, which can envelop large numbers of objects. Before conservation, these corroded objects are x-rayed to reveal what is inside the mass. This shows details that might otherwise be lost or missed completely. If the corrosion were to be removed without the conservator knowing what shape exists inside, he or she may irreparably damage the object .

The process of decay cannot be reversed; it can be only contained, by keeping the object in as stable an atmosphere as possible and controlling the temperature and humidity in which it is stored.

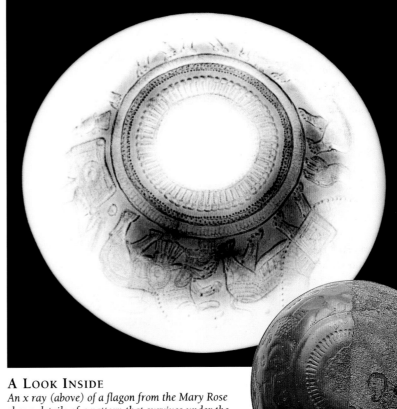

A LOOK INSIDE
An x ray (above) of a flagon from the Mary Rose shows details of a pattern that survives under the layer of corrosion. With the x ray as a guide, the corroded layer can be removed with little damage to the object. On the right, a copper-alloy bowl is shown in its unconserved (right side) and conserved state (left side). The corrosion on the bowl is removed by a scalpel under a microscope.

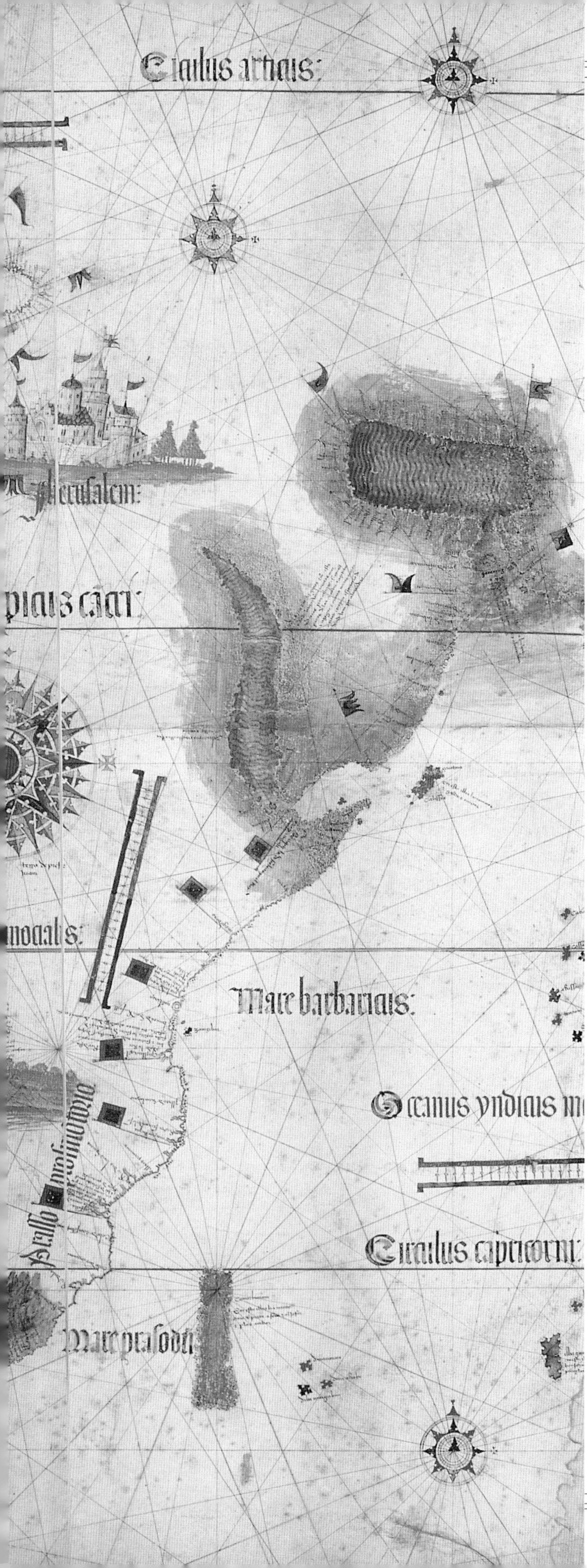

PART TWO
GAZETTEER
SITES

THE SITES FEATURED IN THE
PRECEDING CHAPTERS WILL, WE HOPE,
HAVE INSPIRED YOU TO LEARN MORE
ABOUT ARCHAEOLOGY AND THE WORLD'S
GREAT ARCHAEOLOGICAL SITES. IN THIS
GAZETTEER SECTION, WE HAVE LISTED MORE
THAN A THOUSAND ADDITIONAL SITES BY
GEOGRAPHIC LOCATION, WHICH WILL ENABLE
YOU TO LOCATE ARCHAEOLOGICAL SITES IN
YOUR OWN AREA. COMPARING SITES FROM
DIFFERENT PERIODS AND UNDERSTANDING
HOW THEY RELATE TO ONE ANOTHER
HISTORICALLY, GEOGRAPHICALLY, AND
CULTURALLY IS ONE OF THE MOST SATISFYING
ASPECTS OF AN ARCHAEOLOGIST'S WORK.
THIS GAZETTEER CLEARLY DEMONSTRATES THAT
OUR INTEREST IN THE PAST IS A PASSION SHARED
AROUND THE GLOBE BY PEOPLE OF ALL CULTURES.

MAP OF THE KNOWN WORLD, 1502
*During the late 15th and early 16th centuries, many European
countries sent ships on voyages of discovery around the globe.
This map was commissioned by a Portuguese duke after the
third voyage of Christopher Columbus. It shows, for the first
time, lands discovered in the New World, including the Lesser
Antilles and the coasts of Venezuela and Brazil. Illustrated
details include birds, castles, and African settlements.*

GAZETTEER

THE FOLLOWING PAGES (136–57) contain the 14 gazetteer maps, on which are plotted more than 1,200 archaeological sites. The map below shows how the globe is divided. On pages 158–97, gazetteer listings give details of each site. Every site is given a number, which corresponds to a point on the relevant map. There is a map reference, indicated by a letter and a number. A date is given, which in most cases shows first use or occupation of the site. In some cases, where the exact date is not certain or a site has been occupied over a long period, a more general date is given. The type of site follows, as well as

specific details about the location and any structures and artifacts excavated. The maps and listings have been researched to include archaeological sites from all over the world. Inevitably, some areas of the world have a higher concentration of sites than others. For example, the wealth of archaeology in Peru and the long history of research in the British Isles mean that both have their own maps and listings. The listings are extensive and aim to cover as many different types of archaeological site as possible, ranging from Inuit Arctic settlements and medieval abbeys, to ancient Egyptian temples.

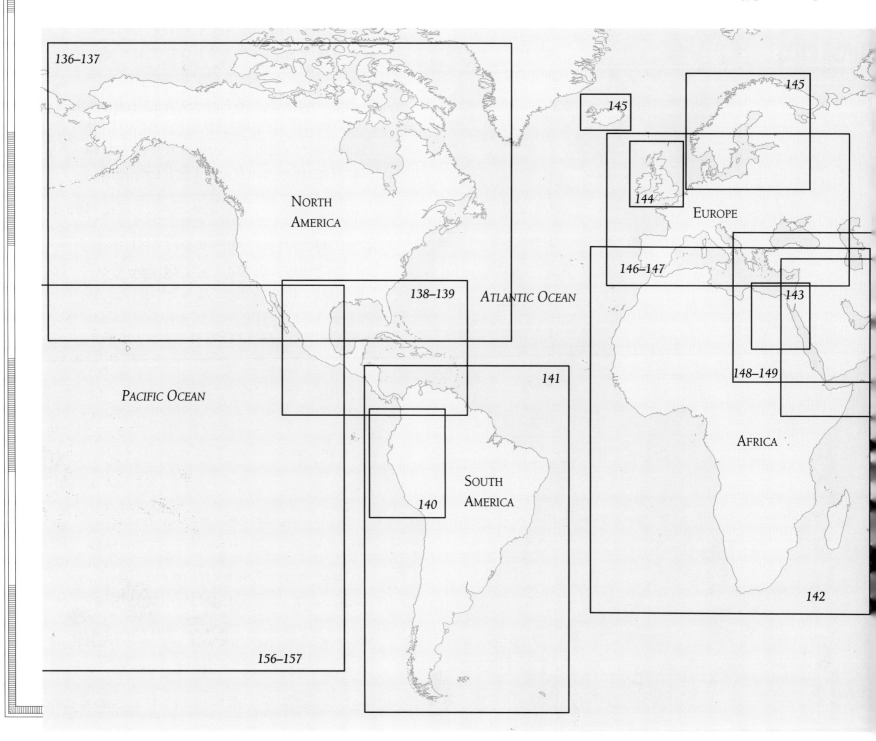

THE MAP SECTION

The world map below shows the area covered by each of the 14 maps in this section of the book. Each box refers to a map that is spread over one or two pages. The text box in the bottom right-hand corner of this page gives the title of every map and the pages on which it can be found with the relevant listings. Each map carries the following information:

• A globe at the side of each introductory paragraph shows the area covered by the map in relation to the whole globe.

• Arrows at the sides of the maps indicate the page numbers of adjacent geographic areas covered by other maps in the section.

• The historical periods when the sites were used or built are indicated by the color of the circle. A key to the colors is given on each double page.

• A short paragraph of text introduces the historical and archaeological background of each of the regions. Page numbers at the end of each paragraph refer to the relevant listings pages.

THE LISTINGS SECTION

For each map, there is a corresponding list. This list provides information about the sites plotted on the maps. These listings start on page 158: the text box below shows the pages on which particular listings can be found. Sites covered in detail in Part One are named only in the listings, although the reader will find that these sites are plotted on the gazetteer maps. The listings give the following information on the sites:

• The name of the site, its location, its type, and its approximate or exact date of use.

• A grid reference to locate the site on the relevant map.

• Site details noting any important structures or artifacts found.

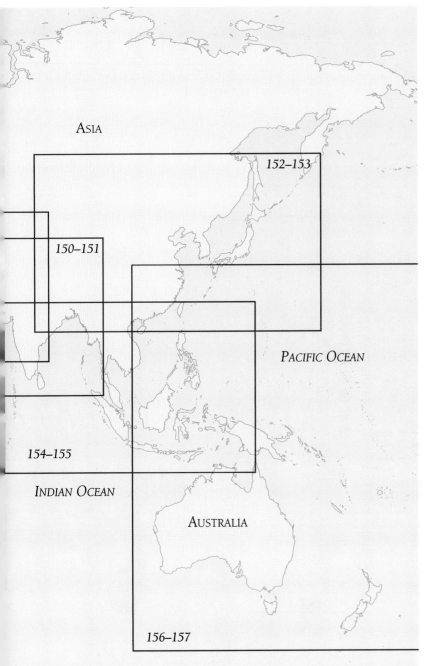

ASIA

150–151

152–153

PACIFIC OCEAN

154–155

INDIAN OCEAN

AUSTRALIA

156–157

KEY TO MAPS AND LISTINGS

GEOGRAPHIC AREA	MAPS	LISTINGS
NORTH AMERICA	136–137	158–160
CENTRAL AMERICA	138–139	161–163
PERU	140	164–165
SOUTH AMERICA	141	165–166
AFRICA	142	167–170
EGYPT	143	170–171
UNITED KINGDOM & IRELAND	144	172–174
SCANDINAVIA & THE BALTIC STATES	145	174–175
CONTINENTAL EUROPE	146–147	176–181
WESTERN ASIA	148–149	182–185
CENTRAL ASIA	150–151	186–188
EASTERN ASIA	152–153	189–192
SOUTHEAST ASIA	154–155	192–194
THE PACIFIC & AUSTRALASIA	156–157	195–197

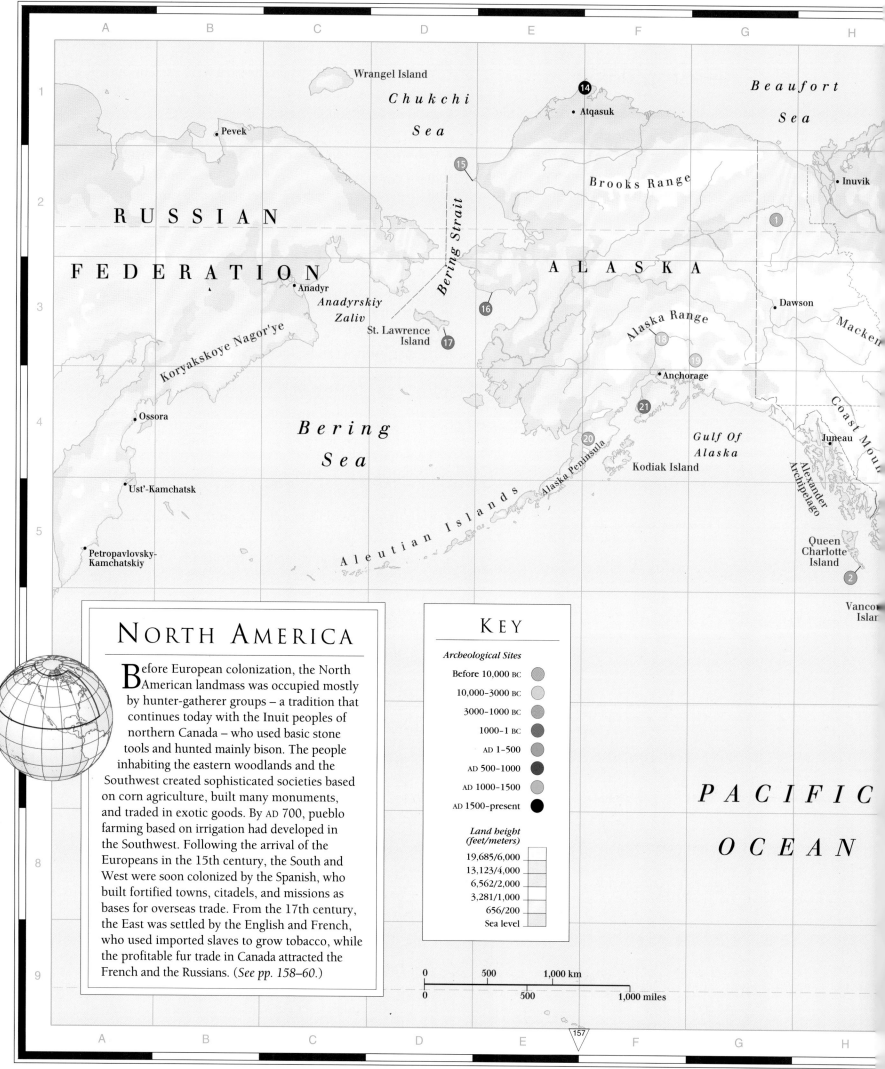

Wrangel Island

Chukchi Sea

14

•Atqasuk

Beaufort Sea

•Pevek

15

•Inuvik

RUSSIAN

Bering Strait

Brooks Range

FEDERATION

1

•Anadyr

ALASKA

Anadyrskiy Zaliv

16

•Dawson

Koryakskoye Nagor'ye

St. Lawrence Island

17

Alaska Range

18

Macken

•Ossora

19

•Anchorage

Coast Moun

•Juneau

21

Bering Sea

20

Gulf Of Alaska

Kodiak Island

Alexander Archipelago

•Ust'-Kamchatsk

Alaska Peninsula

Queen Charlotte Island

2

•Petropavlovsky-Kamchatskiy

Aleutian Islands

Vanco Islan

NORTH AMERICA

Before European colonization, the North American landmass was occupied mostly by hunter-gatherer groups – a tradition that continues today with the Inuit peoples of northern Canada – who used basic stone tools and hunted mainly bison. The people inhabiting the eastern woodlands and the Southwest created sophisticated societies based on corn agriculture, built many monuments, and traded in exotic goods. By AD 700, pueblo farming based on irrigation had developed in the Southwest. Following the arrival of the Europeans in the 15th century, the South and West were soon colonized by the Spanish, who built fortified towns, citadels, and missions as bases for overseas trade. From the 17th century, the East was settled by the English and French, who used imported slaves to grow tobacco, while the profitable fur trade in Canada attracted the French and the Russians. (*See pp. 158–60.*)

KEY

Archeological Sites

Before 10,000 BC

10,000–3000 BC

3000–1000 BC

1000–1 BC

AD 1–500

AD 500–1000

AD 1000–1500

AD 1500–present

Land height (feet/meters)

19,685/6,000
13,123/4,000
6,562/2,000
3,281/1,000
656/200
Sea level

PACIFIC

OCEAN

| 0 | 500 | 1,000 km |
| 0 | 500 | 1,000 miles |

157

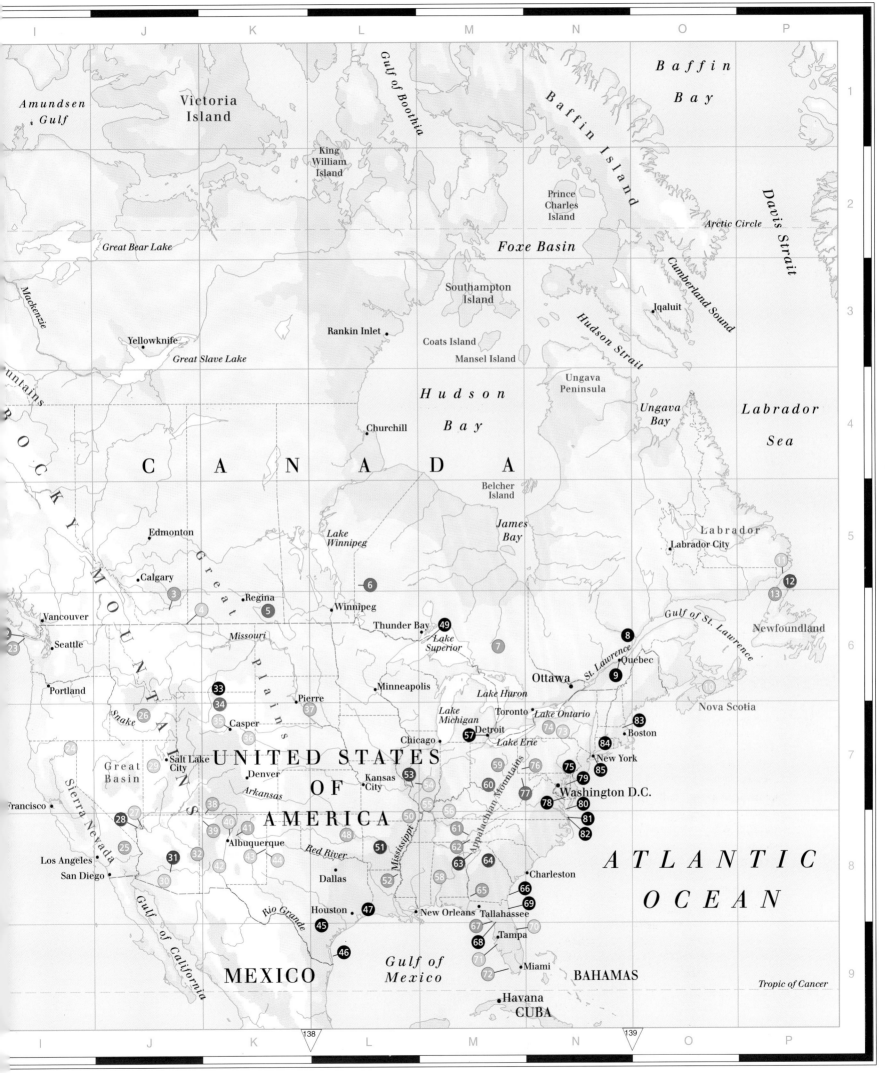

I J K L M N O P

Amundsen Gulf

Victoria Island

Gulf of Boothia

Baffin Bay

Baffin Island

King William Island

Prince Charles Island

Arctic Circle

Davis Strait

Great Bear Lake

Foxe Basin

Cumberland Sound

Iqaluit

Southampton Island

Yellowknife

Great Slave Lake

Rankin Inlet

Coats Island

Mansel Island

Hudson Strait

Ungava Peninsula

Ungava Bay

Labrador Sea

Mackenzie

Hudson Bay

Churchill

Belcher Island

Labrador

Labrador City

C A N A D A

James Bay

Edmonton

Lake Winnipeg

1

2

3

4

5

6

7

8

9

Calgary

Great Plains

Regina 5

Winnipeg

3

4

Thunder Bay 49

Lake Superior

6

7

8 8

9 9

Quebec

Gulf of St. Lawrence

1

12

13

Newfoundland

Vancouver

23

Seattle

Portland

Missouri

Minneapolis

St. Lawrence

Ottawa

10

Nova Scotia

Snake

26

33

34

35

Casper

Pierre

37

36

Lake Huron

Toronto

Detroit 57

Lake Michigan

Lake Ontario

74

73

83

84

Boston

New York

85

24

Great Basin

Salt Lake City

29

Denver

Kansas City

53

54

59

60

76

Appalachian Mountains

75

79

77

78

80

Washington D.C.

Francisco

Arkansas

UNITED STATES OF AMERICA

Sierra Nevada

28

27

38

39 40 41

Albuquerque

Red River

50

55

56

48

51

61

62

63

64

81

82

A T L A N T I C

25

31

32

42

43 44

Los Angeles

San Diego

30

Rio Grande

Dallas

52

58

65

Charleston

66

69

O C E A N

Houston 47

New Orleans

Tallahassee

67

68

70

Tampa

45

46

MEXICO

Gulf of California

Gulf of Mexico

71

72

Miami

BAHAMAS

Tropic of Cancer

Havana

CUBA

138

139

I J K L M N O P

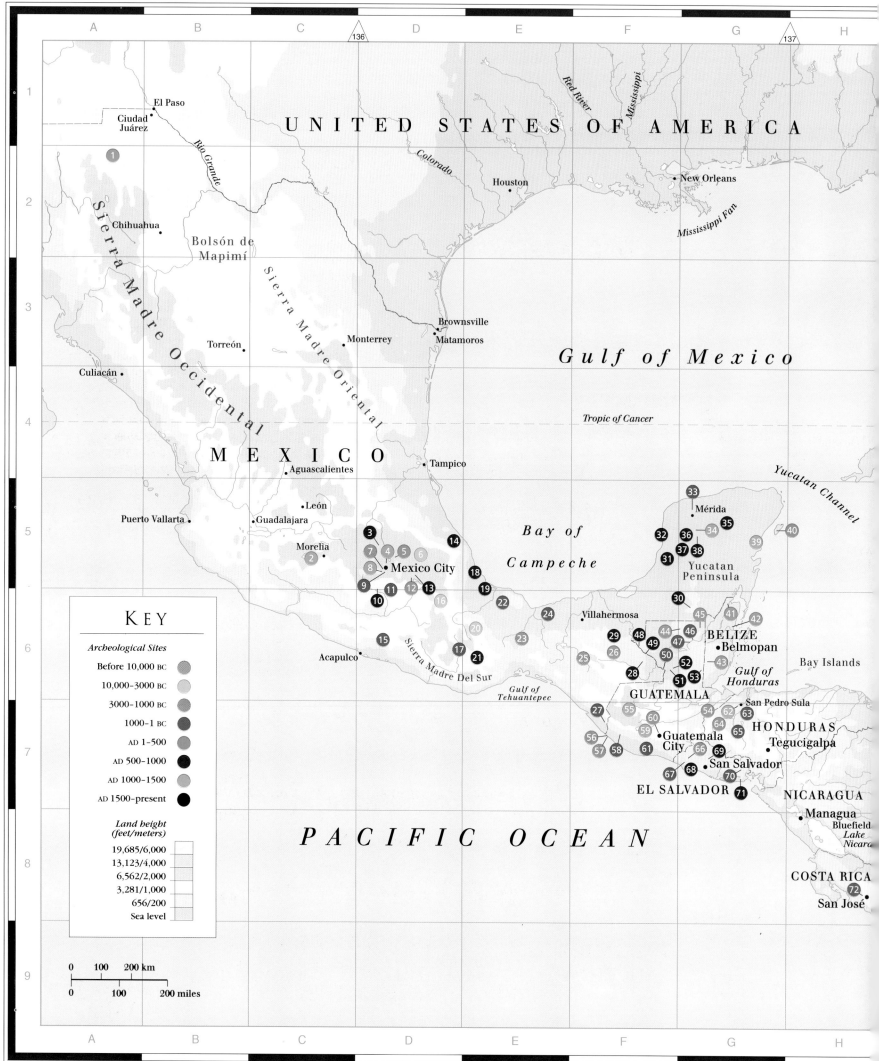

KEY

Archeological Sites

Before 10,000 BC
10,000–3000 BC
3000–1000 BC
1000–1 BC
AD 1–500
AD 500–1000
AD 1000–1500
AD 1500–present

Land height
(feet/meters)

19,685/6,000
13,123/4,000
6,562/2,000
3,281/1,000
656/200
Sea level

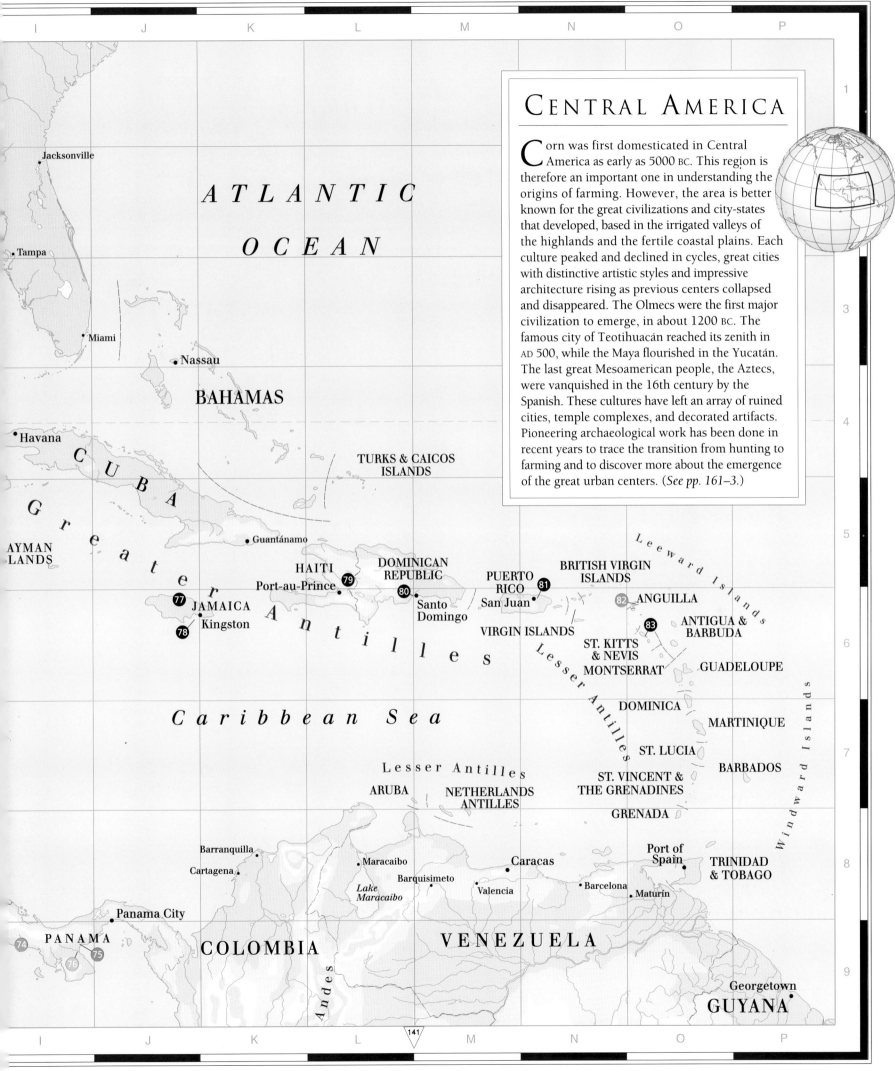

ATLANTIC

OCEAN

CENTRAL AMERICA

Corn was first domesticated in Central America as early as 5000 BC. This region is therefore an important one in understanding the origins of farming. However, the area is better known for the great civilizations and city-states that developed, based in the irrigated valleys of the highlands and the fertile coastal plains. Each culture peaked and declined in cycles, great cities with distinctive artistic styles and impressive architecture rising as previous centers collapsed and disappeared. The Olmecs were the first major civilization to emerge, in about 1200 BC. The famous city of Teotihuacán reached its zenith in AD 500, while the Maya flourished in the Yucatán. The last great Mesoamerican people, the Aztecs, were vanquished in the 16th century by the Spanish. These cultures have left an array of ruined cities, temple complexes, and decorated artifacts. Pioneering archaeological work has been done in recent years to trace the transition from hunting to farming and to discover more about the emergence of the great urban centers. (*See pp. 161–3.*)

Jacksonville

Tampa

Miami

Nassau

BAHAMAS

Havana

TURKS & CAICOS
ISLANDS

CUBA

Greater

AYMAN
LANDS

Guantánamo

Antilles

HAITI
Port-au-Prince

DOMINICAN
REPUBLIC

Santo
Domingo

PUERTO
RICO
San Juan

BRITISH VIRGIN
ISLANDS

Leeward Islands

ANGUILLA

79

80

81

82

ANTIGUA &
BARBUDA

83

77

JAMAICA

78

Kingston

VIRGIN ISLANDS

ST. KITTS
& NEVIS
MONTSERRAT

GUADELOUPE

Lesser Antilles

DOMINICA

MARTINIQUE

Caribbean   Sea

ST. LUCIA

BARBADOS

Lesser Antilles

ARUBA

NETHERLANDS
ANTILLES

ST. VINCENT &
THE GRENADINES

GRENADA

Windward Islands

Barranquilla

Cartagena

Maracaibo

Barquisimeto

Lake
Maracaibo

Caracas

Valencia

Barcelona

Maturín

Port of
Spain

TRINIDAD
& TOBAGO

Panama City

74

PANAMA

76

75

COLOMBIA

Andes

VENEZUELA

Georgetown

GUYANA

141

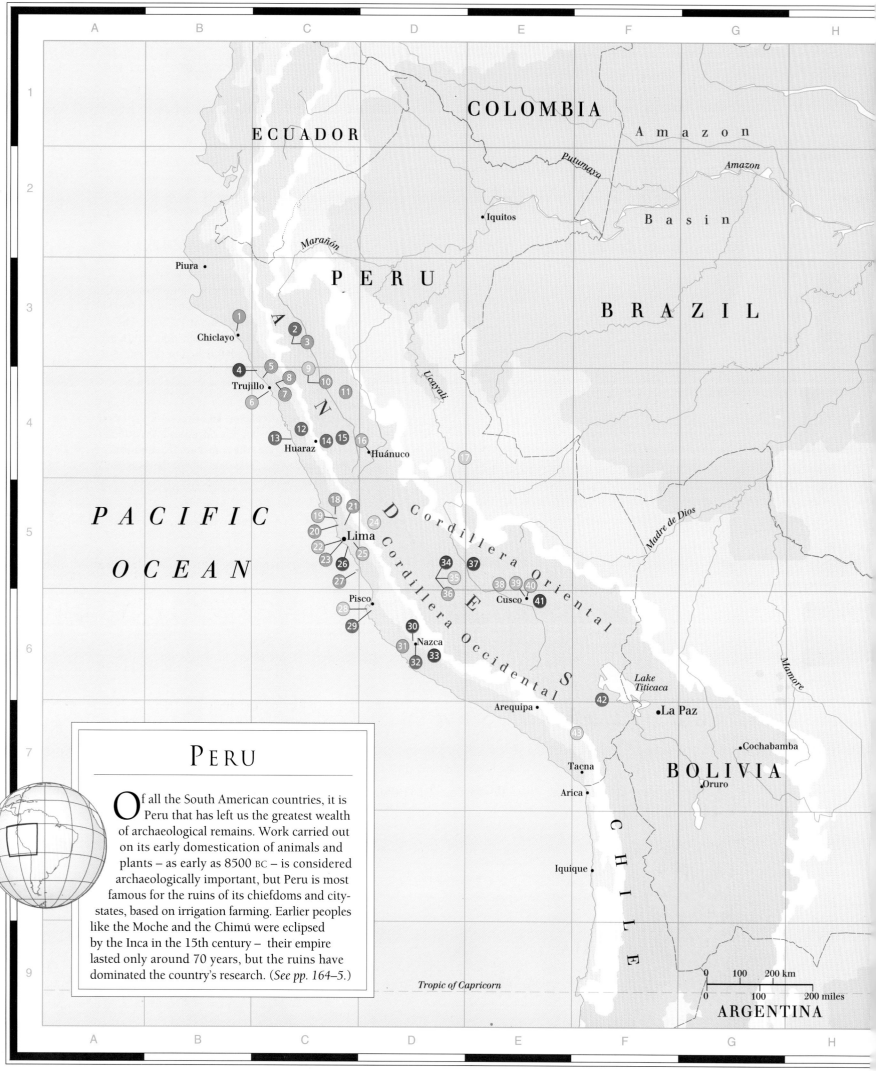

COLOMBIA

ECUADOR

Amazon

Putumayo

Amazon

• Iquitos

B a s i n

PERU

Marañón

Piura •

BRAZIL

Chiclayo •

Ucayali

① ② ③

④ ⑤ ⑨
⑧
Trujillo • ⑦ ⑩
⑥ ⑪

⑫ ⑬ ⑭ ⑮ ⑯
Huaraz   Huánuco •

⑰

Madre de Dios

PACIFIC

⑱ ㉑
⑲ ㉔
⑳ Lima
㉒ ㉓ ㉕ ㉖
㉗

OCEAN

Cordillera Oriental

Cordillera Occidental

㉞ ㊲
㉟
㊱ ㊳㊴㊵
Cusco • ㊶

Pisco •
㉘
㉙

㉚ Nazca
㉛ ㉝
㉜

Lake Titicaca

㊷

Arequipa •   • La Paz

㊸   • Cochabamba

Tacna •   BOLIVIA
• Oruro

Arica •

CHILE

PERU

Of all the South American countries, it is
Peru that has left us the greatest wealth
of archaeological remains. Work carried out
on its early domestication of animals and
plants – as early as 8500 BC – is considered
archaeologically important, but Peru is most
famous for the ruins of its chiefdoms and city-
states, based on irrigation farming. Earlier peoples
like the Moche and the Chimú were eclipsed
by the Inca in the 15th century – their empire
lasted only around 70 years, but the ruins have
dominated the country's research. (*See pp. 164–5.*)

Iquique •

Tropic of Capricorn

0   100   200 km

0   100   200 miles

ARGENTINA

NICARAGUA
Managua
San José
COSTA
RICA
Panama
City
PANAMA

Caribbean Sea

Trujillo

VENEZUELA
Caracas

Orinoco

Georgetown
GUYANA   Paramaribo
SURINAME   Cayenne
FRENCH
GUIANA

Magdalena

Bogotá
COLOMBIA

Equator

Quito
ECUADOR

Putumayo

Negro

Amazon Basin

Amazon

Guiana Highlands

Fortaleza

Galapagos
Islands

Marañón

Madeira

Tapajós

Xingu

Araguaia

Tocantins

Recife

PERU

B R A Z I L

São Francisco

Salvador

Lima

ANDES

PACIFIC

OCEAN

BOLIVIA
La Paz

Mato Grosso
Plateau

Brasilia

Brazilian
Highlands

Tropic of Capricorn

PARAGUAY

Gran Chaco

Paraguay

Rio de Janeiro

São Paulo

Paraná

Asunción

ATLANTIC

OCEAN

Atacama Desert

Porto
Alegre

Santiago
ANDES   ARGENTINA   URUGUAY

Pampas

Buenos
Aires   Montevideo

Colorado

CHILE

Isla de
Chiloé

Patagonia

Deseado

Isla Wellington

Rio Galegros

Falkland
Islands

Tierra del
Fuego

Cape Horn

SOUTH AMERICA

South America has an archaeological
tradition and a historical development
distinct from the rest of the world. The
many hunter-gatherer sites, often with long
periods of occupation, show the transition
to farming, while the later empires of the
Andes established a network of roads and
cities that lasted until Spanish colonization in
the 16th century. Because much of South America
is relatively inaccessible, much archaeological
work is needed before the continent's history can
be said to be fully understood. (*See pp. 165–6.*)

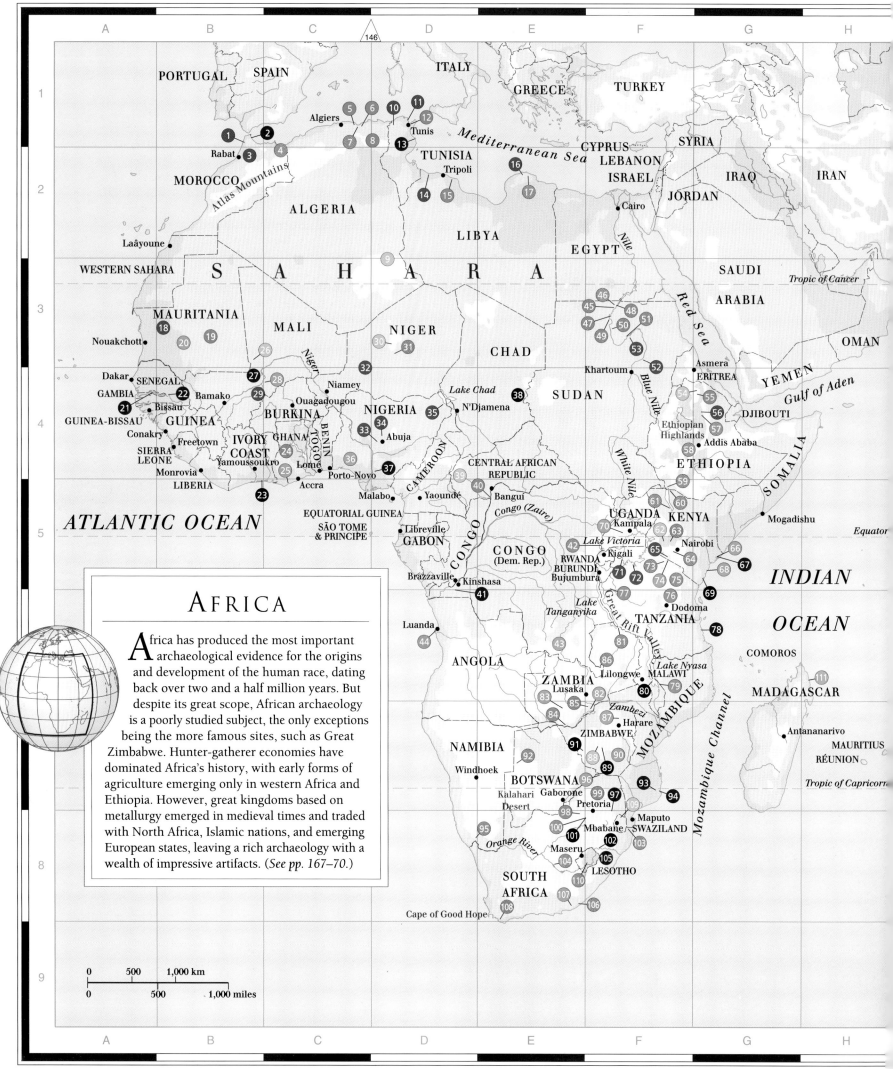

AFRICA

Africa has produced the most important archaeological evidence for the origins and development of the human race, dating back over two and a half million years. But despite its great scope, African archaeology is a poorly studied subject, the only exceptions being the more famous sites, such as Great Zimbabwe. Hunter-gatherer economies have dominated Africa's history, with early forms of agriculture emerging only in western Africa and Ethiopia. However, great kingdoms based on metallurgy emerged in medieval times and traded with North Africa, Islamic nations, and emerging European states, leaving a rich archaeology with a wealth of impressive artifacts. (*See pp. 167–70.*)

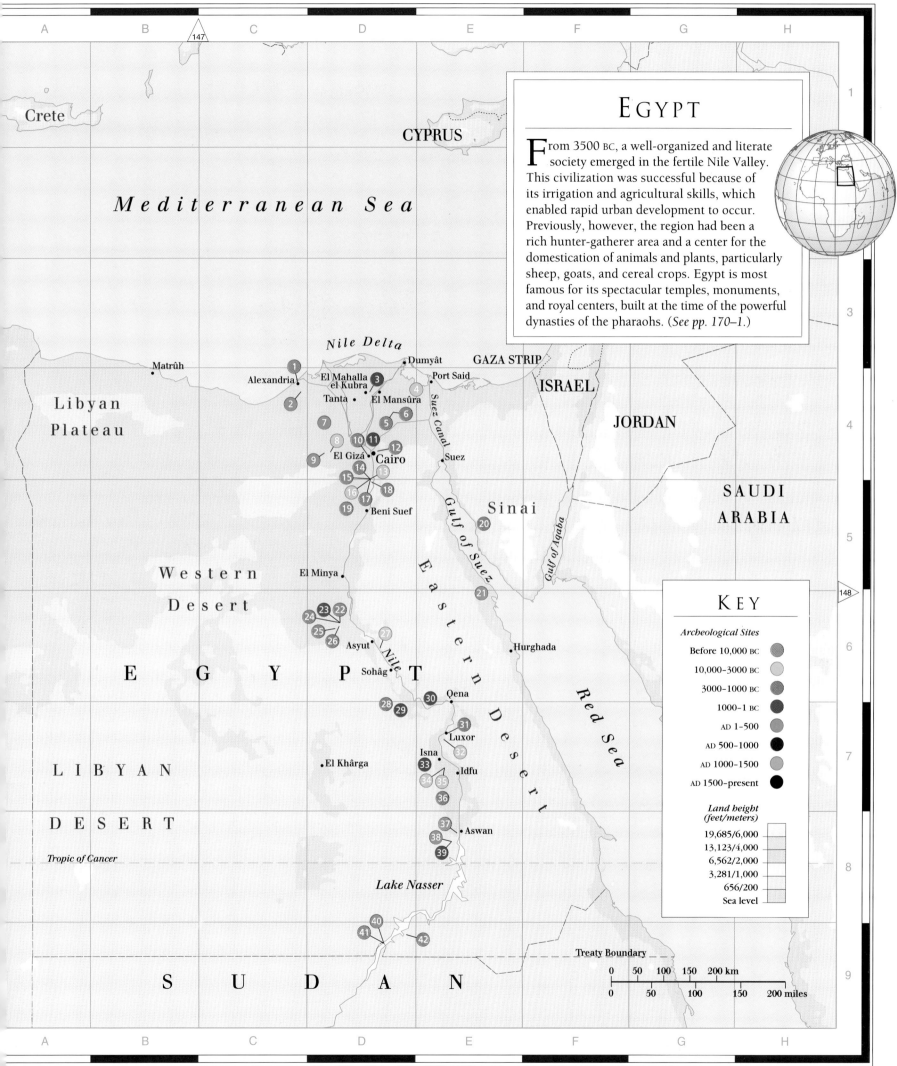

EGYPT

From 3500 BC, a well-organized and literate society emerged in the fertile Nile Valley. This civilization was successful because of its irrigation and agricultural skills, which enabled rapid urban development to occur. Previously, however, the region had been a rich hunter-gatherer area and a center for the domestication of animals and plants, particularly sheep, goats, and cereal crops. Egypt is most famous for its spectacular temples, monuments, and royal centers, built at the time of the powerful dynasties of the pharaohs. *(See pp. 170–1.)*

Crete

CYPRUS

Mediterranean Sea

Nile Delta

Matrûh

Dumyât

GAZA STRIP

Alexandria

Port Said

ISRAEL

El Mahalla el Kubra

Tanta

El Mansûra

Suez Canal

JORDAN

Libyan Plateau

Suez

El Gizá

Cairo

SAUDI ARABIA

Beni Suef

Sinai

Western Desert

El Minya

Gulf of Suez

Gulf of Aqaba

Eastern Desert

Hurghada

Asyut

Nile

Sohâg

Red Sea

Qena

Luxor

El Khârga

Isna

Idfu

E G Y P T

Aswan

Tropic of Cancer

Lake Nasser

L I B Y A N   D E S E R T

S U D A N

KEY

Archeological Sites

Before 10,000 BC

10,000–3000 BC

3000–1000 BC

1000–1 BC

AD 1–500

AD 500–1000

AD 1000–1500

AD 1500–present

Land height (feet/meters)

19,685/6,000

13,123/4,000

6,562/2,000

3,281/1,000

656/200

Sea level

Treaty Boundary

| 0 | 50 | 100 | 150 | 200 km |
| 0 | 50 | 100 | 150 | 200 miles |

UNITED KINGDOM & IRELAND

An immense number of archaeological sites exist in the British Isles. First settled in the Upper Paleolithic period, the islands contain a vast array of sites that range from prehistoric burial chambers and ritual monuments to the many remains of the Roman occupation, and from an array of medieval landscapes to the world's first industrial factories. In part, the wealth of archaeological evidence reflects the long-lived and dense settlement of these small islands, but it is also a result of a long history of archaeological research in these countries. As well as the multitude of sites that continue to be discovered below the soil, many locations, particularly those of later periods, survive as standing monuments or continue to flourish as modern settlements. (*See pp. 172–4.*)

KEY

Archeological Sites

- Before 10,000 BC
- 10,000–3000 BC
- 3000–1000 BC
- 1000–1 BC
- AD 1–500
- AD 500–1000
- AD 1000–1500
- AD 1500–present

Land height (feet/meters)

- 19,685/6,000
- 13,123/4,000
- 6,562/2,000
- 3,281/1,000
- 656/200
- Sea level

0   50   100 km
0   25   50   75   100 miles

Shetland Islands

0   50 km
0   50 miles

Arctic Circle

ICELAND

Reykjavík

0   100 km
0   100 miles

Tromsø

NORWEGIAN
SEA

Arctic Circle

Lapland

Kiruna

SCANDINAVIA &
THE BALTIC STATES

The most prominent archeological sites in this region are associated with the Vikings of the 7th to 10th centuries AD. Included among the many remains are their towns, some deserted and preserved; forts; and many burial sites, some containing remarkable finds of ships and treasures. However, there are also rich prehistoric finds, from Mesolithic sites to the famous Iron Age "bog bodies." Many of the archaeological remains owe their spectacular quality to the high number of waterlogged sites that have been left in a very good state of preservation. Medieval Scandinavia contained a number of flourishing towns and villages that developed and prospered well into the Industrial Age and still contain a wealth of information for archaeologists. (*See pp. 174–5.*)

White Sea

FINLAND

Vaasa

Petrozavodsk

Lake
Onega

Lake
Ladoga

Trondheim

N
O
R
W
A
Y

S
W
E
D
E
N

Gulf of Bothnia

Turku

Helsinki

St. Petersburg

Bergen

Oslo

Örebro

Uppsala

Stockholm

Tallinn

ESTONIA

RUSSIAN

FEDERATION

Vänern

Vättern

Kristiansand

Gothenburg

Gotland

NORTH
SEA

DENMARK

Ålborg

Copenhagen

B
a
l
t
i
c
　
S
e
a

Riga

LATVIA

Moscow

Dvina

LITHUANIA

North European Plain

Kaliningrad

Vilnius

Gdansk

Minsk

0   100   200 km
0   100   200 miles

GERMANY

Hamburg

POLAND

BELARUS

CONTINENTAL EUROPE

Europe, like the British Isles, has a rich archaeological heritage, reflecting its long history of research. It also has a long, complex, and diverse history of human occupation, ranging from the Paleolithic cave sites of Spain and France to the Iron Age towns and hill forts that can be found across the region. The prehistoric period is also well represented by burial monuments, ceremonial sites, and human remains. The first large-scale political power to spread across to the east of Europe was the Greek Empire. It was followed by the Roman Empire, which controlled more than half of Europe and traded with the East. The rich material cultures of the peoples of the so-called Dark Ages developed into the medieval world, characterized by typical European features such as cathedrals, monasteries, and castles. Even the emerging modern world has left a rich archaeology of agricultural landscapes and towns, including the distinctive remains of the industrial revolution. (*See pp. 176–81.*)

KEY

Archeological Sites

Before 10,000 BC
10,000–3000 BC
3000–1000 BC
1000–1 BC
AD 1–500
AD 500–1000
AD 1000–1500
AD 1500–present

Land height (feet/meters)

19,685/6,000
13,123/4,000
6,562/2,000
3,281/1,000
656/200
Sea level

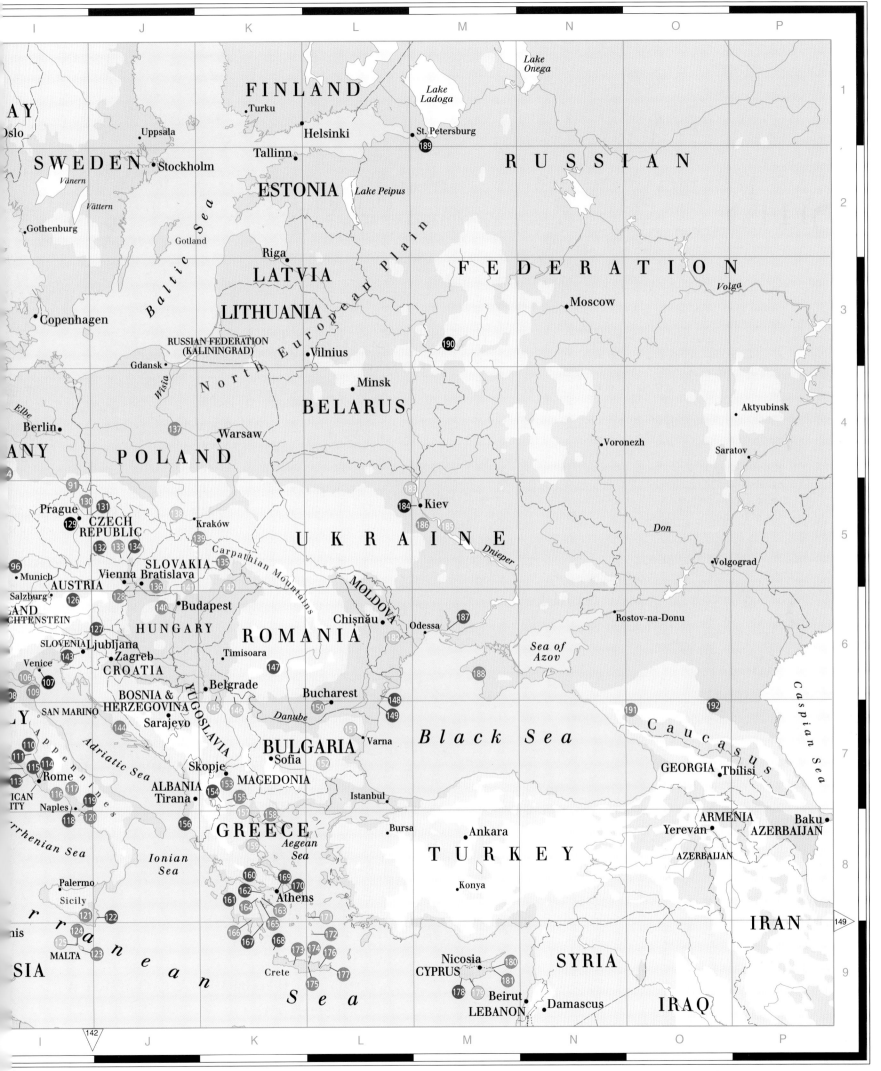

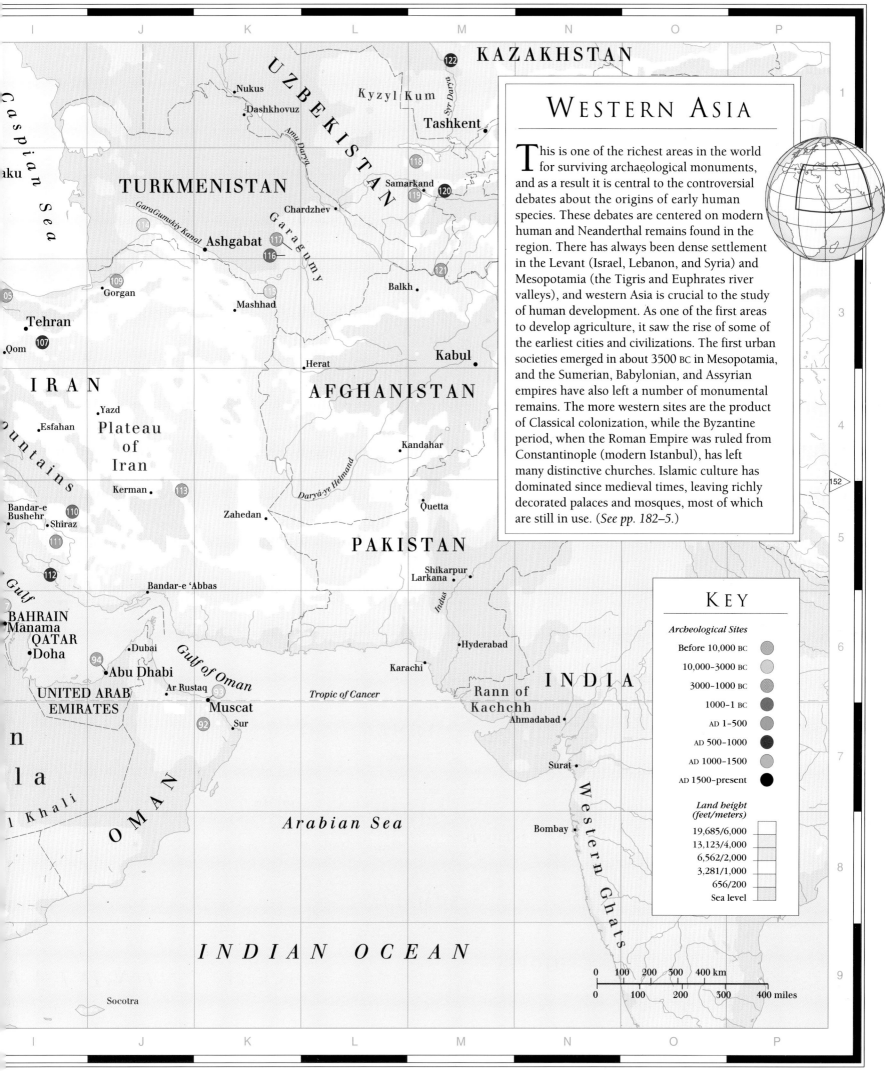

KAZAKHSTAN

UZBEKISTAN

TURKMENISTAN

Nukus
Dashkhovuz
Tashkent
Kyzyl Kum
Syr Darya
Samarkand
Chardzhev
Amu Darya
GaraGumskiy Kanal
Ashgabat
Garagumy
Balkh
Gorgan
Mashhad
Tehran
Qom
Herat
Kabul
AFGHANISTAN
Yazd
Esfahan
Plateau
of
Iran
Kandahar
Kerman
Zahedan
Daryā-ye Helmand
Bandar-e
Bushehr
Shiraz
Quetta
Mountains
IRAN
Bandar-e 'Abbas
PAKISTAN
Shikarpur
Larkana
Indus
BAHRAIN
Manama
QATAR
Doha
Dubai
Abu Dhabi
Ar Rustaq
UNITED ARAB
EMIRATES
Muscat
Sur
Gulf of Oman
Tropic of Cancer
Hyderabad
Karachi
INDIA
Rann of
Kachchh
Ahmadabad
OMAN
al Khali
Arabian Sea
Surat
Western Ghats
INDIAN OCEAN
Bombay
Socotra

Caspian Sea
aku
05
Gulf

WESTERN ASIA

This is one of the richest areas in the world for surviving archaeological monuments, and as a result it is central to the controversial debates about the origins of early human species. These debates are centered on modern human and Neanderthal remains found in the region. There has always been dense settlement in the Levant (Israel, Lebanon, and Syria) and Mesopotamia (the Tigris and Euphrates river valleys), and western Asia is crucial to the study of human development. As one of the first areas to develop agriculture, it saw the rise of some of the earliest cities and civilizations. The first urban societies emerged in about 3500 BC in Mesopotamia, and the Sumerian, Babylonian, and Assyrian empires have also left a number of monumental remains. The more western sites are the product of Classical colonization, while the Byzantine period, when the Roman Empire was ruled from Constantinople (modern Istanbul), has left many distinctive churches. Islamic culture has dominated since medieval times, leaving richly decorated palaces and mosques, most of which are still in use. (*See pp. 182–5.*)

152

KEY

Archeological Sites

Before 10,000 BC
10,000–3000 BC
3000–1000 BC
1000–1 BC
AD 1–500
AD 500–1000
AD 1000–1500
AD 1500–present

*Land height
(feet/meters)*

19,685/6,000
13,123/4,000
6,562/2,000
3,281/1,000
656/200
Sea level

0   100   200   300   400 km
0   100   200   300   400 miles

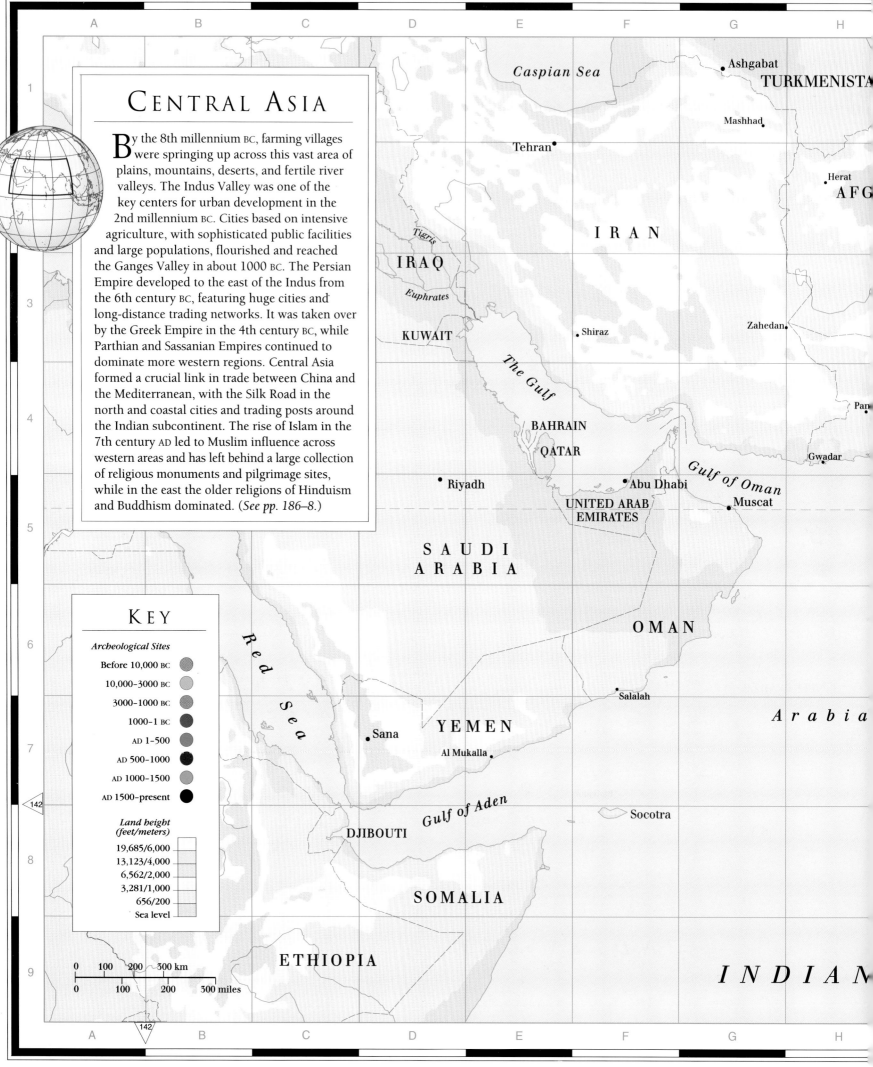

CENTRAL ASIA

By the 8th millennium BC, farming villages were springing up across this vast area of plains, mountains, deserts, and fertile river valleys. The Indus Valley was one of the key centers for urban development in the 2nd millennium BC. Cities based on intensive agriculture, with sophisticated public facilities and large populations, flourished and reached the Ganges Valley in about 1000 BC. The Persian Empire developed to the east of the Indus from the 6th century BC, featuring huge cities and long-distance trading networks. It was taken over by the Greek Empire in the 4th century BC, while Parthian and Sassanian Empires continued to dominate more western regions. Central Asia formed a crucial link in trade between China and the Mediterranean, with the Silk Road in the north and coastal cities and trading posts around the Indian subcontinent. The rise of Islam in the 7th century AD led to Muslim influence across western areas and has left behind a large collection of religious monuments and pilgrimage sites, while in the east the older religions of Hinduism and Buddhism dominated. (*See pp. 186–8.*)

KEY

Archeological Sites

Before 10,000 BC
10,000–3000 BC
3000–1000 BC
1000–1 BC
AD 1–500
AD 500–1000
AD 1000–1500
AD 1500–present

Land height (feet/meters)

19,685/6,000
13,123/4,000
6,562/2,000
3,281/1,000
656/200
Sea level

0   100   200   300 km
0   100   200   300 miles

Caspian Sea
Ashgabat
TURKMENISTA
Mashhad
Tehran
Herat
AFG
IRAN
Tigris
IRAQ
Euphrates
Zahedan
Shiraz
KUWAIT
The Gulf
Pan
BAHRAIN
QATAR
Gwadar
Gulf of Oman
Riyadh
Abu Dhabi
UNITED ARAB
EMIRATES
Muscat
SAUDI
ARABIA
OMAN
Red Sea
Salalah
Arabia
YEMEN
Sana
Al Mukalla
Gulf of Aden
Socotra
DJIBOUTI
SOMALIA
ETHIOPIA
INDIAN

142

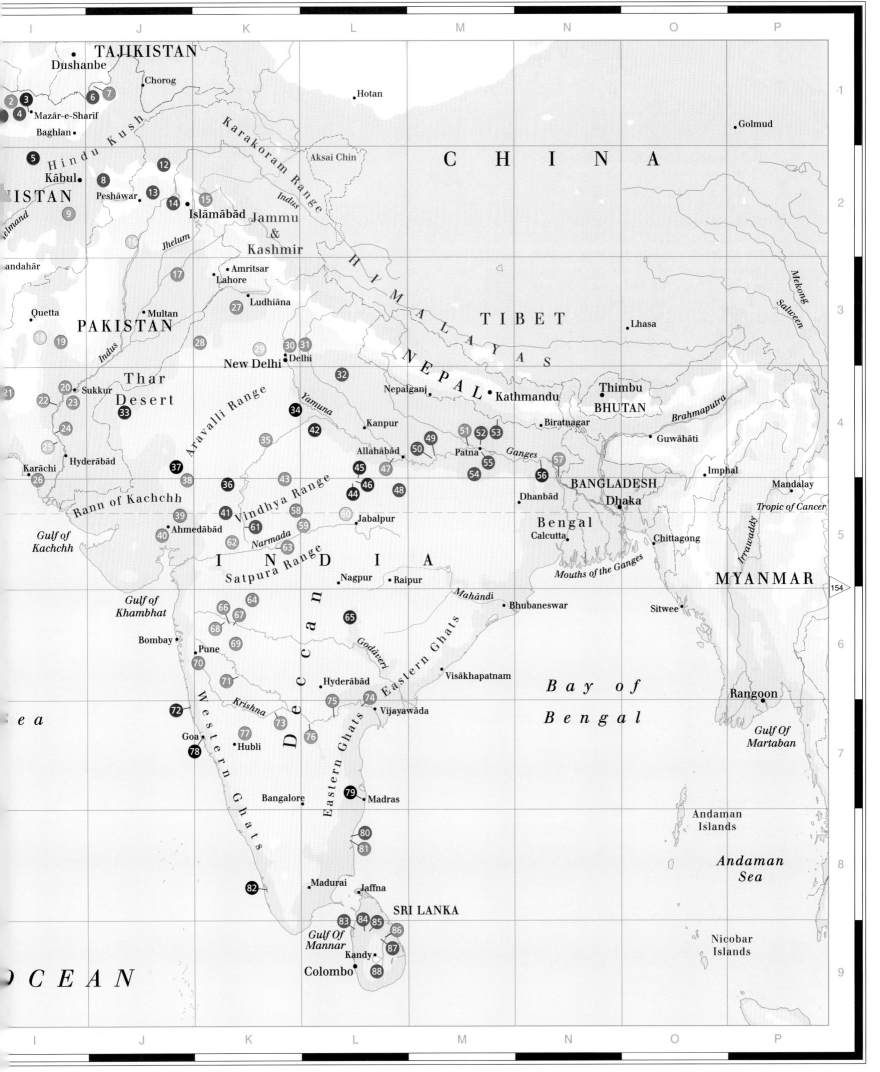

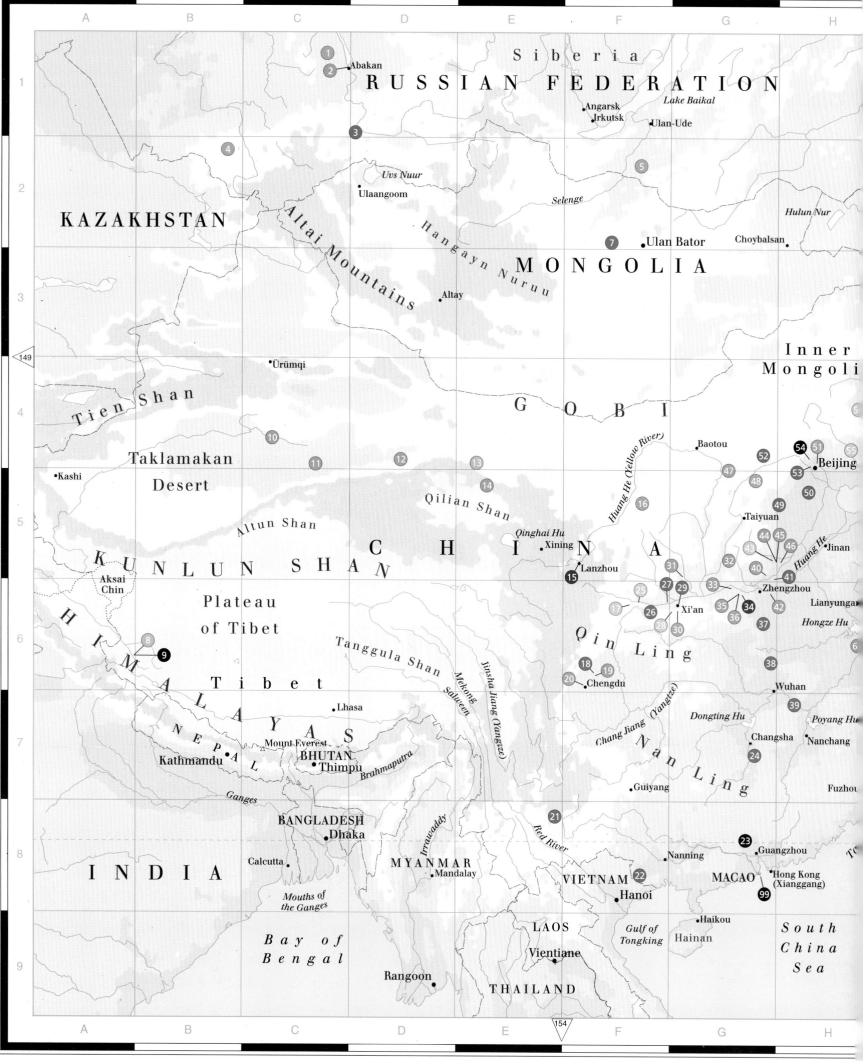

Siberia

RUSSIAN FEDERATION

Lake Baikal

Angarsk
Irkutsk
Ulan-Ude

①
② Abakan

③

④

Uvs Nuur

⑤

Selenge

Hulun Nur

KAZAKHSTAN

Ulaangoom

Altai Mountains

Hangayn Nuruu

⑦ Ulan Bator

Choybalsan

MONGOLIA

Altay

149

Inner
Mongoli

Ürümqi

G O B I

Tien Shan

Huang He (Yellow River)

Baotou

⑤

Taklamakan

⑩

⑪   ⑫   ⑬

⑭

Qilian Shan

Desert

Kashi

⑯

⑤② ⑤④ ⑤① ⑤⑤
Beijing
⑤③
⑤⓪
⑤⑨
Taiyuan

Altun Shan

Qinghai Hu

Xining

C H I N **A**

⑰

⑤⑦

④⑦
④⑧
④④ ④⑤
④③   ④⑥ Jinan
④⓪   Huang He
④①
Zhengzhou
Lianyunga
⑥

K U N L U N   S H A N

Aksai
Chin

Lanzhou
⑮

③①

⑤②

③③

Xi'an
③⑤ ③④
③⑥

③②

③③

④②

Hongze Hu

②⑤
②⑦ ②⑨

⑧
⑨

Plateau
of Tibet

②⑥

②⑧ ③⓪

Qin Ling

③⑦

③⑧

H I M A L A Y A S

Tanggula Shan

Mekong

Jinsha Jiang (Yangtze)

Salween

⑱
②⓪ ⑲ Chengdu

Wuhan

③⑨

Tibet

Lhasa

N E P A L

Mount Everest
BHUTAN
Kathmandu   Thimpu

Brahmaputra

Chang Jiang (Yangtze)

Dongting Hu

Changsha

Poyang Hu

Nanchang

②④

Nan Ling

Ganges

BANGLADESH
Dhaka

②①

Guiyang

Fuzhou

I N D I A

Calcutta

Irrawaddy

MYANMAR
Mandalay

Red River

②③
Nanning

②②

VIETNAM
Hanoi

MACAO

⑨⑨

Guangzhou

Hong Kong
(Xianggang)

Mouths of
the Ganges

Bay of
Bengal

Rangoon

LAOS

Vientiane

THAILAND

Gulf of
Tongking

Haikou

Hainan

South
China
Sea

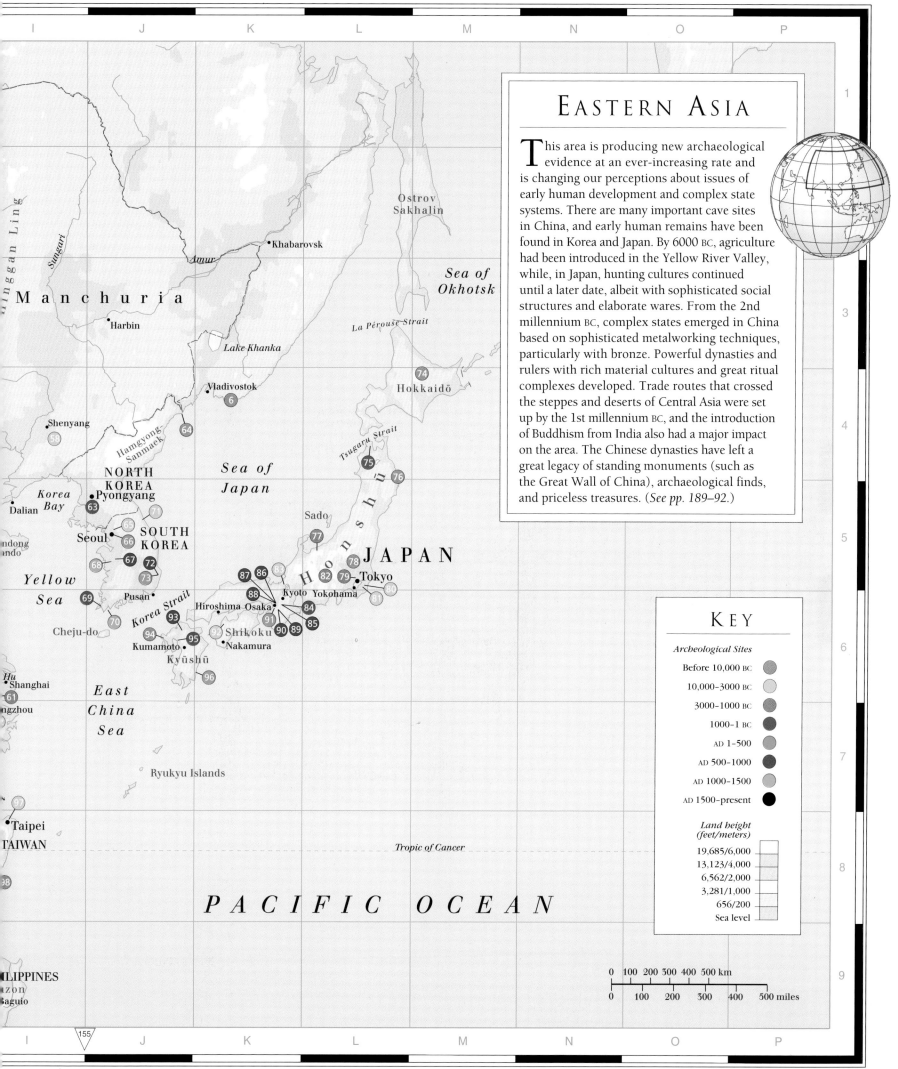

EASTERN ASIA

This area is producing new archaeological evidence at an ever-increasing rate and is changing our perceptions about issues of early human development and complex state systems. There are many important cave sites in China, and early human remains have been found in Korea and Japan. By 6000 BC, agriculture had been introduced in the Yellow River Valley, while, in Japan, hunting cultures continued until a later date, albeit with sophisticated social structures and elaborate wares. From the 2nd millennium BC, complex states emerged in China based on sophisticated metalworking techniques, particularly with bronze. Powerful dynasties and rulers with rich material cultures and great ritual complexes developed. Trade routes that crossed the steppes and deserts of Central Asia were set up by the 1st millennium BC, and the introduction of Buddhism from India also had a major impact on the area. The Chinese dynasties have left a great legacy of standing monuments (such as the Great Wall of China), archaeological finds, and priceless treasures. (*See pp. 189–92.*)

KEY

Archeological Sites

Before 10,000 BC
10,000–3000 BC
3000–1000 BC
1000–1 BC
AD 1–500
AD 500–1000
AD 1000–1500
AD 1500–present

Land height (feet/meters)

19,685/6,000
13,123/4,000
6,562/2,000
3,281/1,000
656/200
Sea level

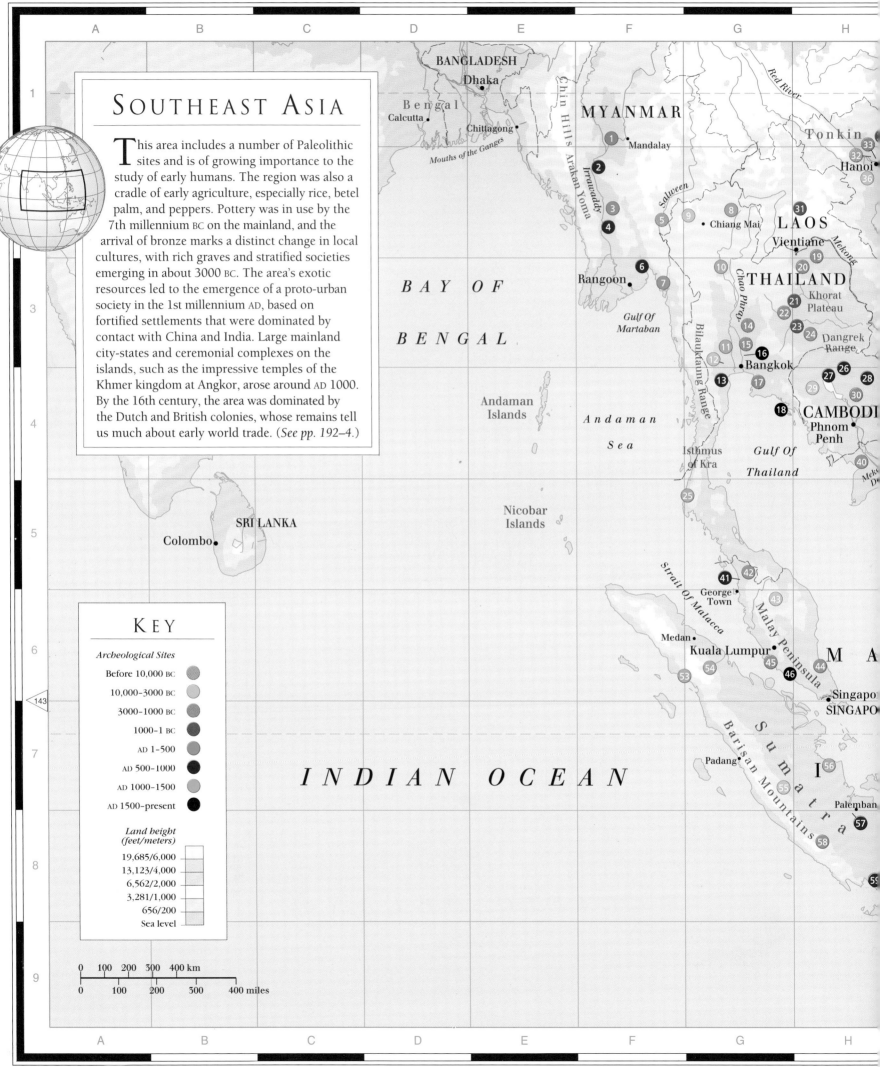

SOUTHEAST ASIA

This area includes a number of Paleolithic sites and is of growing importance to the study of early humans. The region was also a cradle of early agriculture, especially rice, betel palm, and peppers. Pottery was in use by the 7th millennium BC on the mainland, and the arrival of bronze marks a distinct change in local cultures, with rich graves and stratified societies emerging in about 3000 BC. The area's exotic resources led to the emergence of a proto-urban society in the 1st millennium AD, based on fortified settlements that were dominated by contact with China and India. Large mainland city-states and ceremonial complexes on the islands, such as the impressive temples of the Khmer kingdom at Angkor, arose around AD 1000. By the 16th century, the area was dominated by the Dutch and British colonies, whose remains tell us much about early world trade. (*See pp. 192–4.*)

KEY

Archeological Sites

Before 10,000 BC
10,000–3000 BC
3000–1000 BC
1000–1 BC
AD 1–500
AD 500–1000
AD 1000–1500
AD 1500–present

Land height (feet/meters)

19,685/6,000
13,123/4,000
6,562/2,000
3,281/1,000
656/200
Sea level

0 100 200 300 400 km

0 100 200 300 400 miles

CHINA

Nanning

Guangzhou

Hong Kong

MACAO

Taiwan Strait

TAIWAN

Tropic of Cancer

PACIFIC OCEAN

Haikou

Hainan

Gulf Of Tonkin

Da Nang

VIETNAM

Chi Minh City

SOUTH CHINA SEA

Luzon

Manila

PHILIPPINES

PHILIPPINE SEA

Palawan

Sulu Sea

Mindanao

Davao

PALAU

Sabah

Bandar Seri Begawan

BRUNEI

Celebes Sea

AYSIA

Sarawak

Iran Mountains

Borneo

Schwaner Mountains

DONESIA

Molucca Sea

Celebes

Moluccas

Seram Sea

Irian Jaya

Maoke Mountains

Equator

Java Sea

arta

Bandung

Surabaya

Lesser Sunda Islands

Sumbawa

Flores

Bali

Lombok

Sumba

Timor

Arafura Sea

ava

Timor

Timor Sea

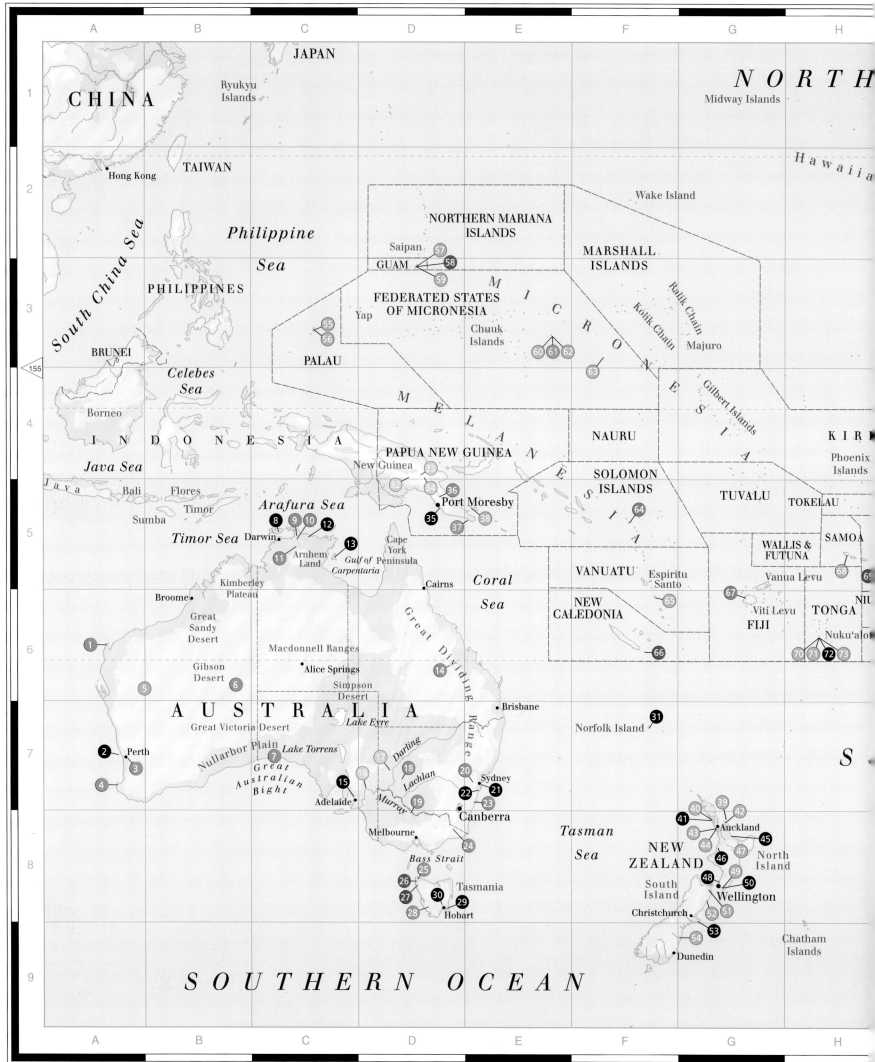

CHINA

JAPAN

Ryukyu
Islands

NORTH

Hong Kong

TAIWAN

Midway Islands

Hawaii

Philippine

Sea

South China Sea

PHILIPPINES

Saipan

GUAM

NORTHERN MARIANA
ISLANDS

Wake Island

MARSHALL
ISLANDS

57

58

59

BRUNEI

Celebes
Sea

155

Borneo

INDONESIA

PALAU

55

56

Yap

FEDERATED STATES
OF MICRONESIA

M
I
C
R
O
N
E
S
I
A

Chuuk
Islands

Ralik Chain

Kolik Chain

Majuro

60 61 62

63

Java Sea

Java

Bali
Sumba

Flores
Timor

MELANE

PAPUA NEW GUINEA

NAURU

Gilbert Islands

KIRI

Phoenix
Islands

Arafura Sea

New Guinea

33

SOLOMON
ISLANDS

TUVALU

TOKELAU

Sumba

32

34 36

Port Moresby

SI

64

SAMOA

Timor Sea Darwin

8 9 10 12

35

37 38

A

WALLIS &
FUTUNA

13

VANUATU

Espiritu
Santo

Vanua Levu

68

69

11 Arnhem
Land

Gulf of
Carpentaria

Cape
York
Peninsula

Cairns

Coral

Sea

NEW
CALEDONIA

67

Viti Levu

TONGA

NIU

Kimberley
Plateau

Broome

Great
Sandy
Desert

1

Gibson
Desert

5

6

Macdonnell Ranges

Alice Springs

Simpson
Desert

55

Nuku'alo

66

70 71 72 73

14

Great Dividing Range

Brisbane

Norfolk Island

31

S

AUSTRALIA

Great Victoria Desert

Lake Eyre

2 Perth

3

Nullarbor Plain *Lake Torrens*

7

Great
Australian
Bight

17

Darling

18

Lachlan

20

Sydney

21

Tasman
Sea

39

4

16

15

19

22

23

41 40

42

Adelaide

Murray

Canberra

24

NEW
ZEALAND

43

44

Auckland

46

47

45

49

North
Island

Melbourne

25

26

Bass Strait

27

30

Tasmania

48

Wellington

50

28

29

Hobart

Christchurch

52 51

Chatham
Islands

54

53

9

SOUTHERN OCEAN

Dunedin

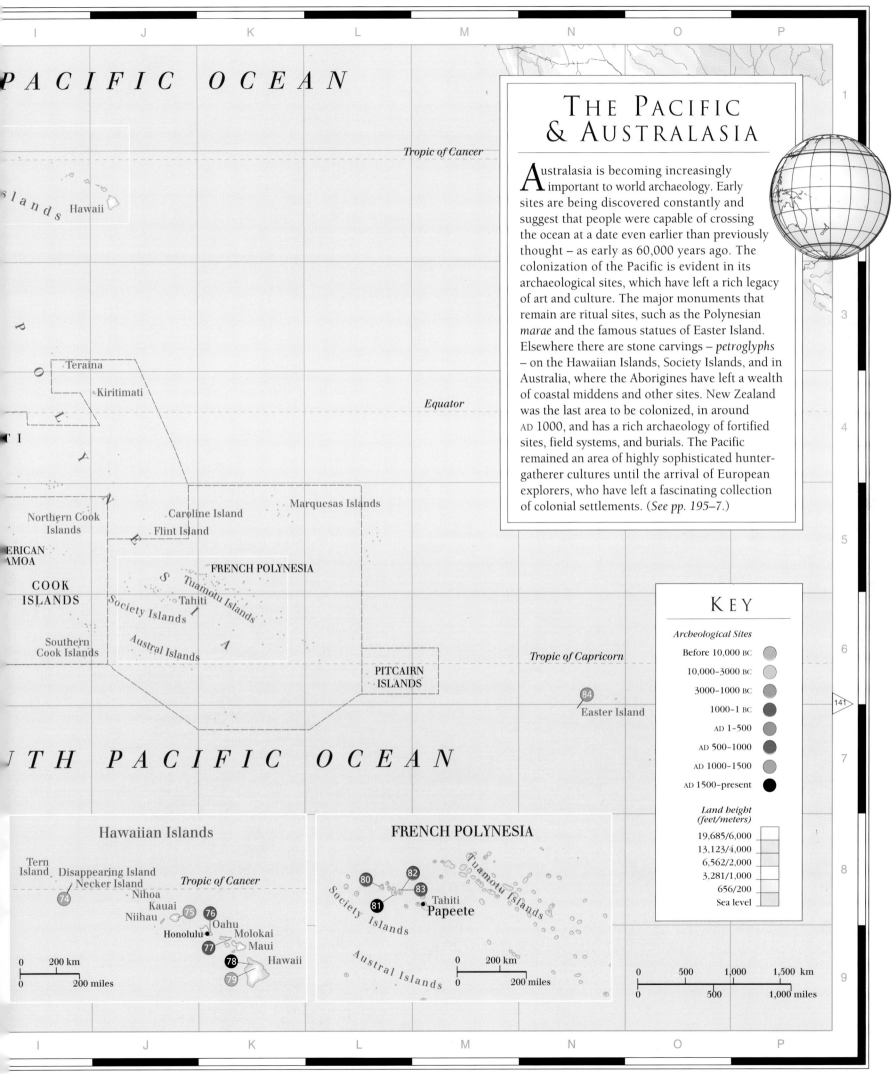

PACIFIC OCEAN

Tropic of Cancer

Hawaii

Teraina

Kiritimati

Equator

Northern Cook
Islands

ERICAN
MOA

COOK
ISLANDS

Caroline Island

Flint Island

Marquesas Islands

FRENCH POLYNESIA

Tuamotu Islands

Tahiti

Society Islands

Southern
Cook Islands

Austral Islands

Tropic of Capricorn

PITCAIRN
ISLANDS

84
Easter Island

TH PACIFIC OCEAN

THE PACIFIC & AUSTRALASIA

Australasia is becoming increasingly important to world archaeology. Early sites are being discovered constantly and suggest that people were capable of crossing the ocean at a date even earlier than previously thought – as early as 60,000 years ago. The colonization of the Pacific is evident in its archaeological sites, which have left a rich legacy of art and culture. The major monuments that remain are ritual sites, such as the Polynesian *marae* and the famous statues of Easter Island. Elsewhere there are stone carvings – *petroglyphs* – on the Hawaiian Islands, Society Islands, and in Australia, where the Aborigines have left a wealth of coastal middens and other sites. New Zealand was the last area to be colonized, in around AD 1000, and has a rich archaeology of fortified sites, field systems, and burials. The Pacific remained an area of highly sophisticated hunter-gatherer cultures until the arrival of European explorers, who have left a fascinating collection of colonial settlements. (*See pp. 195–7.*)

KEY

Archeological Sites

Before 10,000 BC

10,000–3000 BC

3000–1000 BC

1000–1 BC

AD 1–500

AD 500–1000

AD 1000–1500

AD 1500–present

Land height
(feet/meters)

19,685/6,000
13,123/4,000
6,562/2,000
3,281/1,000
656/200
Sea level

Hawaiian Islands

Tern
Island
Disappearing Island
Necker Island *Tropic of Cancer*

74
Nihoa
Kauai
Niihau
75 76
Oahu
Honolulu
Molokai
77
Maui
78 Hawaii
79

0 200 km
0 200 miles

FRENCH POLYNESIA

Tuamotu Islands

80 82
83
81 Tahiti
Papeete

Society Islands

Austral Islands

0 200 km
0 200 miles

0 500 1,000 1,500 km
0 500 1,000 miles

GAZETTEER LISTINGS

The listings in this section correspond to sites plotted on the maps of the preceding pages (pp. *136–57*). The page number of the relevant map is given under the title of each section. After each listing's identifying number, its map reference is given.

NORTH AMERICA

The map for these listings is on pp. 136–7.

1 G2 BLUEFISH CAVES
25,000–10,000 YEARS AGO
LOCATION: Yukon, Canada
TYPE OF SITE: Hunter-gatherer
SITE DETAILS: Caves that are considered some of the oldest archaeological sites in North America – they contain deposits from the Late Glacial period that were part of a bone industry, using mammoth, sheep, and reindeer bones. Later finds include wedge-shaped microliths and burins – a chipped-edge tool that was used for chiseling.

2 H5 NINSTINTS
4TH–19TH CENTURIES AD
LOCATION: British Columbia, Canada
TYPE OF SITE: Early settlement
SITE DETAILS: Impressive village of the Haida people, widely considered to contain some of the best examples of Northwest Coast art. There are still a number of superstructures and mortuary poles from later periods.

3 J6 MAJORVILLE
c. 2500 BC
LOCATION: Alberta, Canada
TYPE OF SITE: Cairn monument
SITE DETAILS: A medicine wheel – a stone ritual monument in the shape of a wheel – consisting of a central cairn with radiating spokes. There are many medicine wheels on the North American plains, yet their exact purpose is not known.

4 J6 HEAD-SMASHED-IN
c. 3700 BC
LOCATION: Alberta, Canada
TYPE OF SITE: Hunter-gatherer
SITE DETAILS: A kill site where hunters regularly made bison stampede over a cliff and used the bodies for food and clothing.

5 K6 MOOSE MOUNTAIN
6TH CENTURY BC
LOCATION: Saskatchewan, Canada
TYPE OF SITE: Cairn monument
SITE DETAILS: Site of a medicine wheel with five rows of stones radiating from the central cairn. Small cairns are found at the end of each of the spokes.

6 L5 BANNOCK POINT
500 BC–AD 800
LOCATION: Ontario, Canada
TYPE OF SITE: Rock art
SITE DETAILS: Site that has a large number of petroglyphs – carvings made on rock faces – depicting fish, snakes, birds, and other animal forms.

7 M6 RICE LAKE
3RD CENTURY AD
LOCATION: Ontario, Canada
TYPE OF SITE: Burial mounds
SITE DETAILS: Hopewell mounds where the leaders of the settlement were buried with a number of goods. Rice Lake is also the site of petroglyphs.

8 N6 CHICOUTIMI
19TH CENTURY AD
LOCATION: Quebec, Canada
TYPE OF SITE: Industrial complex
SITE DETAILS: Pulp mill consisting of five main buildings and an engineering workshop, still in excellent condition.

9 N6 QUEBEC
17TH CENTURY AD
LOCATION: Quebec, Canada
TYPE OF SITE: Colonial settlement
SITE DETAILS: The oldest city in Canada was also one of its earliest forts, founded in AD 1608. The walls of the citadel remain, as do 17th-century buildings, many of them churches. (*See p. 100.*)

10 O6 DEBERT
MID-9TH MILLENNIUM BC
LOCATION: Nova Scotia, Canada
TYPE OF SITE: Hunter-gatherer
SITE DETAILS: A seasonal camp occupied by 15 to 50 hunter-gatherers, who survived on caribou and sea mammals. The site may have been used as base camp and has some finds of fluted points and hearth remains.

11 P5 L'ANSE-AMOUR
5TH MILLENNIUM BC
LOCATION: Labrador, Canada
TYPE OF SITE: Burial mound
SITE DETAILS: A burial that contained grave goods, such as a bone pendant, a walrus tusk, and an antler harpoon.

12 P5 L'ANSE AUX MEADOWS
9TH CENTURY AD
LOCATION: Newfoundland, Canada
TYPE OF SITE: Viking settlement
SITE DETAILS: *See pp. 80, 87.*

13 P6 PORT AUX CHOIX
3RD–2ND MILLENNIA BC
LOCATION: Newfoundland, Canada
TYPE OF SITE: Burial mound
SITE DETAILS: A cemetery of the Arctic Dorset culture. Burial goods included ocher, bone harpoon heads, beads, combs, knives, and scrapers.

14 F1 UTGIAVIK
16TH CENTURY AD
LOCATION: Alaska, US
TYPE OF SITE: Inuit settlement
SITE DETAILS: Remains of 60 Inuit houses, one containing five bodies trapped by a winter ice surge. Archaeologists have deduced from two remaining bodies that the people had a meat-rich diet and suffered from smoke inhalation. There were finds of clothing and hunting tools.

15 D2 IPIUTAK
100 BC–AD 800
LOCATION: Alaska, US
TYPE OF SITE: Inuit settlement
SITE DETAILS: Site of 500 rectangular houses with rounded corners and underground floors, found over several beach ridges. There was a cemetery of log coffins, some with grave goods (linked chains, snow goggles, and carved ivory animals). Other objects from the site include harpoons, stone points, and adze blades.

16 E3 SAFETY SOUND
500 BC–AD 800
LOCATION: Alaska, US
TYPE OF SITE: Early settlement
SITE DETAILS: About 400 dwellings of the Norton culture, most of which had central hearths and pole-and-sod roofs.

17 D3 OKVIK
c. 300 BC
LOCATION: Alaska, US
TYPE OF SITE: Open-air
SITE DETAILS: An area with distinctive ivory artifacts, the most famous being the Okvik "madonna-with-child."

18 F3 DRY CREEK
9TH–4TH MILLENNIA BC
LOCATION: Alaska, US
TYPE OF SITE: Hunter-gatherer
SITE DETAILS: A stratified open-air site of three periods, with the earliest level containing cobble and flake tools, broken blades, and bifacial knives. The second level is from 5,000 years later, with finds of microblades and other artifacts.

19 G3 HEALY LAKE
9TH MILLENNIUM BC
LOCATION: Alaska, US
TYPE OF SITE: Hunter-gatherer
SITE DETAILS: Stratified open-air site where artifacts, including points and blades, were found. (*See p. 12.*)

20 F4 BROOKS RIVER
15TH CENTURY BC
LOCATION: Alaska, US
TYPE OF SITE: Early settlement
SITE DETAILS: The remains of 14 houses, where the inhabitants dug square pits and added a sloped entrance, a sod roof, and a central hearth.

21 F4 KACHEMAK BAY
8TH CENTURY BC
LOCATION: Alaska, US
TYPE OF SITE: Inuit settlement
SITE DETAILS: Villages that consisted of partially underground houses. The people, who subsisted by hunting and fishing, practiced woodworking and weaving, and used stone oil lamps, slate blades, and barbed bone darts.

22 I6 HOKO RIVER
10TH–3RD CENTURIES BC
LOCATION: Washington, US
TYPE OF SITE: Early settlement
SITE DETAILS: Area of three related sites: one waterlogged site, one dry campsite, and a river-mouth site within a large rock shelter. Artifacts unearthed include 400 fishing hooks and wooden racks.

23 I6 OZETTE
15TH–20TH CENTURIES AD
LOCATION: Washington, US
TYPE OF SITE: Hunting village
SITE DETAILS: Site occupied by the Makah people until the 1930s, with a series of settlements built on natural terraces above sea level. A mudslide preserved six 16th-century cedar longhouses.

24 I7 SURPRISE VALLEY
5TH MILLENNIUM BC
LOCATION: California, US
TYPE OF SITE: Hunter-gatherer
SITE DETAILS: A settlement where the inhabitants built small semisubterranean pithouses with frameworks of poles, brush, and soil.

25 J8 CALICO MOUNTAINS
200,000 YEARS AGO
LOCATION: California, US
TYPE OF SITE: Hunter-gatherer
SITE DETAILS: An open-air assemblage of lithics was found that, although unlikely, could prove this to be the earliest archaeological site in North America.

26 J7 WILSON BUTTE CAVE
14,500–10,000 YEARS AGO
LOCATION: Idaho, US
TYPE OF SITE: Archaic settlement
SITE DETAILS: Site that may be one of the

oldest occupations in North America, with finds that included retouched blades and flakes and the remains of camels, horses, and ground sloths.

27 J8 TULE SPRINGS
10TH–7TH MILLENNIA BC
LOCATION: Nevada, US
TYPE OF SITE: Archaic settlement
SITE DETAILS: When hearths and artifacts were excavated here, they were first thought to be 25,000 years old, making them the oldest in North America. However, further examination proved the finds to be of a more recent date.

28 J8 LOST CITY
8TH CENTURY AD
LOCATION: Nevada, US
TYPE OF SITE: Village settlement
SITE DETAILS: The site of a large Anasazi village, where the walls of its structures were made of sandstone blocks with mud-plaster facing.

29 J7 DANGER CAVE
9TH MILLENNIUM BC
LOCATION: Utah, US
TYPE OF SITE: Archaic settlement
SITE DETAILS: See pp. 12, 23.

30 J8 VENTANA CAVE
9000 BC–1ST MILLENNIUM AD
LOCATION: Arizona, US
TYPE OF SITE: Archaic settlement
SITE DETAILS: A rock shelter first used by hunter-gatherers, who hunted horses, bison, and sloths. It was more recently used by Hohokam cultures.

31 J8 SNAKETOWN
10TH–12TH CENTURIES AD
LOCATION: Arizona, US
TYPE OF SITE: Farming village
SITE DETAILS: One of the most impressive Hohokam village sites, where Mexican influence is seen in its ritual ballcourts. The foundations of platform mounds (earth mounds surmounted by ritual structures) have also been uncovered.

32 J8 BROKEN K PUEBLO
13TH CENTURY AD
LOCATION: Arizona, US
TYPE OF SITE: Pueblo settlement
SITE DETAILS: Large 95-house village with three categories of room: the largest were used for daily activities, the smallest for storage, and the rest were kivas – underground chambers used for ritualistic purposes.

33 K6 LITTLE BIGHORN
AD 1876
LOCATION: Wyoming, US
TYPE OF SITE: Battleground
SITE DETAILS: Battle site where General Custer, his men, and Sioux warriors were killed. Archaeologists reconstructed the battle by analyzing the positions of the spent cartridges and evidence showing where soldiers and Native Americans fell.

34 K7 BIG HORN
LATE 1ST MILLENNIUM BC
LOCATION: Wyoming, US
TYPE OF SITE: Cairn monument
SITE DETAILS: A D-shaped medicine wheel with 28 spokes radiating from its center, surrounded by six smaller cairns.

35 K7 CASPER
c. 6000 BC
LOCATION: Wyoming, US
TYPE OF SITE: Hunter-gatherer
SITE DETAILS: The site of a large bison kill where hunters drove 75 of the animals into a sand dune, killing them with spears and stripping them of their skins and meat.

36 K7 HELL GAP
9TH–7TH MILLENNIA BC
LOCATION: Wyoming, US
TYPE OF SITE: Hunter-gatherer
SITE DETAILS: Deeply stratified site that provides a complete Paleo-Indian sequence and a characterizing type of projectile point.

37 K7 CROW CREEK
EARLY 2ND MILLENNIUM AD
LOCATION: South Dakota, US
TYPE OF SITE: Burial mound
SITE DETAILS: Plains Indian village site with 500 skeletons of people who had been massacred, mutilated, and buried in a mass grave.

38 K7 MESA VERDE
11TH–13TH CENTURIES AD
LOCATION: Colorado, US
TYPE OF SITE: Pueblo settlement
SITE DETAILS: Large Anasazi cliff-top farming village of rock-cut apartments, including the impressive Cliff Palace, with its 200 rooms and 23 kivas.

39 K8 CHACO CANYON
10TH–13TH CENTURIES AD
LOCATION: New Mexico, US
TYPE OF SITE: Pueblo settlement
SITE DETAILS: A cluster of farming villages, including the impressive D-shaped Pueblo Bonito. Among the extensive ruins are irrigation canals, road systems, signal stations, and large pueblo buildings, many including kivas.

40 K8 BANDALIER
4000 BC–AD 1500
LOCATION: New Mexico, US
TYPE OF SITE: Rock shelter
SITE DETAILS: A national monument that contains thousands of archaeological sites. Rock shelters date from 4000 BC. The Anasazi occupied the site from the 11th century AD for about 400 years .

41 K8 PECOS
15TH CENTURY AD
LOCATION: New Mexico, US
TYPE OF SITE: Pueblo village
SITE DETAILS: A site consisting of a multi-story quadrangle, with 660 kivas and

living and storage rooms surrounding a central plaza. There is also a ruined 17th-century Spanish mission.

42 K8 BAT CAVE
8000 BC–AD 1200
LOCATION: New Mexico, US
TYPE OF SITE: Rock shelter
SITE DETAILS: A stratified site first used as a shelter for hunting groups. Hearths were built in the 3rd millennium BC. Ceramic levels (pottery from different periods) date from up to 800 years ago.

43 K8 CLOVIS
10TH MILLENNIUM BC
LOCATION: New Mexico, US
TYPE OF SITE: Hunter-gatherer
SITE DETAILS: Type-site for the fluted projectile points of this particular period, made for hunting mammoth. The points are characterized by their symmetrical appearance.

44 K8 BLACKWATER DRAW
7TH MILLENNIUM BC
LOCATION: New Mexico, US
TYPE OF SITE: Hunter-gatherer
SITE DETAILS: Clovis and Folsom points have been excavated, as have the remains of camels, horses, and mammoths and some Paleo-Indian remains. (See p. 12.)

45 L9 ALAMO
19TH CENTURY AD
LOCATION: Texas, US
TYPE OF SITE: Colonial settlement
SITE DETAILS: The San Antonio de Valero mission, which was the site of the battle of the Alamo (1836) on the frontier, still stands. A number of relics of that period have been found here, as well as other colonial artifacts.

46 L9 PADRE ISLAND
16TH CENTURY AD
LOCATION: Texas, US
TYPE OF SITE: Shipwreck
SITE DETAILS: A number of Spanish ships have been wrecked here, including the *Espíritu Santo*, the *San Esteban*, and the *Santa María de Yciar*. The last's cargo included silver coins, bullion, amber, sugar, wool, and cowhides.

47 L8 BEAUMONT
20TH CENTURY AD
LOCATION: Texas, US
TYPE OF SITE: Industrial
SITE DETAILS: The site of an oil boom town that expanded at the beginning of the 20th century. (See p. 110.)

48 L8 SPIRO MOUNDS
11TH–15TH CENTURIES AD
LOCATION: Oklahoma, US
TYPE OF SITE: Ceremonial center
SITE DETAILS: Eight mounds used for ceremonial and burial purposes where a large number of artifacts were found.

49 M6 ISLE ROYALE
19TH CENTURY AD
LOCATION: Michigan, US
TYPE OF SITE: Shipwreck
SITE DETAILS: Sunken ships that have been preserved in extremely good condition due to the freshwater of Lake Superior.

50 M8 POWERS FORT
13TH CENTURY AD
LOCATION: Missouri, US
TYPE OF SITE: Ceremonial center
SITE DETAILS: A fortified religious center that had a central plaza with four adjacent mounds.

51 L8 TOLTEC
10TH–11TH CENTURIES AD
LOCATION: Arkansas, US
TYPE OF SITE: Ceremonial center
SITE DETAILS: A complex settlement built by the Plum Bayou culture. It was an important religious center, consisting of several burial and platform mounds.

52 L8 POVERTY POINT
13TH–5TH CENTURIES BC
LOCATION: Louisiana, US
TYPE OF SITE: Archaic settlement
SITE DETAILS: Large earthworks, one of which is in the shape of a bird. Beads, stone tools, microliths, stone carvings, thousands of clay balls, bird effigies, and other clay objects were excavated.

53 L7 CAHOKIA
8TH–15TH CENTURIES AD
LOCATION: Illinois, US
TYPE OF SITE: Ceremonial center
SITE DETAILS: A number of mounds accredited to the Middle Mississippian culture. It is believed that the population once reached 38,000. Some mounds were defensive, while others contained burials. Arrowheads, polished stone, and a number of skeletons were unearthed in one mound. (See p. 34.)

54 L7 KOSTER
7500 BC–AD 1200
LOCATION: Illinois, US
TYPE OF SITE: Archaic settlement
SITE DETAILS: A stratified site that acts as the defining midwestern site of the Archaic period. It has revealed evidence about the subsistence patterns and wild-plant cultivation.

55 M7 BLACK EARTH SITE
4TH MILLENNIUM BC
LOCATION: Illinois, US
TYPE OF SITE: Archaic settlement
SITE DETAILS: A midden that yielded food remains and storage and processing pits. There were over 150 burials; one male burial contained a bag of eagle talons, projectile points, and other artifacts.

56 M7 INDIAN KNOLL
4TH MILLENNIUM BC
LOCATION: Kentucky, US
TYPE OF SITE: Early settlement

SITE DETAILS: A shell mound and floors of habitation where the remains of around 1,100 burials with grave goods have been found.

57 M7 DETROIT
20TH CENTURY AD
LOCATION: Michigan, US
TYPE OF SITE: Industrial
SITE DETAILS: Founded as a trading post in 1701, it became the site of Henry Ford's pioneering car production line. There are large amounts of industrial machinery. The remains give important evidence about the development of the 20th-century American factory. (*See p. 110.*)

58 M8 MOUNDVILLE
14TH CENTURY AD
LOCATION: Alabama, US
TYPE OF SITE: Ceremonial mound
SITE DETAILS: Site of 20 Mississippian temple platforms, where goods were decorated with a hand-eye motifs. Over 3,000 burials have been analyzed, revealing that the rich elite were buried lavishly with a number of goods.

59 M7 NEWARK MOUNDS
2ND–4TH CENTURIES AD
LOCATION: Ohio, US
TYPE OF SITE: Ceremonial center
SITE DETAILS: A number of mounds built by the Hopewell people that were probably used for ritual purposes.

60 M7 SERPENT MOUND
200 BC–AD 400
LOCATION: Ohio, US
TYPE OF SITE: Ritual earthwork
SITE DETAILS: A long curling mound in the form of a serpent reaching a length of 1,312 ft (400 m), with an oval object in its mouth. (*See p. 34.*)

61 M8 MUD GLYPH CAVE
5TH–18TH CENTURIES AD
LOCATION: Tennessee, US
TYPE OF SITE: Cave art
SITE DETAILS: A series of galleries that contain a rich collection of art, made from incisions cut into the natural mud of the cave.

62 M8 ICEHOUSE BOTTOM VILLAGE
3RD CENTURY AD
LOCATION: Tennessee, US
TYPE OF SITE: Early settlement
SITE DETAILS: Village site situated in a river valley with circular pole-and-thatch buildings. Each of the structures housed about 20 people.

63 M8 ETOWAH
LATE 1ST MILLENNIUM AD
LOCATION: Georgia, US
TYPE OF SITE: Ceremonial center
SITE DETAILS: A South Appalachian Mississippian site of three mounds that together make up one large temple mound. Artifacts include statuettes.

64 M8 ROCK EAGLE MOUND
10TH CENTURY AD
LOCATION: Georgia, US
TYPE OF SITE: Ritual monument
SITE DETAILS: A large effigy mound in the form of an eagle. Hundreds of thousands of rocks were transported here to create the figure.

65 M8 KOLOMOKI
1ST CENTURY AD
LOCATION: Georgia, US
TYPE OF SITE: Early settlement
SITE DETAILS: Large area with a large number of burial mounds and a platform mound. Each of them was a village center for around 2,000 people.

66 M8 ST. CATHERINE'S ISLAND
17TH CENTURY AD
LOCATION: Georgia, US
TYPE OF SITE: Colonial settlement
SITE DETAILS: Built on a site that had been inhabited for 4,000 years, the Santa Catalina de Guale mission, which served 25,000 Native Americans, consisted of a friary and church in a fortified compound. Nave foundations have been uncovered, with 400 Guale Native American burials underneath. Finds include majolica vessels, glass trade beads, and other artifacts.

67 M8 CRYSTAL RIVER MOUNDS
EARLY 1ST MILLENNIUM AD
LOCATION: Florida, US
TYPE OF SITE: Ceremonial mound
SITE DETAILS: Excavations of six mounds have revealed a number of burials.

68 M8 TATHAM
16TH CENTURY AD
LOCATION: Florida, US
TYPE OF SITE: Burial mound
SITE DETAILS: The remains of 70 Native Americans who may have been the victim of an epidemic, although some of the bodies have sword cuts. Spanish artifacts, including metal, glass, beads, and armor fragments, have been found.

69 M8 ST. AUGUSTINE
16TH–18TH CENTURIES AD
LOCATION: Florida, US
TYPE OF SITE: Colonial settlement
SITE DETAILS: Founded in AD 1585; the remains of the 17th-century Castillo de San Marcos still exist. An 18th-century house has been found. (*See p.100.*)

70 M9 WINDOVER
c. 5400 BC
LOCATION: Florida, US
TYPE OF SITE: Burial
SITE DETAILS: Waterlogged site that revealed America's earliest mass burial when excavated. Some 168 individuals were found, from which several complete brains were exhumed.

71 M9 LITTLE SALT SPRING
10TH–4TH MILLENNIA BC
LOCATION: Florida, US
TYPE OF SITE: Hunter-gatherer
SITE DETAILS: Stratified site that has unearthed a hearth, an oak boomerang, and an antler projectile point. Later Archaic remains include a mass burial.

72 M9 KEY MARCO
EARLY 2ND MILLENNIUM BC
LOCATION: Florida, US
TYPE OF SITE: Early settlement
SITE DETAILS: Thatch houses built on stilts preserved by lagoon swamps. The people fished and hunted, using weapons including shark-teeth swords. There have also been finds of unusual wooden sculptures with naturalistic animal carvings, stools, trays, bowls, and wooden dugout canoes.

73 N7 LAMOKA LAKE
8TH–5TH MILLENNIA BC
LOCATION: New York, US
TYPE OF SITE: Early settlement
SITE DETAILS: Probably a base camp, suggested by traces of 27 rectangular houses. Floors were covered with bark, skins, and mats. There were also storage pits and meat-drying mats.

74 N7 SACKETT
12TH CENTURY AD
LOCATION: New York, US
TYPE OF SITE: Fortified settlement
SITE DETAILS: Village site that was protected by raised earthworks. It was linked to the Owasco culture.

75 N7 HOPEWELL VILLAGE
19TH CENTURY AD
LOCATION: Pennsylvania, US
TYPE OF SITE: Industrial
SITE DETAILS: Prime example of an industrial ironworking site in North America that has now been restored. Goods produced ranged from cannons and shot to a distinctive range of stoves.

76 M7 MEADOWCROFT
17,000 YEARS AGO
LOCATION: Pennsylvania, US
TYPE OF SITE: Stratified rock shelter
SITE DETAILS: A cave inhabited for very long, intermittent periods, representing all the major periods of human occupation in North American history. Artifacts found here include flakes, blades, unifaces, and bifaces.

77 M7 GRAVE CREEK MOUND
3RD–2ND CENTURIES BC
LOCATION: West Virginia, US
TYPE OF SITE: Burial mound
SITE DETAILS: The largest conical mound in North America, 62 ft (19 m) high and 240 ft (73 m) in diameter. It was built over a high-status burial and was surrounded by a moat. The site was the center of the Adena people.

78 N7 ST. MARY'S CITY
17TH CENTURY AD
LOCATION: Maryland, US
TYPE OF SITE: Colonial settlement
SITE DETAILS: See pp. 100, 102–5.

79 N7 ANNAPOLIS
17TH CENTURY AD
LOCATION: Maryland, US
TYPE OF SITE: Colonial settlement
SITE DETAILS: City that has a high level of preservation from its colonial era. Among the structures investigated have been an 18th-century tavern, a print shop, private residences, and some gardens.

80 N7 WOLSTENHOLME TOWNE
AD 1616–22
LOCATION: Virginia, US
TYPE OF SITE: Colonial settlement
SITE DETAILS: An English settlement destroyed by a Native American attack. There are remains of stores, dwellings, a timber fort, British arms and armor, and the graves of the massacred victims.

81 N8 JAMESTOWN
17TH CENTURY AD
LOCATION: Virginia, US
TYPE OF SITE: Colonial settlement
SITE DETAILS: Site of the first permanent English settlement in America, founded in AD 1607. It contained 17 houses, surrounded by a defensive wall, with battle stations on three corners.

82 N8 MARTIN'S HUNDRED
AD 1619
LOCATION: Virginia, US
TYPE OF SITE: Colonial settlement
SITE DETAILS: A pilgrim plantation site where the dwellers were attacked by Native Americans and 58 people were killed. Much has been preserved.

83 N7 LOWELL
19TH CENTURY AD
LOCATION: Massachusetts, US
TYPE OF SITE: Industrial
SITE DETAILS: A textile mill set up by Francis Lowell on the Merrimack River, where the buildings are preserved. (*See p. 110.*)

84 N7 HARTFORD
17TH–19TH CENTURIES AD
LOCATION: Connecticut, US
TYPE OF SITE: Colonial settlement
SITE DETAILS: Originally the site of a Dutch fort, later settled by the English, it is important for the industrial remains found at the Colt gun factory, built in AD 1855. (*See p. 110.*)

85 N7 NEW YORK
17TH CENTURY AD
LOCATION: New York, US
TYPE OF SITE: Colonial settlement
SITE DETAILS: See pp. 100, 106–7.

CENTRAL AMERICA

The map for these listings is on pp. 138–9.

① A2 CASAS GRANDES
11TH–14TH CENTURIES AD
LOCATION: Mexico
TYPE OF SITE: Trading settlement
SITE DETAILS: Influenced by the Toltec culture, there are the remains of a number of pyramids, including the Mound of the Cross, the Mound of the Heroes, and a ritual ballcourt.

② C5 TZINTZUNTZAN
14TH–15TH CENTURIES AD
LOCATION: Central Highlands, Mexico
TYPE OF SITE: Pre-Columbian settlement
SITE DETAILS: Remains include terraced circular platforms of five Purépecha temples. There are also the remains of dwellings, believed to be those of high priests. Tombs have yielded huge deposits of human bones.

③ D5 TULA
9TH–12TH CENTURIES AD
LOCATION: Hidalgo, Mexico
TYPE OF SITE: Pre-Columbian town
SITE DETAILS: Site of a capital with several pyramids and *chac mools* – altars in the shape of reclining human figures, used as receptacles for ritual offerings. The surfaces are carved with figures of jaguars, coyotes, and eagles. It was also a later Aztec town.

④ D5 TENOCHTITLAN
14TH–16TH CENTURIES AD
LOCATION: Mexico City, Mexico
TYPE OF SITE: Aztec capital
SITE DETAILS: Discovered in the center of Mexico City, the Aztec capital city had a major ceremonial precinct with a twin-temple pyramid called the Templo Mayor. It was built on artificial islands, with canals running through the city. There were also remains of a market-place, ballcourts, and other structures.

⑤ D5 TEOTIHUACAN
1ST–8TH CENTURIES AD
LOCATION: Mexico
TYPE OF SITE: Ceremonial city
SITE DETAILS: See pp. 34, 43.

⑥ D5 SANTA ISABEL IZTAPAN
8TH MILLENNIUM BC
LOCATION: Chiapas, Mexico
TYPE OF SITE: Hunter-gatherer
SITE DETAILS: Site of two mammoth kills where finds included straight-based and slightly stemmed points, obsidian blades, and side-scrapers.

⑦ D5 TLATILCO
12TH CENTURY BC
LOCATION: Mexico
TYPE OF SITE: Cemetery
SITE DETAILS: Lavish burial site with white Olmec-like figurines and "hollow-babies," which suggest some form of link to a fertility cult.

⑧ D5 CALIXTLAHUACA
13TH–16TH CENTURIES AD
LOCATION: Mexico
TYPE OF SITE: Pre-Columbian town
SITE DETAILS: Occupied by the Matlatzinca people, who were enemies of the Aztecs. It is best known for its circular Temple of Quetzalcoatl. Later, it became an Aztec town.

⑨ D5 CUICUILCO
600 BC–AD 200
LOCATION: Mexico
TYPE OF SITE: Pre-Classic center
SITE DETAILS: A ceremonial site with an impressive circular temple platform that had been covered in lava and once had four tiers and ramps. The ceremonial hub was over 390 ft (119 m) long and was faced with a number of stone slabs.

⑩ D5 XOCHICALCO
6TH–10TH CENTURIES AD
LOCATION: Morelos, Mexico
TYPE OF SITE: Fortified settlement
SITE DETAILS: Important Late Classic center situated on an artificially terraced hill. It is believed to have been the retreat of the elite of Teotihuacán. It includes the Pyramid of the Plumed Serpent, a ceremonial ballcourt, stelae, and underground passages.

⑪ D5 CHALCATZINGO
9TH–41H CENTURIES BC
LOCATION: Puebla, Mexico
TYPE OF SITE: Pre-Classic settlement
SITE DETAILS: An Olmec site where a number of carved heads have been found. The main relief shows a richly dressed Olmec woman seated within the mouth of a cave, beneath clouds of pouring rain.

⑫ D5 CHOLULA
2ND–12TH CENTURIES AD
LOCATION: Puebla, Mexico
TYPE OF SITE: Early settlement
SITE DETAILS: An Olmec center, with an impressive pyramid mound built over four periods, which reached a height of 180 ft (55 m). Some impressively decorated pottery has been found here.

⑬ D5 CACATXLA
8TH–9TH CENTURIES AD
LOCATION: Tlaxcala, Mexico
TYPE OF SITE: Late Classic settlement
SITE DETAILS: Site with a number of stelae and buildings painted with scenes of warfare. One shows the victory of a jaguar over a birdlike figure.

⑭ D5 EL TAJIN
7TH–11TH CENTURIES AD
LOCATION: Tamaulipas, Mexico
TYPE OF SITE: Ceremonial center
SITE DETAILS: Classic site linked to many of the period's dominant cultures. The Pyramid of the Niches has 365 niches that held an idol for each day of the year. A previous structure was discovered underneath. Other remains include 11 ballcourts and the Building of the Columns.

⑮ D6 JUXTLAHUACA
9TH–7TH CENTURIES BC
LOCATION: Guerrero, Mexico
TYPE OF SITE: Cave art
SITE DETAILS: A cave decorated with art motifs that combine humans and animals with rain-fertility and origin-myth iconography in its two chambers.

⑯ D6 TEHUACAN VALLEY
7000 BC–AD 1520
LOCATION: Puebla, Mexico
TYPE OF SITE: Early farming area
SITE DETAILS: The extensive site where Mesoamerican agriculture originated, reflecting a transition from a hunter-gatherer culture to a reliance on domestic foods. (See p. 24.)

⑰ D6 MONTE ALBAN
500 BC–AD 750
LOCATION: Mexico
TYPE OF SITE: Early settlement
SITE DETAILS: Site of four distinct periods, although its Classic stage is considered the most important. Plazas, terraces, carved stelae, temple platforms and an unexcavated mound have been found here. Burials were accompanied by terra-cotta urns.

⑱ E5 CEMPOALA
7TH–15TH CENTURIES AD
LOCATION: Veracruz, Mexico
TYPE OF SITE: Classic settlement
SITE DETAILS: A site with distinctive carvings and ballcourts. It became an Aztec outpost in the 13th century.

⑲ E5 CERRO DE LAS MERAS
6TH CENTURY AD
LOCATION: Mexico
TYPE OF SITE: Early settlement
SITE DETAILS: The site of earth mounds and 15 stelae where the burial goods link the site to Teotihuacán. A large number of Olmec jades were also discovered here.

⑳ E6 GUILA NAQUITZ
9TH–7TH MILLENNIA BC
LOCATION: Oaxaca, Mexico
TYPE OF SITE: Rock shelter
SITE DETAILS: A cave occupied by hunters and farmers intermittently over a long period of time. The remains of six living levels were found; pathways, hearths, storage areas, and cooking areas have been excavated. (See p. 12.)

㉑ E6 MITLA
9TH–12TH CENTURIES AD
LOCATION: Oaxaca, Mexico
TYPE OF SITE: Pre-Columbian town
SITE DETAILS: An early Zapotec religious center, with a shrine and plaza flanked by pyramids. It later became a Mixtec settlement containing five groups of palaces of long buildings. There have also been finds of earlier ceramics.

㉒ E6 TRES ZAPOTES
10TH–1ST CENTURIES BC
LOCATION: Veracruz, Mexico
TYPE OF SITE: Ceremonial center
SITE DETAILS: An Olmec site famous for its stela dating to 32 BC. There were 50 earth mounds discovered here, with finds of pottery and clay figurines.

㉓ E6 SAN LORENZO
12TH CENTURY BC
LOCATION: Mexico
TYPE OF SITE: Early village
SITE DETAILS: Olmec center where monuments and sculptures include eight large and distinctive Olmec heads. Iron-ore beads, a stela, an altar, earth platforms, and around 200 house mounds were excavated.

㉔ E6 LA VENTA
EARLY 1ST MILLENNIUM BC
LOCATION: Mexico
TYPE OF SITE: Ceremonial center
SITE DETAILS: An Olmec site with evidence of elite housing and public buildings. One burial revealed a cache of votive figurines, arranged to show a ritual activity. (See p. 46.)

㉕ F6 CHIAPA DE CORZO
15TH–9TH CENTURIES BC
LOCATION: Chiapas, Mexico
TYPE OF SITE: Early settlement
SITE DETAILS: A farming site with evidence of corn grinding on milling stones and pottery production. Its stratified layers make it especially interesting to archaeologists. It later became a Maya town.

㉖ F6 TONINA
4TH CENTURY AD
LOCATION: Chiapas, Mexico
TYPE OF SITE: Maya settlement
SITE DETAILS: Currently under excavation. By 1997, a pyramid mound, two ritual ballcourts, temple ruins, sculptures, and some impressive reliefs had been discovered.

㉗ F7 IZAPA
100 BC–AD 100
LOCATION: Chiapas, Mexico
TYPE OF SITE: Maya settlement
SITE DETAILS: About 80 pyramid mounds,

large stelae, and altars have been excavated at this Maya center.

28 F6 BONAMPAK

9TH CENTURY AD

LOCATION: Chiapas, Mexico
TYPE OF SITE: Maya settlement
SITE DETAILS: Probably a Classic Maya palace, with monuments and paintings that show images of sacrifice and other Maya rituals.

29 F6 PALENQUE

6TH–9TH CENTURIES AD

LOCATION: Chiapas, Mexico
TYPE OF SITE: Maya settlement
SITE DETAILS: A city most prominent in the Late Classic period under the leadership of Lord Pacal, who was buried inside the Temple of the Inscriptions. Other structures include the Temple of the Cross, the Temple of the Foliated Cross, and the Temple of the Sun, which are all intricately carved inside.

30 G6 RIO BEC

7TH–9TH CENTURIES AD

LOCATION: Mexico
TYPE OF SITE: Maya settlement
SITE DETAILS: A pyramid-palace where the structures include towering temples, most of which have ornamented exteriors.

31 F5 EDZNA

7TH CENTURY AD

LOCATION: Campeche, Mexico
TYPE OF SITE: Maya settlement
SITE DETAILS: Site of a ritual ballcourt and the Templo de los Cinco, which was used for habitation as well as ritual.

32 F5 JAINA ISLAND

7TH–9TH CENTURIES AD

LOCATION: Mexico
TYPE OF SITE: Maya settlement
SITE DETAILS: A number of remarkable clay figurines, often hollow and fitted with whistles, have been discovered. This is likely to have been an elite burial site.

33 G5 DZIBILCHALTUN

500 BC–AD 1200

LOCATION: Yucatán, Mexico
TYPE OF SITE: Maya settlement
SITE DETAILS: A ceremonial center – over 6,000 sacrificial offerings have been found in the Temple of the Seven Dolls, named after the figurines that were discovered there. In all, 8,400 structures have been recorded.

34 G5 MAYAPAN

13TH–14TH CENTURIES AD

LOCATION: Yucatán, Mexico
TYPE OF SITE: Pre-Columbian city
SITE DETAILS: Founded by the Itzá with many shrines and finds of potsherds, some depicting gods. Its castillo was virtually a copy of the one found at Chichén Itzá. The population varied between 6,000 and 15,000.

35 G5 CHICHEN ITZA

8TH–10TH CENTURIES AD

LOCATION: Yucatán, Mexico
TYPE OF SITE: Maya settlement
SITE DETAILS: Famous Maya site with a number of ceremonial and religious complexes. It includes the Temple of the Warriors, the Court of One Thousand Warriors, an impressive ballcourt, and the Castillo. There are a number of chac mools, and jade objects, wooden idols, copper and cast-gold bells, and human remains were found.

36 G5 UXMAL

7TH–10TH CENTURIES AD

LOCATION: Yucatán, Mexico
TYPE OF SITE: Maya settlement
SITE DETAILS: Large city whose main structure was the Pyramid of the Magician, carved with reliefs of sky serpents and divine and human figures. Also found here are the Platform of the Stelae and the Great Pyramid. Many of the structures are impressively carved.

37 G5 KABAH

8TH–10TH CENTURIES AD

LOCATION: Yucatán, Mexico
TYPE OF SITE: Maya settlement
SITE DETAILS: Late Classic center that was mysteriously abandoned; it is connected to Uxmal by a causeway.

38 G5 SAYIL

9TH CENTURY AD

LOCATION: Yucatán, Mexico
TYPE OF SITE: Maya settlement
SITE DETAILS: Late Classic center with finds of a palace, residences, plazas, and a number of temple pyramids. Its most impressive monument is a three-story palace, which contained 94 rooms.

39 G5 TULUM

11TH–12TH CENTURIES AD

LOCATION: Yucatán, Mexico
TYPE OF SITE: Maya settlement
SITE DETAILS: Ceremonial center with concentric patterns of residences surrounding it. Structures include the Temple of the Frescoes, with the shrine encased in a larger building, and the Temple of the Diving God.

40 G5 COZUMEL

1ST–14TH CENTURIES AD

LOCATION: Mexico
TYPE OF SITE: Pre-Columbian town
SITE DETAILS: Based on an island, this was an important port and pilgrimage site for the cult of Ix Chel (the Maya goddess of women), reaching its peak around AD 1400. Long causeways lead to inland storage platforms.

41 G6 CERROS

1ST–9TH CENTURIES AD

LOCATION: Belize
TYPE OF SITE: Maya settlement
SITE DETAILS: Originally a small fishing village, major renovations turned it into an important trading town – its ruins include temples and remains of other structures decorated with mask façades. It has two ballcourts and the remains of fields surrounded by a canal.

42 G6 ALTUN HA

3RD–9TH CENTURIES AD

LOCATION: Belize
TYPE OF SITE: Maya center
SITE DETAILS: Classic city with remains of tombs and temples, dominated by the Temple of the Masonry Altars. A number of jade artifacts were found.

43 G6 CARACOL

EARLY 1ST MILLENNIUM AD

LOCATION: Belize
TYPE OF SITE: Maya city
SITE DETAILS: Currently under excavation – an acropolis, a ritual ballcourt, and the Temple of the Wooden Lintel have been excavated. At its height, the city held between 30,000 and 60,000 people. There are a number of raised causeways, known as sacbe, leading to smaller sites.

44 G6 EL MIRADOR

4TH–6TH CENTURIES AD

LOCATION: Petén, Guatemala
TYPE OF SITE: Maya settlement
SITE DETAILS: An early urban center that has one of the culture's largest temple pyramids. A causeway joins it to another site, Nakbe.

45 G6 RIO AZUL

4TH–9TH CENTURIES AD

LOCATION: Petén, Guatemala
TYPE OF SITE: Early settlement
SITE DETAILS: A minor center with a number of Early Classic palace structures. There are also a number of tombs, many of which are painted.

46 G6 UAXACTUN

3RD–6TH CENTURIES AD

LOCATION: Petén, Guatemala
TYPE OF SITE: Maya settlement
SITE DETAILS: Late Pre-Classic site with pyramids and temples; the foundations of one temple may be 4,000 years old.

47 G6 TIKAL

800 BC–AD 900

LOCATION: Petén, Guatemala
TYPE OF SITE: Maya ceremonial center
SITE DETAILS: Around 3,000 structures situated in an area of 6 sq mi (15 sq km), including five huge temple pyramids with rich burials. There is a great plaza with a number of altars and stelae. The Temple of the Jaguar rises over 140 ft (43 m). A 212-ft (65-m) mound remains unexcavated and is likely to be the largest pyramid of the Maya.

48 F6 PIEDRAS NEGRAS

6TH CENTURY AD

LOCATION: Guatemala
TYPE OF SITE: Maya settlement
SITE DETAILS: Late Classic site with a number of stelae, mounds, and pyramids. It is especially known for its sculpture.

49 F6 YAXCHILAN

6TH–8TH CENTURIES AD

LOCATION: Guatemala
TYPE OF SITE: Maya settlement
SITE DETAILS: Following the natural contours of the river, this Maya site is noted for its low-relief carving on the stelae and the lintels of its structures.

50 F6 ALTAR DE SACRIFICIOS

800 BC–AD 900

LOCATION: Guatemala
TYPE OF SITE: Maya settlement
SITE DETAILS: Site with finds of Fine Orange pottery, different from the Classic Maya types, which indicates that the area was invaded in the 9th century AD. Most of the ruins come from the Classic period, when platforms and pyramids were arranged around central plazas.

51 F6 DOS PILAS

7TH–8TH CENTURIES AD

LOCATION: Guatemala
TYPE OF SITE: Maya city
SITE DETAILS: A large center whose most impressive site is the tomb of a Maya ruler, with an impressive headdress discovered at the base of a temple.

52 F6 SEIBAL

9TH CENTURY AD

LOCATION: Guatemala
TYPE OF SITE: Maya settlement
SITE DETAILS: A center with stelae and traces of round buildings linked to the worship of the god Quetzalcoatl.

53 G6 NAJ TUNICH

8TH CENTURY AD

LOCATION: Guatemala
TYPE OF SITE: Cave art
SITE DETAILS: A Maya cave art site containing 100 paintings, some with hieroglyphic text and human figures.

54 G7 QUIRIGUA

4TH–12TH CENTURIES AD

LOCATION: Guatemala
TYPE OF SITE: Maya settlement
SITE DETAILS: A small center that contains a large number of zoomorphs – objects resembling animal forms. It is also the site of the largest Maya stela, whose carvings depict a successful ruler. The site was later occupied by the Toltecs.

55 F7 ZACULEU

EARLY 2ND MILLENNIUM AD

LOCATION: Guatemala
TYPE OF SITE: Maya settlement
SITE DETAILS: A Classic and Post-Classic religious center with ballcourts and temples. The site has revealed much about the burial practices of the Maya through its tombs. The cremated bodies were placed in urns, along with fine slate-backed plaques and jade carvings.

56 F7 OCOS
LATE 2ND MILLENNIUM BC
LOCATION: Guatemala
TYPE OF SITE: Open-air
SITE DETAILS: Site containing some of the earliest pottery in Mesoamerica. Often shell-marked and sometimes painted, it was characteristic of the Ocós culture. Their pottery often consisted of *tecomates* – neckless jars and flat-bottomed bowls.

57 F7 LA VICTORIA
MID-2ND MILLENNIUM BC
LOCATION: Guatemala
TYPE OF SITE: Early settlement
SITE DETAILS: Village of the Ocós culture, with a group of a dozen or so houses on low platforms, positioned next to a silted tidal estuary.

58 F7 ABAJ TAKALIK
1ST CENTURY BC
LOCATION: Guatemala
TYPE OF SITE: Maya settlement
SITE DETAILS: Site still under excavation. Finds include Olmec-type heads, one of which has a were-jaguar beard.

59 F7 IXIMCHE
12TH CENTURY AD
LOCATION: Guatemala
TYPE OF SITE: Maya center
SITE DETAILS: A Cakchiquel Maya capital with four ceremonial plazas, temples, and ritual ballcourts, similar in style to Monte Albán.

60 F7 KAMINALJUYU
5TH–6TH CENTURIES AD
LOCATION: Guatemala
TYPE OF SITE: Maya center
SITE DETAILS: Linked with Teotihuacán, it was a religious and trading center of a culture known as Las Charcas. There were finds of over 200 earth-and-clay mounds, which contained richly filled log tombs, one with 340 objects.

61 F7 MONTE ALTO
1ST MILLENNIUM BC
LOCATION: Guatemala
TYPE OF SITE: Early settlement
SITE DETAILS: A Late Formative site with finds of heads, a jade mask, and many reliefs, linked to the Izapa culture. It was a later Maya settlement.

62 G7 NACO
11TH–16TH CENTURIES AD
LOCATION: Honduras
TYPE OF SITE: Maya settlement
SITE DETAILS: Large coastal trading site that may have had a population of up to 10,000. There are remains of small residential mounds, elite residences, public buildings, and a ballcourt. There were also finds of distinctive ceramics.

63 G7 PLAYA DE LOS MUERTOS
7TH–3RD CENTURIES BC
LOCATION: Honduras
TYPE OF SITE: Early settlement

SITE DETAILS: Site mostly salvaged from river waters, with at least four periods of habitation. The people lived in pole-and-thatch-roofed structures and buried their dead in cemeteries with rich offerings – there have been finds of 100 figurines, most individually carved.

64 G7 COPAN
4TH–9TH CENTURIES AD
LOCATION: Honduras
TYPE OF SITE: Maya center
SITE DETAILS: A large site with a number of funerary pyramids and other structures. Structure 16, a nine-level pyramid, is the largest, and the Hieroglyphic Stairway is a pyramid with 2,200 glyphs. There are also stelae, ritual ballcourts, and an acropolis.

65 G7 LOS NARANJOS
8TH–5TH CENTURIES BC
LOCATION: Honduras
TYPE OF SITE: Fortified settlement
SITE DETAILS: Site of a fairly complex fortified site with a 20-ft- (6-m-) tall platform that yielded an elite burial with jade ornaments, including an ax.

66 G7 CHALCHUAPA
12TH–3RD CENTURIES BC
LOCATION: El Salvador
TYPE OF SITE: Pre-Classic settlement
SITE DETAILS: Earliest evidence of human occupation in the country, with finds of shards, figurines, and chipped and ground stone. It is the site of later obsidian manufacture. It grew into a sizable community, with a 66-ft- (20-m-) high pyramid and a large platform-plaza. It was probably linked to the Olmec culture.

67 G7 SANTA LETICIA
400 BC–AD 200
LOCATION: El Salvador
TYPE OF SITE: Early settlement
SITE DETAILS: A site that contained a large artificial terrace and a number of structures. Artifacts found at the site included several pot-bellied sculptures, carved stone masks, and bell-shaped pots that still contained corn and sunflower seeds.

68 G7 CEREN
6TH CENTURY AD
LOCATION: El Salvador
TYPE OF SITE: Maya settlement
SITE DETAILS: Village site preserved after it was covered in ash by a nearby volcano. There are finds of wooden digging sticks, pestles, and dishes of food left on hearths. The fields were well-preserved and have revealed much about the daily life of the Maya.

69 G7 CIHAUTAN
8TH CENTURY AD
LOCATION: El Salvador
TYPE OF SITE: Early settlement
SITE DETAILS: A center that is dominated by a central pyramid plaza-ballcourt

complex, with finds of ceramics, wheeled figures, and *chac mools*.

70 G7 QUELEPA
500 BC–AD 625
LOCATION: El Salvador
TYPE OF SITE: Early settlement
SITE DETAILS: Site with evidence of Mesoamerican blade technology. A platform has the Jaguar Altar in its center, flanked by serpents and stylized jaguars. Large-scale ritual pyramids were built later on artificially constructed platforms. Finds include small carved jade beads, legged *metates* used as ritual vessels, and stone balls.

71 G7 LOS LLANITOS
LATE 1ST MILLENNIUM AD
LOCATION: El Salvador
TYPE OF SITE: Early settlement
SITE DETAILS: Ritual and residential center with ten small structures and a ballcourt arranged around two plazas.

72 H8 BARRIAL DE HEREDIA
500 BC–AD 1300
LOCATION: Costa Rica
TYPE OF SITE: Pre-Columbian settlement
SITE DETAILS: A center of many periods whose most interesting remains date from the 8th–11th centuries AD. Eight cobble-foundation houses have been excavated – half were rectangular, with burials under the floor containing imported Nicoya pottery, and four were circular, with hearths, stone tools, and flakes.

73 H8 TORRALBA
8TH MILLENNIUM BC
LOCATION: Costa Rica
TYPE OF SITE: Hunter-gatherer
SITE DETAILS: A stratified site with finds of many chipped-stone artifacts, including 18 fluted points, scrapers, and blades.

74 I9 CHIRIQUI
12TH–16TH CENTURIES AD
LOCATION: Panama
TYPE OF SITE: Pre-Columbian settlement
SITE DETAILS: An area with a number of deep graves, where gold objects and pottery, sometimes painted, have been excavated.

75 I9 SITIO CONTE
3RD–13TH CENTURIES AD
LOCATION: Coclé Province, Panama
TYPE OF SITE: Burial mound
SITE DETAILS: A cemetery containing 60 graves filled with gold, regalia, carved stone, bone ornaments, and painted ceramics. All items were decorated with elaborate scrolls and animal designs.

76 I9 CERRO MANGOTE
6TH MILLENNIUM BC
LOCATION: Panama
TYPE OF SITE: Shell midden
SITE DETAILS: Site of a shell mound

where chopping and edge-grinding tools were excavated.

77 J6 SEVILLA NUEVA
16TH CENTURY AD
LOCATION: Jamaica
TYPE OF SITE: Colonial settlement
SITE DETAILS: Ruins of a Spanish fort built in 1509, where excavations have been made on two caravel ships in its harbor area.

78 J6 PORT ROYAL
17TH CENTURY AD
LOCATION: Jamaica
TYPE OF SITE: Colonial settlement
SITE DETAILS: A thriving port submerged by an earthquake in 1692. It remains in an excellent state of preservation. Objects salvaged include cannonballs, guns, clay pipes, bottles of rum, and even a watch that stopped at the exact time of the earthquake.

79 L5 SANS SOUCI PALACE
19TH CENTURY AD
LOCATION: Milot, Haiti
TYPE OF SITE: Colonial palace
SITE DETAILS: The site of a large royal palace ruined by an earthquake. It is still largely intact, with many galleries, paneled rows, and paintings.

80 M6 LA ISABELA LUPERON
16TH CENTURY AD
LOCATION: Dominican Republic
TYPE OF SITE: Colonial settlement
SITE DETAILS: The site of Columbus's first base, which has been excavated and restored.

81 N6 CASTILLO DE SAN FELIPE DEL MORRO
17TH CENTURY AD
LOCATION: San Juan, Puerto Rico
TYPE OF SITE: Fortified castle
SITE DETAILS: Site of the remains of a large fortress castle that rises 145 ft (44 m) above the ocean.

82 O6 ANGUILLA
2ND MILLENNIUM BC
LOCATION: Anguilla
TYPE OF SITE: Open-air
SITE DETAILS: This tiny island has the earliest-known indigenous sites, where finds have included a number of tools and other artifacts. Also on the island is a national park containing a number of petroglyphs.

83 O6 FORT THOMAS
17TH CENTURY AD
LOCATION: St. Kitts
TYPE OF SITE: Early settlement
SITE DETAILS: Site of the ruins of a colonial fort, where there are still a number of artifacts from the period, including weapons, clothing, and domestic wares.

PERU

The map for these listings is on p. 140.

1 B3 SIPAN

EARLY 1ST MILLENNIUM AD
LOCATION: Lambayeque, Peru
TYPE OF SITE: Burial mound
SITE DETAILS: Site of a number of lavish burials. The most important is the tomb of the Lord of Sipán, accompanied by four other burials and some pottery.

2 C3 EL CUMBEMAYO

3RD CENTURY BC
LOCATION: Cajamarca, Peru
TYPE OF SITE: Artificial water channel
SITE DETAILS: A cleverly engineered water channel built long before any by the Inca. It is so complex that it can have been used only for ritual purposes.

3 C3 HUACALOMA

1400 BC–EARLY 1ST MILLENNIUM AD
LOCATION: Cajamarca, Peru
TYPE OF SITE: Early Horizon settlement
SITE DETAILS: The earliest structure contained a central hearth. Later structures include six platforms built on a hillside. It reached its peak as a center in the early centuries AD.

4 C4 PACATNAMU

7TH–15TH CENTURIES AD
LOCATION: Jequetepeque Valley, Peru
TYPE OF SITE: Middle Horizon settlement
SITE DETAILS: Over 50 truncated pyramid complexes, with mounds, courtyards, and plazas excavated. Inhabited briefly before the arrival of the Inca.

5 C4 HUACA PRIETA

c. 2300 BC
LOCATION: Chicama, Peru
TYPE OF SITE: Farming settlement
SITE DETAILS: A preceramic site and an agricultural settlement with finds of elaborately twined cotton textiles that depict birds and other images.

6 C4 CHAN CHAN

12TH CENTURY AD
LOCATION: Trujillo, Peru
TYPE OF SITE: Chimú center
SITE DETAILS: The capital of the Chimú empire with three main sectors, the Tschudi temple-citadel being the largest. The area also contains two main *huacas* (temple buildings) – Huaca Esmerelda and Arco Iris.

7 C4 ALTO SALAVERY

18TH–10TH CENTURIES BC
LOCATION: Moche Valley, Peru
TYPE OF SITE: Early settlement
SITE DETAILS: Remains of religious and domestic architecture linked to a preceramic culture. It has a sunken ceremonial room, typical of the Initial period, when ceramics were first used in South America.

8 C4 CABALLO MUERTO

19TH–1ST CENTURIES BC
LOCATION: Moche Valley, Peru
TYPE OF SITE: Ceremonial center
SITE DETAILS: Pre-Chavín mounds – the most fascinating is the U-shaped, multi-level Huaca de los Reyes, which has red, yellow, and white reliefs.

9 C4 GUITARRERO CAVE

8TH MILLENNIUM BC
LOCATION: Callejón de Huaylas, Peru
TYPE OF SITE: Rock shelter
SITE DETAILS: The first traces of domestic beans were found here, dating back 10,000 years. It may have been occupied only during the wet season.

10 C4 HUARICOTO

2ND MILLENNIUM BC
LOCATION: Callejón de Huaylas, Peru
TYPE OF SITE: Ceremonial center
SITE DETAILS: A small artificial mound occupied over a long period that consists of 13 structures that contain ritual chambers and hearths.

11 C4 LA GALGADA

EARLY 2ND MILLENNIUM BC
LOCATION: Peru
TYPE OF SITE: Ceremonial center
SITE DETAILS: Temple structures that contained central fire pits, with the remains of chili pepper seeds. Once each temple was filled in, sometimes with a burial; the next was built over it. There is evidence that it supported a substantial residential population.

12 C4 MOXEKE

9TH CENTURY BC
LOCATION: Casma Valley, Peru
TYPE OF SITE: Ceremonial center
SITE DETAILS: Site of a terraced platform pyramid mound consisting of eight terraces. Most structures are covered in clay relief rather than stones.

13 C4 CERRO SECHIN

9TH–7TH CENTURIES BC
LOCATION: Ancash Coast, Peru
TYPE OF SITE: Ceremonial center
SITE DETAILS: Temple platforms, dwellings, and cemeteries have been excavated. The central platform mound has 90 stone monolithic slabs, which represent warriors and their trophy heads.

14 C4 CHAVIN DE HUANTAR

9TH–3RD CENTURIES BC
LOCATION: Peru
TYPE OF SITE: Ceremonial center
SITE DETAILS: At its peak, it was one of the largest religious centers in the world, with around 3,000 priests and temple attendants. The south wing of the main temple is known as the Castillo, built of dressed stone with carved gargoyles. A carved monolith, the Lanzón, is found inside. Excavation has revealed a canal system in the temple.

15 C4 KOTOSH

8TH CENTURY BC
LOCATION: Peru
TYPE OF SITE: Ceremonial center
SITE DETAILS: Important Chavín religious site, with a number of temples built on its hillside, including the Temple of the Crossed Hands.

16 D4 HUANUCO PAMPA

15TH–16TH CENTURIES AD
LOCATION: Peru
TYPE OF SITE: Inca city
SITE DETAILS: Administrative center based around a central plaza. Excavation of high-status residences and a compound show a socially stratified society. Hundreds of storage buildings can be found overlooking the town.

17 D4 LAURICOCHA

8TH–5TH MILLENNIA BC
LOCATION: Peru
TYPE OF SITE: Stratified rock shelter
SITE DETAILS: Caves that served as seasonal hunting camp; the remains of deer were found, along with some projectile points.

18 C5 ASPERO

c. 2600 BC
LOCATION: Peru
TYPE OF SITE: Early settlement
SITE DETAILS: A center with six platform mounds; the largest was the Huaca de Los Idolos, where a child wrapped in textiles was found buried. Other finds include 135 carved wooden sticks, sugarcane, cotton, plant material, clay figurines, and feathers.

19 C5 CHANCAY

11TH–15TH CENTURIES AD
LOCATION: Peru
TYPE OF SITE: Pre-Inca cemetery
SITE DETAILS: A burial site that is important for its distinctive ware. The pottery was often elongated, with a face painted on the neck and black geometric designs on the body It seems, from nearby burials, that the culture was socially stratified. It was later conquered by the Inca.

20 C5 ANCON

3000 BC–AD 1500
LOCATION: Peru
TYPE OF SITE: Early settlement
SITE DETAILS: A number of artifacts have been found in burials, including pottery, textiles, and an ancient archer's bow. It was a site later occupied by the Chimú and the Inca.

21 C5 EL PARAISO

MID-2ND MILLENNIUM BC
LOCATION: Chillón Valley, Peru
TYPE OF SITE: Ceremonial center
SITE DETAILS: An important preceramic religious site of U-shaped mounds, where excavation has revealed that the people subsisted on fishing and plants.

22 C5 HUACA PUQUIANA

15TH CENTURY AD
LOCATION: Lima, Peru
TYPE OF SITE: Ceremonial mound
SITE DETAILS: A huge pre-Inca mound with a hollow core, which some archaeologists postulate may have been frog-shaped – the symbol of a rain god.

23 C5 GARAGAY

18TH–10TH CENTURIES BC
LOCATION: near Lima, Peru
TYPE OF SITE: Ceremonial center
SITE DETAILS: Initial Period site where a large U-shaped temple complex has been found. The architectural style isa local variation of Chavín art.

24 D5 TELARMACHAY

7TH–4TH MILLENNIA BC
LOCATION: Peru
TYPE OF SITE: Rock shelter
SITE DETAILS: A seasonal shelter with evidence of camelid domestication, and where the transition from hunting to herding took place in around 3800 BC.

25 C5 TAMBO COLORADO

15TH CENTURY AD
LOCATION: Peru
TYPE OF SITE: Inca center
SITE DETAILS: An administrative center and a roadside depot with an excellently preserved adobe (dried clay) structure. It still retains red-painted niched walls.

26 C5 PACHACAMAC

7TH–15TH CENTURIES AD
LOCATION: Peru
TYPE OF SITE: Ceremonial center
SITE DETAILS: A settlement later used by the Inca. Structures include the Temple of the Moon and the Convent of the Sun Virgins. Small cotton bags containing pigments, reeds, and human hair, and distinctive pottery were excavated.

27 C5 ASIA

3RD MILLENNIUM BC
LOCATION: Peru
TYPE OF SITE: Preceramic settlement
SITE DETAILS: An early agricultural settlement. Burials suggest a socially stratified society, some of the bodies showing signs of trephination – the drilling of holes into the skull.

28 D6 CABEZAS LARGAS

4TH MILLENNIUM BC
LOCATION: Peru
TYPE OF SITE: Early cemetery
SITE DETAILS: Up to 60 mummies were

found in one grave at this burial site. They were accompanied by a number of personal objects.

29 D6 PARACAS
10TH–2ND CENTURIES BC
LOCATION: Peru
TYPE OF SITE: Ceremonial center
SITE DETAILS: A Chavín-influenced religious center where a necropolis revealed some fine prehistoric textiles, preserved by the desert climate. The pottery was geometrically incised and is seen as a precursor to the Nazca style.

30 D6 PACHECO
7TH–10TH CENTURIES AD
LOCATION: Nazca Valley, Peru
TYPE OF SITE: Early settlement
SITE DETAILS: A site linked to the Huari culture – an empire of the Middle Horizon period – with a number of structures. Polychrome (multicolored) vessels were found in a pit.

31 D6 CAHUACHI
3RD–5TH CENTURIES AD
LOCATION: Nazca Valley, Peru
TYPE OF SITE: Ceremonial center
SITE DETAILS: Main religious site of the Nazca culture, with 40 adobe mounds, pyramids, and extensive cemeteries. (See p. 46.)

32 D6 NAZCA
200 BC–AD 600
LOCATION: Nazca Valley, Peru
TYPE OF SITE: Pre-Inca center
SITE DETAILS: The Nazca culture is most famous for its mysteriously carved lines in the desert, which show geometric designs of abstract beings and animals It is also noted for its distinctive forms of polychrome-painted pottery.

33 D6 ESTAQUERIA
10TH CENTURY AD
LOCATION: Nazca Valley, Peru
TYPE OF SITE: Middle Horizon settlement
SITE DETAILS: A site incorporating a large rectangular area with 12 rows of posts, an adobe compound, and cemeteries.

34 D5 CONCHOPATA
6TH–7TH CENTURIES AD
LOCATION: Ayacucho, Peru
TYPE OF SITE: Middle Horizon cemetery
SITE DETAILS: Offerings of Huari pottery, often smashed, have been found in burial pits. The shards are similar to ware found at Tiahuanaco.

35 D5 WICHQUANA
LATE 4TH MILLENNIUM BC
LOCATION: Peru
TYPE OF SITE: Agricultural settlement
SITE DETAILS: Site of an early village that shows evidence of subsistence farming.

36 D5 PACHAMACHAY CAVE
11TH–1ST CENTURIES BC
LOCATION: Peru
TYPE OF SITE: Hunter-gatherer

SITE DETAILS: The oldest-known occupied place in the Andes. Herding developed here in about 1600 BC.

37 E5 TRES CRUCES
6TH CENTURY AD
LOCATION: Peru
TYPE OF SITE: Ceremonial center
SITE DETAILS: A celebrated earthwork that was also a site of pilgrimage, with finds of Late Nazcan pottery.

38 E5 MACHU PICCHU
15TH–16TH CENTURIES AD
LOCATION: Peru
TYPE OF SITE: Inca center
SITE DETAILS: The most impressive Inca site, situated on a virtually inaccessible ridge. Structures include the Temple of the Three Windows, the Royal Palace, and the Hitching Post of the Sun. North of the main plaza were the residential and agricultural areas.

39 E6 CUZCO
13TH–16TH CENTURIES AD
LOCATION: Peru
TYPE OF SITE: Inca city
SITE DETAILS: The Inca capital was built in the shape of a puma and divided into two. It was centered on a U-shaped plaza, and surrounded by buildings made of extremely carefully cut and tight-fitting stones.

40 E6 PISAC
15TH CENTURY AD
LOCATION: Cuzco Valley, Peru
TYPE OF SITE: Inca center
SITE DETAILS: A settlement with a Temple of the Sun built on an outcrop of rock. Much of the area is still unexcavated, with several burial sites in the vicinity.

41 E6 PIKILLACTA
7TH–10TH CENTURIES AD
LOCATION: near Cuzco, Peru
TYPE OF SITE: Huari center
SITE DETAILS: An administrative center with small structures thought to be used as both dwellings and storehouses.

42 F6 PUCARA
3RD CENTURY BC
LOCATION: Peru
TYPE OF SITE: Early settlement
SITE DETAILS: Site of a monumental sunken court and stone-slab walled structures, with finds of carved Paracas ceramics. It later became an Inca center.

43 F7 ASANA
8TH MILLENNIUM BC
LOCATION: Peru
TYPE OF SITE: Preceramic settlement
SITE DETAILS: Circular and ovoid domestic structures have been uncovered. There was a later religious complex with two structures that may well be altars.

SOUTH AMERICA
The map for these listings is on p. 141.

1 C1 PUERTO HORMIGA
c. 3200 BC
LOCATION: Colombia
TYPE OF SITE: Early settlement
SITE DETAILS: Finds of very old ceramics with evidence that vegetable fibers were incorporated into the pottery.

2 C1 MONSU
EARLY 4TH MILLENNIUM BC
LOCATION: Colombia
TYPE OF SITE: Open-air
SITE DETAILS: Most likely to be the site of the oldest ceramics in the Americas distinguishable by their broad incisions.

3 C1 PUEBLITO
EARLY 2ND MILLENNIUM AD
LOCATION: Colombia
TYPE OF SITE: Early settlement
SITE DETAILS: Site of the prehistoric Tairona culture that included around 3,000 houses and public buildings.

4 C1 BURITACA
14TH CENTURY AD
LOCATION: Colombia
TYPE OF SITE: Early settlement
SITE DETAILS: A large Tairona center whose remains include house platforms, stone stairways, canal and drainage systems, ceremonial plazas, and roads.

5 C2 ZIPA
EARLY 2ND MILLENNIUM AD
LOCATION: Colombia
TYPE OF SITE: Muisca center
SITE DETAILS: The capital of the Muisca people that stretched nearly 2 mi (3 km) along a lagoon.

6 C2 QUIMBAYA VALLEY
4TH–16TH CENTURIES AD
LOCATION: Colombia
TYPE OF SITE: Open-air
SITE DETAILS: An important site where excavations have revealed some of the best metallurgic techniques in the Americas, especially in the use of gold. There were also finds of ceramics.

7 C2 TIERRADENTRO
8TH–9TH CENTURIES AD
LOCATION: Colombia
TYPE OF SITE: Cemetery
SITE DETAILS: Underground shaft tombs, many of which are painted with black, orange, and red figures.

The artifacts found here are thought to be Amazonian in origin.

8 C2 SAN AGUSTIN
1ST MILLENNIUM AD
LOCATION: Colombia
TYPE OF SITE: Burial mound
SITE DETAILS: Megalithic mounds covering a huge area, many with stone-lined chambers inside. There are also around 300 sculptures and monumental figures.

9 C2 LA TOLITA
300 BC–AD 350
LOCATION: Ecuador
TYPE OF SITE: Burial mound
SITE DETAILS: Coastal cemetery with burial mounds laid around a plaza. Postholes show that structures were once situated on top of the mounds. The inhabitants made also clay and gold objects.

10 C2 COTOCOLLAO
c. 540 BC
LOCATION: Quito Valley, Ecuador
TYPE OF SITE: Early cemetery
SITE DETAILS: Site of a large burial where up to 400 inhumations and decorated stone bowls were found.

11 C2 EL INGA
10TH–2ND MILLENNIA BC
LOCATION: Ecuador
TYPE OF SITE: Hunter-gatherer
SITE DETAILS: Fluted-stemmed fishtail Clovis and Folsom points have been found here.

12 C2 MACHALILLA
c. 2000 BC
LOCATION: Ecuador
TYPE OF SITE: Early settlement
SITE DETAILS: Site of an early agricultural village that has unearthed fire-incised pottery, probably a type of the Valdivian ware.

13 C2 LA PLATA ISLAND
2500 BC–AD 1500
LOCATION: Ecuador
TYPE OF SITE: Early settlement
SITE DETAILS: A market for spondylus shells, which were used for ritual. It was also a sanctuary that has revealed hundreds of figurines.

14 C2 VALDIVIA
c. 3200 BC
LOCATION: Ecuador
TYPE OF SITE: Formative settlement
SITE DETAILS: Some of the earliest pottery in the Americas was found here, along with figurines and textiles. (See p. 24.)

15 C2 REAL ALTO
1ST MILLENNIUM BC
LOCATION: Ecuador
TYPE OF SITE: Early village
SITE DETAILS: Valdivian settlement with an open plaza surrounded by mounds. Large wood-and-thatch houses and elite burials have been excavated.

16 C2 INCA PIRCA

AD 1463–71

LOCATION: Ecuador

TYPE OF SITE: Inca center

SITE DETAILS: Ruins that may have been a garrison. The Temple of the Sun contains impressive Inca stonework.

17 D1 RANCHO PELUDO

2800 BC–AD 1600

LOCATION: Venezuela

TYPE OF SITE: Early settlement

SITE DETAILS: Multioccupational river settlement that had its own coarse grit-tempered ware.

18 D1 EL JOBO

c. 10,000 BC

LOCATION: Venezuela

TYPE OF SITE: Early settlement

SITE DETAILS: Artifacts include a distinctive form of leaf-shaped spear point known as the Jobo point, which have been found at a number of nearby mammoth kill sites.

19 D1 TAIMA-TAIMA

15,000–12,000 YEARS AGO

LOCATION: Venezuela

TYPE OF SITE: Early settlement

SITE DETAILS: Finds include mastadon bones and stone tools. (See p. 12.)

20 E2 PACOVAL

11TH CENTURY AD

LOCATION: Brazil

TYPE OF SITE: Early burial

SITE DETAILS: Site of cemeteries with burial urns arranged in straight lines or semicircles and finds of ceramics.

21 E2 TAPERINHA

6TH MILLENNIUM BC

LOCATION: Brazil

TYPE OF SITE: Shell midden

SITE DETAILS: Deposits of mollusk shells have been excavated, and there is some evidence of ancient wares. (See p. 12.)

22 F2 ATERRO DOS BICHOS

400 BC–AD 1300

LOCATION: Marajó Island, Brazil

TYPE OF SITE: Early settlement

SITE DETAILS: Mounds have revealed the foundations of 20 houses arranged around an open space and surrounded by a large earth wall. The people subsisted on corn, seeds, and fish.

23 F2 OS CAMUTINS

400 BC–AD 1300

LOCATION: Marajó Island, Brazil

TYPE OF SITE: Early settlement

SITE DETAILS: A village site of about 40 artificial mounds built on platforms.

24 G3 PEDRA FURADA

25,000 YEARS AGO

LOCATION: Brazil

TYPE OF SITE: Early settlement

SITE DETAILS: Cave with the oldest-known signs of human habitation in South America. Some wall paintings were done 10,000 years later.

25 G3 TOCO DE ESPERANZA

295,000 YEARS AGO

LOCATION: Brazil

TYPE OF SITE: Hunter-gatherer

SITE DETAILS: Though of extremely unlikely date, finds include fauna and basic stone tools.

26 F4 SANTANA DO RIACHO

8TH–2ND MILLENNIA BC

LOCATION: Minas Gerais, Brazil

TYPE OF SITE: Hunter-gatherer

SITE DETAILS: Rock paintings and a cemetery that has unearthed small flakes, axes, and bone projectile points.

27 F5 PRAIA DA TAPEIRA

6TH–10TH CENTURIES AD

LOCATION: Brazil

TYPE OF SITE: Early village

SITE DETAILS: Shell mounds and houses, 49 ft (15 m) in diameter with pole walls, have been excavated.

28 D4 MOUND VELARDE

7TH CENTURY AD

LOCATION: Bolivia

TYPE OF SITE: Burial mound

SITE DETAILS: Ceramics of different periods have been found, as have urn burials, bowls, bones, and shell remains.

29 D4 ISKANWAYA

10TH–15TH CENTURIES AD

LOCATION: Bolivia

TYPE OF SITE: Early settlement

SITE DETAILS: A Mollo center where the houses were built on platforms and terraces. Some of the houses were painted with red ocher, with children buried beneath the floors; adults were placed in *chullpas* – burial towers.

30 D4 CHIRIPA

13TH–6TH CENTURIES BC

LOCATION: Bolivia

TYPE OF SITE: Ceremonial mound

SITE DETAILS: Centered on an artificial temple platform, this is the type-site for one of the country's oldest cultures. There are a few *chullpas* nearby.

31 D4 TIAHUANACO

4TH–7TH CENTURIES AD

LOCATION: Bolivia

TYPE OF SITE: Early city

SITE DETAILS: Kalasasaya was the first major structure built here, when the city supported a population of between 100,000 and 400,000. The Akapana step pyramid was the city's central focus, while the monumental Gateway of the Sun is decorated with reliefs. (See p. 34.)

32 D4 WAKALLANI

4TH–8TH CENTURIES AD

LOCATION: Bolivia

TYPE OF SITE: Early settlement

SITE DETAILS: Site with the remains of three burial towers and two stelae.

33 D4 CHULLPA PAMPA

EARLY 1ST MILLENNIUM AD

LOCATION: Bolivia

TYPE OF SITE: Open-air

SITE DETAILS: Site of a culture that immediately preceded that at Tiahuanaco, where burial urns and distinctive gray and red flat-bottomed beakers have been excavated.

34 D4 WANKARANI

c. 1200 BC

LOCATION: Bolivia

TYPE OF SITE: Early settlement

SITE DETAILS: Remains of round adobe houses with the dead buried under their clay floors. Stone heads of llamas and alpacas were also found.

35 D4 POTOSI

16TH CENTURY AD

LOCATION: Bolivia

TYPE OF SITE: Colonial settlement

SITE DETAILS: Once the biggest industrial complex in the world, with a number of mills and industrial monuments. The colonial town has a church and the *barrios mitayos* – the workers' living quarters. (See p. 100.)

36 D4 CHINCHORRO

3RD–1ST MILLENNIA BC

LOCATION: Chile

TYPE OF SITE: Burial mound

SITE DETAILS: Site of the earliest mummies in America, which were painted red, skinned, and then covered with clay.

37 D4 QUIANI

4TH MILLENNIUM BC

LOCATION: Chile

TYPE OF SITE: Shell midden

SITE DETAILS: The earliest finds are characterized by fishhooks and leaf points. There are later finds of *bolas* (rounded, faceted stones) and choppers from around 250 BC.

38 D4 AZAPA

1ST MILLENNIUM BC

LOCATION: Chile

TYPE OF SITE: Early settlement

SITE DETAILS: An Alto Ramirez culture burial containing silver and copper ornaments and textiles.

39 D4 CAMARONES

3RD–1ST MILLENNIA BC

LOCATION: Chile

TYPE OF SITE: Early settlement

SITE DETAILS: A Chinchorro site with a cemetery containing around 150 burials.

40 D4 TILIVICHE

c. 3500 BC

LOCATION: Chile

TYPE OF SITE: Agricultural settlement

SITE DETAILS: An oval-shaped site with the remains of 90 houses. There is evidence of early corn agriculture and finds of mortars, pestles, stone projectiles, and scrapers.

41 D5 LAS CONCHAS

c. 7500 BC

LOCATION: Chile

TYPE OF SITE: Shell midden

SITE DETAILS: Type-site where artifacts excavated include choppers, flake tools, mortars, bone tools, and distinctive geometric sandstone objects.

42 D6 TAGUA-TAGUA

c. 9400 BC

LOCATION: Chile

TYPE OF SITE: Hunter-gatherer

SITE DETAILS: Site where artifacts including unifacial tools, blades, and mastadon-bone tools were excavated.

43 D7 PADRE LAS CASAS

LATE 1ST MILLENNIUM AD

LOCATION: Chile

TYPE OF SITE: Burial mound

SITE DETAILS: Cemetery with finds of burial urns within funerary canoes. The richer burials have stone beads and copper rings, as well as ceramics.

44 D7 MONTE VERDE

13,000 YEARS AGO

LOCATION: Chile

TYPE OF SITE: Early settlement

SITE DETAILS: See p. 12.

45 D9 FELL'S CAVE

10TH–9TH MILLENNIA BC

LOCATION: Chile

TYPE OF SITE: Early settlement

SITE DETAILS: Preceramic cave containing fluted-stemmed fishtail points and the remains of the predecessor of the llama.

46 D9 MARASSI SHELTER

c. 7640 BC

LOCATION: Chile

TYPE OF SITE: Hunter-gatherer

SITE DETAILS: A seasonally occupied rock shelter whose finds include stone tools, small mammal bones, and two *bolas*.

47 D5 TAFI

200 BC–AD 600

LOCATION: Argentina

TYPE OF SITE: Early settlement

SITE DETAILS: A site where rituals were centered on artificial mounds and carvings of feline figures were found.

48 D8 LOS TOLDOS

9TH MILLENNIUM BC

LOCATION: Argentina

TYPE OF SITE: Early settlement

SITE DETAILS: A cave that unearthed large flake tools, unifacial and bifacial tools, flaked and fishtail points, bone awls, and spatulas. There is also some early South American rock art.

AFRICA

The map for these listings is on p. 142.

1 B1 LIXUS
MID-1ST MILLENNIUM BC
LOCATION: Morocco
TYPE OF SITE: Classical city
SITE DETAILS: Previously a Carthaginian coastal city, its most substantial remains are of a 1st-century AD Roman temple and sanctuaries.

2 B1 BELYUNECH
10TH CENTURY AD
LOCATION: Morocco
TYPE OF SITE: Islamic town
SITE DETAILS: An Islamic site whose palace is cited as the first example of an Andalusian-Moroccan mix of architecture. There were also finds of decorated pottery.

3 B2 VOLUBILIS
3RD CENTURY BC
LOCATION: Morocco
TYPE OF SITE: Roman colony
SITE DETAILS: First settled by the local Mauri kings. Most of the remains date from the Roman period – the Capitolium temple, the Arch of Caracalla, a basilica, law courts, and some impressive mosaics.

4 C2 TAFORALT
20,000–10,000 BC
LOCATION: Morocco
TYPE OF SITE: Hunter-gatherer
SITE DETAILS: A cave site with many sequences, from Mousterian occupation to an Iberomaurasian cemetery, with the remains of 185 people.

5 C1 TIPASA
2ND–4TH CENTURIES AD
LOCATION: Algeria
TYPE OF SITE: Early town
SITE DETAILS: Originally a Phoenician trading post, among the ruins are two Roman temples, mosaics, a well-preserved forum, a Christian basilica, and a chapel.

6 C1 DJEMILA
3RD CENTURY AD
LOCATION: Algeria
TYPE OF SITE: Roman colony
SITE DETAILS: Impressive ruins, including the Temple of Septimien, a number of mosaics, an old forum, a civil basilica, grand baths, temples, baptisteries, and a theater with a capacity of 3,000.

7 C1 AFALOU BOU RHUMMEL
15,000 YEARS AGO
LOCATION: Algeria
TYPE OF SITE: Hunter-gatherer
SITE DETAILS: The site has revealed terra-cotta figurines and has evidence of the emergence of portable art.

8 D1 TIMGAD
1ST CENTURY AD
LOCATION: Algeria
TYPE OF SITE: Roman town
SITE DETAILS: Known as the "Pompeii of Africa" due to its near-perfect condition, its central area is an almost-perfect square of *insulae* (city blocks). Its buildings include a forum, a basilica, a temple, baths, and the Arch of Trajan. A Byzantine fort stood outside the town

9 D2 TASSILI N'AJJER
6TH MILLENNIUM BC
LOCATION: Algeria
TYPE OF SITE: Rock-art
SITE DETAILS: Drawn by a Saharan people with over 15,000 examples of art, with many depictions of human figures.

10 D1 DOUGGA
2ND CENTURY BC
LOCATION: Tunisia
TYPE OF SITE: Roman town
SITE DETAILS: The remains are mostly from its time as a Roman settlement, combining Libyan, Punic, and Roman cultures. There are many temples, dolmens (megalithic tombs), a theater, baths, and houses.

11 D1 CARTHAGE
900 BC–AD 200
LOCATION: Tunisia
TYPE OF SITE: Early city
SITE DETAILS: A Phoenician colony that became a powerful city in its own right. Excavations revealed a sanctuary and traces of a harbor, with some remains of ships. It was destroyed in 146 BC by the Romans, although their subsequent amphitheater and aqueduct remain. (See p. 88.)

12 D1 ZAGHOUAN AQUEDUCT
2ND CENTURY AD
LOCATION: Tunisia
TYPE OF SITE: Roman aqueduct
SITE DETAILS: Site of an impressive aqueduct, part of which stands up to 66 ft (20 m) high. It runs mainly at ground level, while tunnels were dug in some places to maintain the slope.

13 D1 MAHDIA
10TH CENTURY AD
LOCATION: Tunisia
TYPE OF SITE: Islamic town
SITE DETAILS: Site of a fortified Islamic harbor with a special encircling wall.

14 D2 SABRATHA
MID-1ST MILLENNIUM BC
LOCATION: Libya
TYPE OF SITE: Early town
SITE DETAILS: A Phoenician trading post that was taken over by the Romans in the 2nd century AD. More than half of the ancient Roman city has been excavated, including the forum, harbor, baths, temples, fountains, a catacomb, and later Christian churches. (See p. 68.)

15 D2 LEPTIS MAGNA
2ND CENTURY AD
LOCATION: Libya
TYPE OF SITE: Roman town
SITE DETAILS: A Phoenician coastal city conquered by the Romans, whose ruins dominate the site. There are a number of impressive temples, although the best-preserved structure is the Hunting Baths.

16 E2 CYRENE
630 BC–AD 200
LOCATION: Libya
TYPE OF SITE: Greek town
SITE DETAILS: A colony founded by emigrants from Thera. It has a large Doric temple to Apollo, linked by road to an acropolis and *agora* (meeting place). Most impressive are the rock-carved Greek baths. Later Roman ruins include a theater and baths. Thousands of tombs surround the city.

17 E2 HAUA FTEAH CAVE
60,000–7000 YEARS AGO
LOCATION: Libya
TYPE OF SITE: Hunter-gatherer
SITE DETAILS: A cave that acted as a summer camp to exploit coastal resources, with evidence of a number of stone and blade industries. (See p. 13.)

18 B3 AKJOUJT REGION
1ST MILLENNIUM BC
LOCATION: Mauritania
TYPE OF SITE: Open-air
SITE DETAILS: A region with evidence of ancient copper mining, where some arrowheads have been excavated.

19 B3 DAR TICHITT
12TH–4TH CENTURIES BC
LOCATION: Mauritania
TYPE OF SITE: Agricultural settlement
SITE DETAILS: Some of the earliest evidence of farming evolution in the southern Sahara. Village settlements of circular compounds were connected by wide paths. Stone axes, arrowheads, gouges, and pottery were found. Protective walls and fortified entrance gates were added at a later date.

20 B3 AWADAGHOST
11TH CENTURY AD
LOCATION: Mauritania
TYPE OF SITE: Medieval town
SITE DETAILS: The site of a trading center with many links to the Islamic trade. It later became part of the Ghanan empire.

21 A4 SINE-SALOUM
8TH CENTURY AD
LOCATION: Gambia
TYPE OF SITE: Cemetery
SITE DETAILS: A burial site with over 4,000 megalithic tombs, many of which are filled with grave goods.

22 B4 TIEKENE-BOUSSOURA
6TH–8TH CENTURIES AD
LOCATION: Senegal
TYPE OF SITE: Cemetery
SITE DETAILS: Megalithic monuments that make up nine circles of upright stones, with some ceramic finds.

23 B5 NDALOA
17TH CENTURY AD
LOCATION: Ivory Coast
TYPE OF SITE: Ritual monument
SITE DETAILS: Site of the Eotile culture where two terra-cotta heads, perhaps of high-ranking chiefs, were excavated.

24 C4 KINTAMPO
c. 1400 BC
LOCATION: Ghana
TYPE OF SITE: Early settlement
SITE DETAILS: There have been finds of ceramics and flaked-stone tools used by hunter-gatherers, and some remains of goats and bovine animals.

25 C5 ELMINA
16TH CENTURY AD
LOCATION: Ghana
TYPE OF SITE: Colonial settlement
SITE DETAILS: The site of early Portuguese colony where a fortress castle was built. (See p. 101.)

26 B3 KARKARICHINKAT
4TH MILLENNIUM BC
LOCATION: Tilemsi Valley, Mali
TYPE OF SITE: Early settlement
SITE DETAILS: Two mounds that were occupied by herder-fishers.

27 B4 KOUMBI SALEH
LATE 1ST MILLENNIUM AD
LOCATION: Mali
TYPE OF SITE: Islamic city
SITE DETAILS: A large site believed to have been the capital of the Ghanan empire. It includes two cemeteries and the ruins of a large mosque. Groups of narrow-roomed houses were excavated. It was certainly the biggest city of its day, possibly home to 15,000 inhabitants.

28 C4 BANDIAGARA
11TH–16TH CENTURIES AD
LOCATION: Mali
TYPE OF SITE: Rock-cut cemeteries
SITE DETAILS: Large necropolises, some predating the Dogon era. The local Tellem people used to bury their dead in the cliff-side caves. Excavated artifacts included broken pots and skulls.

29 B4 JENNE-JENO
3RD CENTURY BC
LOCATION: Mali
TYPE OF SITE: Ancient city
SITE DETAILS: The oldest city south of the Sahara, it was orginally occupied by an iron-working people. In the 4th century AD the city rapidly expanded, and it was a thriving urban center until its rulers turned to Islam and built a new city. The walls were up to 36 ft (11 m) wide in places. Remains include a residential area, a cemetery, burial urns, and houses.

30 D3 TAGALTAGAL
8TH MILLENNIUM BC
LOCATION: Niger
TYPE OF SITE: Early settlement
SITE DETAILS: Neolithic site where old ceramics, stone tools, arrowheads, and grinding equipment have been found.

31 D3 AFUNFUN
2ND MILLENNIUM BC
LOCATION: Niger
TYPE OF SITE: Iron Age
SITE DETAILS: An early copper-working site with slag and a series of furnaces.

32 C4 AZELIK
MID-1ST MILLENNIUM BC
LOCATION: Niger
TYPE OF SITE: Early settlement
SITE DETAILS: Evidence of copper smelting that shows a brief Copper Age before the discovery of iron.

33 D4 TARUGA
5TH–3RD CENTURIES BC
LOCATION: Nigeria
TYPE OF SITE: Early settlement
SITE DETAILS: Ten furnaces point to an extensive iron-working industry of the Nok culture. Some terra-cotta sculptures have been found.

34 D4 NOK
500 BC–AD 200
LOCATION: Nigeria
TYPE OF SITE: Iron Age settlement
SITE DETAILS: The type-site of an artistic, flourishing culture, famous for its human and animal terra-cotta figures found in waterlogged areas. There is evidence of an iron-working industry.

35 D4 DAIMA
800 BC–AD 400
LOCATION: Nigeria
TYPE OF SITE: Early settlement
SITE DETAILS: A long-standing permanent settlement of cattle herders and fishermen that remained mostly unchanged, despite the technological changes brought with iron-working.

36 C4 IFE
13TH–15TH CENTURIES AD
LOCATION: Nigeria
TYPE OF SITE: Urban center
SITE DETAILS: Settlement famous for its terra-cotta and bronze sculpture, and influenced by the Nok and Benin cultures. Sculpted heads were mounted on wooden effigies and used for funerary rituals. Old town walls enclose the sacred groves where most of the sculptures were found.

37 D4 IGBO UKWU
9TH CENTURY AD
LOCATION: Nigeria
TYPE OF SITE: Early settlement
SITE DETAILS: A village with a shrine full of bronze vessels – the wooden chamber pit contained a major burial with a studded headdress, beads, armlets, anklets, and a copper crown, and five other bodies. A disposal pit with bronze and copper items was excavated.

38 E4 DARFUR
18TH CENTURY AD
LOCATION: Chad
TYPE OF SITE: Islamic palace
SITE DETAILS: The ruins of the great red-brick palace of Sultan Teirab (1752–87)

39 D4 BOUAR
6TH–5TH MILLENNIA BC
LOCATION: Central African Republic
TYPE OF SITE: Burial mound
SITE DETAILS: The site of a large number of megalithic standing stones positioned on oval tumuli.

40 E5 BATALIMO
4TH CENTURY AD
LOCATION: Central African Republic
TYPE OF SITE: Early settlement
SITE DETAILS: Flakes, sidescrapers, dark pottery, and axes have been excavated.

41 D6 SANGA
8TH–18TH CENTURIES AD
LOCATION: Democratic Republic of Congo
TYPE OF SITE: Early settlement
SITE DETAILS: An old town that has an Iron Age cemetery with finds of Kisalian ware, lavish burials, and advanced metallurgy.

42 F5 ISHANGO
20,000–9,000 YEARS AGO
LOCATION: Democratic Republic of Congo
TYPE OF SITE: Hunter-gatherer
SITE DETAILS: The site of a Stone Age midden with stone artifacts, bone industries, and backed microliths.

43 E6 KATOTO
13TH CENTURY AD
LOCATION: Democratic Republic of Congo
TYPE OF SITE: Iron Age cemetery
SITE DETAILS: A burial site with collective burials linked to the Kisalian culture – a rich Iron Age complex.

44 D6 BENFICA
2ND CENTURY AD
LOCATION: Angola
TYPE OF SITE: Shell midden
SITE DETAILS: A coastal village with a number of mounds, where some Iron Age pottery has been excavated.

45 F3 BUHEN
3RD–2ND MILLENNIA BC
LOCATION: Sudan
TYPE OF SITE: Early city
SITE DETAILS: An Egyptian settlement used for mining expeditions and trade. There are remains of some temples to the god Min and goddess Isis. It later became an important fortress site.

46 F3 ARKIN
100,000 YEARS AGO
LOCATION: Sudan
TYPE OF SITE: Hunter-gatherer
SITE DETAILS: A Paleolithic site that has traces of a camp and some Mousterian flakes and *racloir* points (sidescrapers).

47 F3 SAI
2ND MILLENNIUM BC
LOCATION: Sudan
TYPE OF SITE: Early settlement
SITE DETAILS: Site of an island town and fort with remains of a temple. There is also a Kerma cemetery.

48 F3 SEMNA
19TH–18TH CENTURIES BC
LOCATION: Sudan
TYPE OF SITE: Fortified settlement
SITE DETAILS: Two Egyptian mudbrick fortresses, with inscriptions left by Senusert III to proclaim it as the frontier with Nubia. Each fortress contained a temple, one dedicated to Thutmose III and the other to Taharqa.

49 F3 SOLEB
2ND MILLENNIUM BC
LOCATION: Sudan
TYPE OF SITE: Early city
SITE DETAILS: An Upper Nubian site that is home to the Temple of Amenophis III and a New Kingdom cemetery.

50 F3 KERMA
17TH–16TH CENTURIES BC
LOCATION: Sudan
TYPE OF SITE: Early city
SITE DETAILS: A major Kushite settlement with the Western Deffufa – a castle structure with a series of apartments. A nearby cemetery has tumuli graves and a royal necropolis.

51 F3 GEBEL BARKAL
2ND MILLENNIUM BC
LOCATION: Sudan
TYPE OF SITE: Ceremonial center
SITE DETAILS: A religious center for the worship of Amun Ra; a temple was built by Ramses II, and enlarged by later kings. There are a number of other palaces and nearby pyramids.

52 F3 MEROE
3RD CENTURY BC–4TH CENTURY AD
LOCATION: Sudan
TYPE OF SITE: Ancient city
SITE DETAILS: The capital of the Meroe and Kush kingdoms, set up after they withdrew from ruling Egypt. Among the ruins are palaces, temples, and pyramid tombs. It was later destroyed by the Axumite empire.

53 F3 NURI
7TH–3RD CENTURIES BC
LOCATION: Sudan
TYPE OF SITE: Ancient cemetery
SITE DETAILS: Royal pyramid for Napatan kings that included the granite sarcophagus of King Aspelta.

54 G4 GOBEDRA
10TH–3RD MILLENNIA BC
LOCATION: Ethiopia
TYPE OF SITE: Rock shelter
SITE DETAILS: A stratified shelter with finds of pottery in its later periods. There are remains of cultivated crops and domestic animals.

55 G4 AXUM
1ST–7TH CENTURIES AD
LOCATION: Ethiopia
TYPE OF SITE: Early city
SITE DETAILS: The capital of a large Christian empire that has remains of a palace and the impressive 98-ft- (30-m-) high stelae with religious carvings.

56 G4 LALIBELA
500 BC–14TH CENTURY AD
LOCATION: Ethiopia
TYPE OF SITE: Early settlement
SITE DETAILS: A cave site that revealed bones, pottery, and vegetables. The later Christian settlement has a number of impressive rock-cut churches.

57 G4 HADAR
3.4 MILLION YEARS AGO
LOCATION: Ethiopia
TYPE OF SITE: Open-air
SITE DETAILS: The site where the upright hominid "Lucy" was discovered, with 40 percent of the skeleton intact.

58 F4 MELKA KUNTURE
1.7 MILLION YEARS AGO
LOCATION: Ethiopia
TYPE OF SITE: Hunter-gatherer
SITE DETAILS: Large numbers of stone artifacts were discovered, including flakes, cores, choppers, and antelope and hippopotamus bones.

59 F5 OMO VALLEY
2.5 MILLION YEARS AGO
LOCATION: Ethiopia
TYPE OF SITE: Hunter-gatherer
SITE DETAILS: One of the oldest hominid encampments found, with discoveries of flakes and quartz pebbles.

60 F5 KOOBI FORA
2.5–1 MILLION YEARS AGO
LOCATION: Kenya
TYPE OF SITE: Hunter-gatherer
SITE DETAILS: About 150 hominid fossils

and stone artifacts have been found. The processing of food probably took place.

61 F5 LOTHAGAM
5.5 MILLION YEARS AGO
LOCATION: Kenya
TYPE OF SITE: Hunter-gatherer
SITE DETAILS: A site that may have the remains of several hominid species. There is also evidence of a Late Stone Age fishing settlement.

62 F5 GAMBLE'S CAVE
6TH MILLENNIUM BC
LOCATION: Kenya
TYPE OF SITE: Early settlement
SITE DETAILS: The site of an Eburran tool industry, with finds of bone harpoons, decorated pottery, and several skeletons.

63 F5 HYRAX HILL
3RD MILLENNIUM BC
LOCATION: Kenya
TYPE OF SITE: Early settlement
SITE DETAILS: Finds of stone bowls and pottery, and a Neolithic settlement with a nearby cemetery, where 19 burials have been uncovered.

64 F5 OLORGESAILIE
900,000–700,000 YEARS AGO
LOCATION: Kenya
TYPE OF SITE: Hunter-gatherer
SITE DETAILS: A rich collection of artifacts were found at an Acheulian camp. There are remains of animal bones and hand-ax cleavers.

65 F5 NAROSURA
9TH–5TH CENTURIES BC
LOCATION: Kenya
TYPE OF SITE: Early settlement
SITE DETAILS: A pastoral Neolithic site with an industry of obsidian-backed microliths and burins, stone axes, ground bowls, and incised pottery.

66 G5 SHANGA
15TH–16TH CENTURIES AD
LOCATION: Kenya
TYPE OF SITE: Islamic town
SITE DETAILS: A trading post that includes remains of a mosque, palaces, hundreds of tombs, and finds of imported Ming Dynasty pottery.

67 G5 MANDA
9TH CENTURY AD
LOCATION: Kenya
TYPE OF SITE: Islamic town
SITE DETAILS: A coastal Islamic trading post containing a number of ruins. There have been finds of Sasano-Islamic pottery and later Chinese porcelain. There are house remains and evidence of an iron-smelting industry.

68 G5 GEDI
14TH CENTURY AD
LOCATION: Kenya
TYPE OF SITE: Islamic town
SITE DETAILS: Ruins of prosperous Islamic trading town with a palace, six mosques, and three pillar tombs.

69 G5 FORT JESUS
16TH CENTURY AD
LOCATION: Mombasa, Kenya
TYPE OF SITE: Colonial settlement
SITE DETAILS: *See pp. 101, 108.*

70 F5 BIGO
13TH–16TH CENTURIES AD
LOCATION: Uganda
TYPE OF SITE: Early settlement
SITE DETAILS: Some earthworks give credence to an early kingdom in the area, and some ceramics have been found.

71 F5 BUHAYA (KATURUKA)
5TH CENTURY BC
LOCATION: Tanzania
TYPE OF SITE: Iron Age settlement
SITE DETAILS: The site of early iron smelting, one of the earliest in this area of Africa. (*See p. 59.*)

72 F5 KANSYORE
1ST MILLENNIUM BC
LOCATION: Lake Ukerewe, Tanzania
TYPE OF SITE: Shell midden
SITE DETAILS: Site with distinctive pottery remains found in shell mounds.

73 F5 OLDUVAI GORGE
c. 1.9 MILLION–100,000 YEARS AGO
LOCATION: Tanzania
TYPE OF SITE: Hunter-gatherer
SITE DETAILS: *See pp. 13, 20.*

74 F5 ENGARUKA
15TH CENTURY AD
LOCATION: Tanzania
TYPE OF SITE: Early settlement
SITE DETAILS: An Iron Age community that used stone walling and ditching and intensive farming methods.

75 F5 LAETOLI
4–3.5 MILLION YEARS AGO
LOCATION: Tanzania
TYPE OF SITE: Hunter-gatherer
SITE DETAILS: Hominid and animal imprints preserved in hardened mud. They are attributed to *Australopithecus afarensis* or *Australopithecus africanus*.

76 F6 BOMBA KABURI
3RD CENTURY AD
LOCATION: Tanzania
TYPE OF SITE: Early settlement
SITE DETAILS: An Iron Age village site where some distinctive pottery has been excavated.

77 F6 UVINZA
5TH–18TH CENTURIES AD
LOCATION: Tanzania
TYPE OF SITE: Iron Age settlement
SITE DETAILS: Extensive evidence of a salt-working industry. Along with metal, salt was considered an important commodity in eastern Africa at this time.

78 G6 KILWA
9TH CENTURY AD
LOCATION: Tanzania
TYPE OF SITE: Islamic port
SITE DETAILS: Trading center on an island that passed into Portuguese hands. The merchants dealt in gold, silver, pearls, perfumes, crockery, and Chinese and Persian porcelain.

79 F6 NKOPE
EARLY 1ST MILLENNIUM AD
LOCATION: Malawi
TYPE OF SITE: Early settlement
SITE DETAILS: Sites that are linked to the Chifumbaze Iron Age culture.

80 F6 NKUDZI
18TH–19TH CENTURIES AD
LOCATION: Malawi
TYPE OF SITE: Cemetery
SITE DETAILS: Iron Age cemetery whose grave goods emphasize the importance of the trade with the east African coast.

81 F6 KALAMBO FALLS
120,000 YEARS AGO
LOCATION: Zambia
TYPE OF SITE: Hunter-gatherer
SITE DETAILS: Stratified site whose earliest artifacts were from the Middle Stone Age, including well-preserved pollens, charcoal, and vegetable matter. There was also evidence of Acheulian camps. Later artifacts include axes and bored stones (*c.* 1900 BC). The last period (AD 360–1000) has revealed channel-decorated pottery.

82 F7 INGOMBE ILEDE
12TH CENTURY AD
LOCATION: Zambia
TYPE OF SITE: Iron Age town
SITE DETAILS: A center of trade that used advanced metallurgy. Burials with woven fabrics, bangles, and ingots have been excavated.

83 E7 ISAMU PATI
11TH CENTURY AD
LOCATION: Zambia
TYPE OF SITE: Iron Age settlement
SITE DETAILS: A village mound that has unearthed storage pits, lathe-turned ivory, ironwork, and clay figurines. The villagers dealt in long-distance trade.

84 E7 KUMADZULO
5TH–7TH CENTURIES AD
LOCATION: Zambia
TYPE OF SITE: Iron Age settlement
SITE DETAILS: Village remains with finds of grindstones, iron hoes, copper ingots, and glass fragments.

85 F7 GWISHO
3000–1500 BC
LOCATION: Zambia
TYPE OF SITE: Early settlement
SITE DETAILS: Mound occupied by the Stone Age Wilton people. The remains of 35 skeletons were found in graves.

86 F6 NACHIKUFU
22,000–14,000 YEARS AGO
LOCATION: Zambia
TYPE OF SITE: Hunter-gatherer
SITE DETAILS: Cave that gives its name to a microlithic industry. Its distinctive artifacts include pointed, backed blades, scrapers, and small stones.

87 F7 MAXTON
11TH CENTURY AD
LOCATION: Zimbabwe
TYPE OF SITE: Early settlement
SITE DETAILS: Type-site of an Iron Age complex situated on a hilltop and that had distinctive stone walling.

88 F7 POMONGWE
9TH–8TH MILLENNIA BC
LOCATION: Zimbabwe
TYPE OF SITE: Hunter-gatherer
SITE DETAILS: Finds of scrapers that were part of the Oakhurst microlithic complex. The objects are characterized by large quadrilateral flakes. One cave has some fine examples of rock art.

89 F7 DHLO-DHLO
17TH CENTURY AD
LOCATION: Zimbabwe
TYPE OF SITE: Iron Age settlement
SITE DETAILS: Site that succeeded Great Zimbabwe and Khami as a major center.

90 F7 GREAT ZIMBABWE
13TH–15TH CENTURIES AD
LOCATION: Zimbabwe
TYPE OF SITE: Iron Age settlement
SITE DETAILS: *See pp. 59, 64–5.*

91 E7 KHAMI
16TH–17TH CENTURIES AD
LOCATION: Zimbabwe
TYPE OF SITE: Iron Age settlement
SITE DETAILS: Stone ruins of the city that succeeded Great Zimbabwe – it has house platforms and decorated walls.

92 E7 GI
150,000–30,000 YEARS AGO
LOCATION: Botswana
TYPE OF SITE: Hunter-gatherer
SITE DETAILS: Middle Stone Age remains that have yielded a microlithic industry, pits, and an open living site.

93 F7 MANEKWENIGI
9TH–10TH CENTURIES AD
LOCATION: Mozambique
TYPE OF SITE: Iron Age town
SITE DETAILS: Stone enclosure similar to that of Great Zimbabwe and related to the commercial activities of the area.

94 F7 CHIBUENE
8TH CENTURY AD
LOCATION: Mozambique
TYPE OF SITE: Iron Age town
SITE DETAILS: Coastal trade center that became an important link with Great Zimbabwe and the coastal towns. Excavations have unearthed deposits of glazed and unglazed wares.

95 D8 APOLLO II CAVE
26,000–19,000 YEARS AGO
LOCATION: Namibia
TYPE OF SITE: Rock art
SITE DETAILS: Site with the oldest rock paintings in Africa.

96 F7 MAPUNGUBWE
13TH CENTURY AD
LOCATION: South Africa
TYPE OF SITE: Iron Age town
SITE DETAILS: A site that was similar to Great Zimbabwe (see pp. 64–5). There has been debate as to who built the town.

97 F7 LYDENBURG
6TH CENTURY AD
LOCATION: South Africa
TYPE OF SITE: Early settlement
SITE DETAILS: A village where modeled terra-cotta heads, probably used in ritual ceremony, have been excavated. Thick stone walls were built at a later date.

98 E8 BROEDERSTROOM
5TH CENTURY AD
LOCATION: South Africa
TYPE OF SITE: Iron Age settlement
SITE DETAILS: Village with remains of circular huts, iron debris, skeletons, and fauna.

99 F7 MAKAPAN LIMEWORKS
2.5 MILLION YEARS AGO
LOCATION: South Africa
TYPE OF SITE: Hunter-gatherer
SITE DETAILS: Site where an early hominid and animal fossil remains have been excavated. Part of the skull of an *Australopithecus africanus* was found.

100 F8 STERKFONTEIN
2.5 MILLION YEARS AGO
LOCATION: South Africa
TYPE OF SITE: Hunter-gatherer
SITE DETAILS: A cave where a skull named "Mrs. Ples" (*Plesianthropus transvaalensis*), a *Homo habilis*-type skull, and some *Australopithecus* fossils have been excavated.

101 F8 JOHANNESBURG
19TH CENTURY AD
LOCATION: South Africa
TYPE OF SITE: Colonial settlement
SITE DETAILS: One of the first colonies in South Africa, founded as a result of the discovery of gold. (See p. 101.)

102 F8 DRAKENSBURG MOUNTAINS
16TH–19TH CENTURIES AD
LOCATION: South Africa
TYPE OF SITE: Rock art
SITE DETAILS: Bushman art with images based on the eland antelope, reflecting San mythology.

103 F8 BORDER CAVE
130,000–30,000 YEARS AGO
LOCATION: South Africa
TYPE OF SITE: Hunter-gatherer
SITE DETAILS: Middle Stone Age site where full remains of modern humans have been found. The skeleton of a child, thought to be around 100,000 years old, is the only one of its kind in the south of the continent.

104 E8 ORANGIA
120,000–35,000 YEARS AGO
LOCATION: South Africa
TYPE OF SITE: Hunter-gatherer
SITE DETAILS: A type-site for an early lithic industry, where six stone semi-circles have been found that may have been seasonal hunting camps.

105 F8 NDONDONDWANE
8TH CENTURY AD
LOCATION: South Africa
TYPE OF SITE: Iron Age settlement
SITE DETAILS: The remains of a village containing 2,000 fragments, from ivory bangles to large ceramics.

106 E8 MELKOUTBOOM
15,000 YEARS AGO
LOCATION: South Africa
TYPE OF SITE: Hunter-gatherer
SITE DETAILS: Cave with finds of tools and plant remains revealing their importance in the local diet.

107 E8 ZEEKOEI VALLEY
EARLY 1ST MILLENNIUM AD
LOCATION: South Africa
TYPE OF SITE: Hunter-gatherer
SITE DETAILS: A valley that has unearthed many Stone Age sites and their ceramic sequences, which predate the arrival of European settlers in the 1700s.

108 E8 DIE KELDERS
EARLY 2ND MILLENNIUM BC
LOCATION: South Africa
TYPE OF SITE: Early settlement
SITE DETAILS: A site with some evidence of sheep domestication – the first in the area. Pottery has also been found.

109 F8 NGWENYA
7TH MILLENNIUM BC
LOCATION: Swaziland
TYPE OF SITE: Hunter-gatherer
SITE DETAILS: Caves with signs of Late Stone Age mining for hematite, which was used for body painting.

110 F8 SEHONGHONG
70,000 YEARS AGO
LOCATION: Lesotho
TYPE OF SITE: Hunter-gatherer
SITE DETAILS: Rock shelter with a very long sequence of remains, especially from a period about 20,000 years ago.

111 H6 IHARANA
12TH–13TH CENTURIES AD
LOCATION: Madagascar
TYPE OF SITE: Cemetery
SITE DETAILS: Rich burials positioned in an Islamic fashion, accompanied by swords and daggers. Finds included Chinese and Persian plates and bowls, mother-of-pearl spoons, bronze mirrors, silver jewelry, and a bead necklace.

EGYPT

The map for these listings is on p. 143.

1 C4 ALEXANDRIA
3RD–1ST MILLENNIA BC
LOCATION: Egypt
TYPE OF SITE: Greek city
SITE DETAILS: While there is evidence of ancient walls, the classical city was founded by Alexander the Great. The lighthouse was one of the Seven Wonders of the Ancient World. There was also a huge library, famous throughout the Mediterranean region. Other structures included the Serapeum, two obelisks, and catacombs with sculpture and reliefs. (See p. 69.)

2 C4 CHURCH OF ST. MENAS
5TH–6TH CENTURIES AD
LOCATION: Egypt
TYPE OF SITE: Early Christian
SITE DETAILS: The remains of a large church and a popular pilgrimage site, though archaeological records of its earlier periods are inadequate.

3 D4 BEHBEIT EL-HAGAR
4TH CENTURY BC
LOCATION: Nile Delta, Egypt
TYPE OF SITE: Early temple
SITE DETAILS: Remains of one of ancient Egypt's most important temples. Built in hard stone, it was dedicated to the goddess Isis. The temple has now collapsed, although some elements of the original architecture are evident.

4 D4 MINSHAT ABU OMAR
4TH MILLENNIUM BC
LOCATION: Egypt
TYPE OF SITE: Cemetery
SITE DETAILS: The site of an important pre-Dynastic cemetery, with finds of pottery, stone vessels, copper tools, and jewelry, some imported from Upper Egypt and Palestine.

5 D4 TANIS
c. 2300 BC
LOCATION: Egypt
TYPE OF SITE: Ancient city
SITE DETAILS: Excavations have revealed lakes, foundations of the Temple of Amun, and some rich royal tombs.

6 D4 AVARIS
2ND MILLENNIUM BC
LOCATION: Egypt
TYPE OF SITE: Ancient city
SITE DETAILS: Under excavation at present – the site is probably the capital of Hyksos and shows links with Mycenae.

7 D4 KELLIA
4TH–8TH CENTURIES AD
LOCATION: Egypt
TYPE OF SITE: Christian settlement
SITE DETAILS: An area settled by Coptic monks with important finds of Late Roman pottery.

8 D4 MERIMDA
5TH MILLENNIUM BC
LOCATION: Egypt
TYPE OF SITE: Early settlement
SITE DETAILS: Small dwellings that may have held up to 16,000 people. It consists of a 7-ft- (2-m-) high occupation mound with finds including pottery, stone axes, and arrowheads.

9 D4 WADI EL-NATRUN
c. AD 330
LOCATION: Egypt
TYPE OF SITE: Early Christian settlement
SITE DETAILS: Site of an early Coptic monastery currently under excavation.

10 D4 GIZA
3RD MILLENNIUM BC
LOCATION: Egypt
TYPE OF SITE: Burial complex
SITE DETAILS: The site of the three pyramids (those of Khufu, Khafre, and Menkaure), the Sphinx, and the tomb of Hetepheres. Many other structures have been excavated, and thousands of artifacts have been recovered. (See p. 35.)

11 D4 BUBASTIS (TELL BASTA)
10TH–8TH CENTURIES BC
LOCATION: Egypt
TYPE OF SITE: Ancient city
SITE DETAILS: A large settlement, once an early capital, with a number of remains, including the Temple of Bastet and other smaller temples. There is also an animal cemetery, which contains a number of sacred-cat burials.

12 D4 HELIOPOLIS (TELL HISN)
3RD MILLENNIUM BC
LOCATION: Cairo, Egypt
TYPE OF SITE: Ancient city
SITE DETAILS: A capital and a center of the Ra cult, with a number of temples dedicated to the sun god. Due to the plundering of its stone, archaeological work has been difficult, but there are a number of structures left, including statues, obelisks, and reliefs. Tombs of high priests have also been discovered.

13 D4 MA'ADI
LATE 4TH MILLENNIUM BC
LOCATION: Egypt
TYPE OF SITE: Early settlement
SITE DETAILS: A pre-Dynastic site that may have been an important trade center, with remains of houses, pottery,

and storage pits. Cellars full of jars with animal bones, beads, and imported asphalt have been excavated.

14 D4 MEMPHIS

3RD MILLENNIUM BC

LOCATION: Egypt

TYPE OF SITE: Ancient city

SITE DETAILS: Once the capital of ancient Egypt, it has almost completely disappeared, although a huge statue of Ramses II remains. Four chapels containing paintings, caskets, stones, jewels, and statues were excavated.

15 D4 ABUSIR

3RD MILLENNIUM BC

LOCATION: Egypt

TYPE OF SITE: Royal cemetery

SITE DETAILS: A site with a number of temples and pyramids. The sun temple here is the earliest of its type in ancient Egypt, while the 5th Dynasty pyramids include the complexes of Sahure, Neferkirare, and Neuserre. There are also a number of private tombs.

16 D4 EL-OMARI

c. 3200 BC

LOCATION: Egypt

TYPE OF SITE: Early settlement

SITE DETAILS: An area that is the site of about 100 pre-Dynastic dwellings.

17 D4 SAQQARA

3RD MILLENNIUM BC

LOCATION: Egypt

TYPE OF SITE: Burial complex

SITE DETAILS: The necropolis with royal mastaba tombs and 15 pyramids, including a step pyramid built by Imhotep (c. 2650 BC) and the Pyramid of Djoser. Underground vaults were discovered containing "Pyramid Texts."

18 D5 MEIDUM

3RD MILLENNIUM BC

LOCATION: Egypt

TYPE OF SITE: Burial complex

SITE DETAILS: A pyramid that collapsed in the last stages of construction. The final layer of casing stones could not support the accumulated weight and were squeezed out of place. Nevertheless, it was probably the first Egyptian pyramid to be built.

19 D5 EL-LAHUN

2ND MILLENNIUM BC

LOCATION: Egypt

TYPE OF SITE: Early settlement

SITE DETAILS: A Middle Kingdom site consisting of a pyramid and a town. There have been finds of papyri.

20 E5 SERABIT EL-KHADIM

EARLY 2ND MILLENNIUM BC

LOCATION: Sinai, Egypt

TYPE OF SITE: Early settlement

SITE DETAILS: An important site, including the Temple of Hathor, a shrine for Sopdu (the god of the eastern desert), and stelae. The area was important for its turquoise mining.

21 E6 ABU SHAOAR

3RD CENTURY AD

LOCATION: Egypt

TYPE OF SITE: Fortified settlement

SITE DETAILS: Built of limestone, the fort had about a dozen buildings, barracks, storage magazines, administration buildings, and the remains of baths.

22 D6 HERMOPOLIS

c. 2000 BC

LOCATION: Egypt

TYPE OF SITE: Ancient city

SITE DETAILS: An important Middle Kingdom site, including the temples of Thoth and other fragmentary stone structures. There are also some substantial remains of a Roman basilica.

23 D6 TIHNA EL-GEBEL

6TH CENTURY BC

LOCATION: Egypt

TYPE OF SITE: Burial complex

SITE DETAILS: The site of a necropolis with Late Period and Greco-Roman tombs, and where the remains of three temples have been found.

24 D6 BENI HASAN

c. 2000 BC

LOCATION: Egypt

TYPE OF SITE: Early settlement

SITE DETAILS: Site of an ancient village that is most famous for its 39 Old Kingdom rock-cut tombs, many of which were painted.

25 D6 MEIR

3RD MILLENNIUM BC

LOCATION: Egypt

TYPE OF SITE: Burial complex

SITE DETAILS: The site of rock-cut tombs of local *nomarchs* – chief officials – decorated with reliefs.

26 D6 EL-AMARNA

16TH CENTURY BC

LOCATION: Egypt

TYPE OF SITE: Ancient city

SITE DETAILS: Once the capital of the king Akhenaten, it later became an administrative and religious center. There are remains of palaces, houses, and temples. Letters written on cuneiform clay tablets relating to diplomatic relations were excavated.

27 D6 EL-HAMMAMIYA

c. 3700 BC

LOCATION: Egypt

TYPE OF SITE: Early settlement

SITE DETAILS: An early stratified site, including finds of thin Badarian pottery. It is also the site of a nobleman's tomb.

28 D7 ABYDOS

EARLY 3RD MILLENNIUM BC

LOCATION: Egypt

TYPE OF SITE: Religious center

SITE DETAILS: Late Archaic royal tombs and early 19th-Dynasty temples have been found, including the Temple of Osiris. The site was excavated by the Egyptian archaeology pioneer Sir Flinders Petrie.

29 D7 DENDERA

1ST MILLENNIUM BC

LOCATION: Egypt

TYPE OF SITE: Ancient city

SITE DETAILS: Once an important capital whose best-preserved remains are of the Temple of Hathor. There is an Early Dynastic necropolis on the outskirts and some substantial Roman remains – a Ptolemaic birth house, a Christian basilica, and other remains – enclosed by a mudbrick wall.

30 E6 COPTUS

1ST CENTURY BC

LOCATION: Egypt

TYPE OF SITE: Early settlement

SITE DETAILS: The ruins of an emporium on the Nile River connected to the Egyptian Red Sea ports.

31 E7 THEBES

c. 2700–1070 BC

LOCATION: Egypt

TYPE OF SITE: Ancient city

SITE DETAILS: See pp. 35, 44–5.

32 E7 NAQADA

4TH MILLENNIUM BC

LOCATION: Egypt

TYPE OF SITE: Early settlement

SITE DETAILS: The village where the first evidence of Neolithic people in Egypt was discovered. Rock-cut tombs are found to the north. Excavated artifacts include ivory tablets, vase fragments, and clay seals.

33 E7 ESNA

2ND CENTURY BC

LOCATION: Egypt

TYPE OF SITE: Religious settlement

SITE DETAILS: The site of a temple to Khnum decorated by Roman emperors. There is a completely preserved hypostyle hall and columns richly decorated with text. There is also a Middle- to- New Kingdom cemetery.

34 E7 KOM EL-AHMAR

4TH MILLENNIUM BC

LOCATION: Egypt

TYPE OF SITE: Early settlement

SITE DETAILS: A major pre-Dynastic settlement with a series of shrines and a cache of votive objects. There is also a painted tomb from 4000 BC.

35 E7 EL-KAB

6TH MILLENNIUM BC

LOCATION: Egypt

TYPE OF SITE: Early settlement

SITE DETAILS: A microlithic industry has been found here that predates most of the other Egyptian Neolithic cultures.

It was the site of a later Early Dynastic town, surrounded by large brick walls, and contained a Temple of Nekhbet and rock-cut tombs.

36 E7 EDFU

3RD–1ST MILLENNIA BC

LOCATION: Egypt

TYPE OF SITE: Early settlement

SITE DETAILS: A center and tomb complex spanning the Old Kingdom to Greco-Roman periods. The temple, some of whose roof still exist, is one of the best preserved in Egypt. Its inner chapel still has two black granite statues of Horus the falcon.

37 E8 WADI KUBBANIYA

18,250–16,960 YEARS AGO

LOCATION: Egypt

TYPE OF SITE: Hunter-gatherer

SITE DETAILS: A relatively sophisticated Paleolithic site, where remains of carbonized plants, stone grinders, pestles, and mortars have been excavated. There is evidence that there may have been some settled housing.

38 E8 MONASTERY OF ST. SIMEON

EARLY 1ST MILLENNIUM AD

LOCATION: Aswan, Egypt

TYPE OF SITE: Christian settlement

SITE DETAILS: A compound found on two levels and surrounded by a wall.

39 E8 PHILAE

1ST MILLENNIUM BC

LOCATION: Egypt

TYPE OF SITE: Religious center

SITE DETAILS: A complex of temples sacred to Isis, mostly built by the Ptolemaic kings. There are also the ruins of a Coptic church. The temples were moved after the area was flooded.

40 D9 BALLANA

4TH–6TH CENTURIES AD

LOCATION: Egypt

TYPE OF SITE: Burial mounds

SITE DETAILS: The site that houses the royal burials of the X-Group (a Nubian culture dating to c. AD 350–600) – with evidence of human sacrifice.

41 D9 ABU SIMBEL

13TH CENTURY BC

LOCATION: Egypt

TYPE OF SITE: Temple complex

SITE DETAILS: Two hugely impressive temples built by Ramses II. One temple has two large figures of Ramses and Nefertiti flanking its entrance.

42 D9 QASR IBRIM

2ND MILLENNIUM BC

LOCATION: Egypt

TYPE OF SITE: Early settlement

SITE DETAILS: An almost continually occupied site situated on a fortified cliff. It has rock-cut shrines, reliefs, and a stela. There is also some material from the X-Group and Christian cultures.

UNITED KINGDOM & IRELAND

The map for these listings is on p. 144.

1 **D4 MOUNT SANDEL**
7TH MILLENNIUM BC
LOCATION: Northern Ireland
TYPE OF SITE: Mesolithic settlement
SITE DETAILS: Foundations of wooden huts with hearth remains and artifacts including microliths, tranchet axes, and early polished stone axes. The people subsisted on salmon and shellfish.

2 **E5 CARRICKFERGUS CASTLE**
12TH CENTURY AD
LOCATION: Northern Ireland
TYPE OF SITE: Medieval castle
SITE DETAILS: Massive stone castle with a large gatehouse and a keep, added in 1250. It also included stables, a storeroom, a chapel, and a bakery.

3 **D5 DEVENISH ISLAND**
EARLY 2ND MILLENNIUM AD
LOCATION: Northern Ireland
TYPE OF SITE: Medieval monastery
SITE DETAILS: A settlement founded by St. Molaise. Most of the buildings are now ruined, with only Teampall Mor in relatively good condition. There is a 12th-century stone tower, stone carvings, and a carved 15th-century stone cross.

4 **D5 ARMAGH**
5TH CENTURY AD
LOCATION: Northern Ireland
TYPE OF SITE: Christian settlement
SITE DETAILS: After the decline of Navan, Ard Macha, as the town was then known, became a major Celtic center. An important church was set up by St. Patrick in AD 455, inside the banks of an Iron Age hill.fort. The streets of the modern city center follow the ditch that surrounded the original building. There are also a couple of impressive medieval cathedrals and a high cross.

5 **D5 NAVAN**
1ST CENTURY BC
LOCATION: Northern Ireland
TYPE OF SITE: Celtic settlement
SITE DETAILS: See pp. 58, 60–3.

6 **C5 CARROWKEEL**
4TH MILLENNIUM BC
LOCATION: Republic of Ireland
TYPE OF SITE: Neolithic cemetery
SITE DETAILS: Passage tombs and a cemetery with 14 corbeled graves, with a chamber that is illuminated on the day of the summer solstice.

7 **D6 MONASTERBOICE**
5TH–10TH CENTURIES AD
LOCATION: Republic of Ireland
TYPE OF SITE: Monastic settlement
SITE DETAILS: The ruins of an early medieval monastery founded by St. Buite. It includes a roofless round tower and 10th-century high crosses, carved with biblical scenes.

8 **D6 MELLIFONT ABBEY**
11TH–13TH CENTURIES AD
LOCATION: Republic of Ireland
TYPE OF SITE: Medieval monastery
SITE DETAILS: The first Cistercian abbey in the Republic of Ireland, founded by St. Mallachy. It has the remains of a Romanesque cloister and a unique 13th-century *lavabo*, where monks washed their hands before meals.

9 **D6 KNOWTH**
MID-4TH MILLENNIUM BC
LOCATION: Republic of Ireland
TYPE OF SITE: Megalithic tomb
SITE DETAILS: The greatest concentration of megalithic art in Europe, consisting of two richly decorated passage tombs inside a large Neolithic burial mound, 197 ft (60 m) in diameter. There are curbstones and a number of tombs surrounding the monument.

10 **D6 NEWGRANGE**
c. 3100 BC
LOCATION: Republic of Ireland
TYPE OF SITE: Burial mound
SITE DETAILS: A large mound with a surrounding stone circle containing a Neolithic passage grave, with excellent examples of megalithic art. The mound was originally 39 ft (12 m) high, with a hole above the entrance that is aligned with the midwinter sunrise. (See p. 35.)

11 **D6 KELLS**
6TH CENTURY AD
LOCATION: Republic of Ireland
TYPE OF SITE: Medieval monastery
SITE DETAILS: A settlement set up by St. Columba in the 6th century, where the illuminated *Book of Kells* was later written. There are many high crosses and the remains of St. Columba's House.

12 **D6 HILL OF TARA**
300 BC–AD 500
LOCATION: Republic of Ireland
TYPE OF SITE: Celtic settlement
SITE DETAILS: See pp. 58, 67.

13 **D6 DUBLIN**
9TH CENTURY AD
LOCATION: Republic of Ireland
TYPE OF SITE: Viking settlement
SITE DETAILS: A trading center set up by the Vikings in the 9th century. Excavations have revealed the remains of Viking houses, and finds have included brooches, stickpins, and inscribed wooden objects.

14 **D6 CLONMACNOISE**
6TH CENTURY AD
LOCATION: Republic of Ireland
TYPE OF SITE: Early Christian settlement
SITE DETAILS: A monastery founded by St. Ciaran in AD 545 that flourished between the 7th and 12th centuries. Its remains include stone churches, two round towers, three high crosses, and a cathedral.

15 **C6 ARAN ISLANDS**
1ST–15TH CENTURIES AD
LOCATION: Republic of Ireland
TYPE OF SITE: Early settlement
SITE DETAILS: Islands with a number of sites: Dún Aonghasa, an Iron Age promontory fort with four concentric walls and a ring of stone stakes; Clochán na Carraige, a large beehive hut; Na Seacht d'Teampaill, the "Seven Churches" monastic settlement; Dún Eochla, a circular Bronze Age fort; and Teampall Chiaráin, a ruined 12th-century church with several nearby stones inscribed with crosses.

16 **C6 KILMACDUAGH**
7TH CENTURY AD
LOCATION: Republic of Ireland
TYPE OF SITE: Monastic settlement
SITE DETAILS: Founded by St. Colman, the ruins include an 11th-century tower and a pre-Norman cathedral, remodeled in Gothic style with fine tomb carvings.

17 **B7 SKELLIG MICHAEL**
5TH CENTURY AD
LOCATION: Republic of Ireland
TYPE OF SITE: Medieval monastery
SITE DETAILS: An almost inaccessible monastic settlement that has the remains of six beehive cells, two boat-shaped oratories, and a 10th-century monastery.

18 **D7 ROCK OF CASHEL**
5TH CENTURY AD
LOCATION: Republic of Ireland
TYPE OF SITE: Religious settlement
SITE DETAILS: A rocky stronghold that was the seat of the kings of Munster until it was handed over to the church in 1101. In the complex can be found the Hall of the Vicars' Choral and a 92-ft- (28-m-) high round tower.

19 **D6 GLENDALOUGH**
6TH–12TH CENTURIES AD
LOCATION: Republic of Ireland
TYPE OF SITE: Monastic settlement
SITE DETAILS: Structures remaining include the gatehouse to the monastic enclosure, a 110-ft- (33-m-) high round tower, a roofless cathedral, and a number of small priests' dwellings.

20 **D7 WATERFORD**
9TH CENTURY AD
LOCATION: Republic of Ireland
TYPE OF SITE: Medieval town
SITE DETAILS: Originally a Viking settlement; excavations have revealed their early dwellings. It was later fortified by the Normans, including a large tower that still stands. It is also important for the 18th-century factory that manufactures its famous crystal.

21 **(SEE INSET) UNDERHOULL**
10TH CENTURY AD
LOCATION: Shetland, Scotland
TYPE OF SITE: Viking settlement
SITE DETAILS: Site of a Viking farm with remains of a boathouse in a nearby bay. There is an Iron Age *broch* – a circular stone tower – and at one time there was also a watermill here.

22 **(SEE INSET) JARLSHOF**
9TH CENTURY AD
LOCATION: Shetland, Scotland
TYPE OF SITE: Viking settlement
SITE DETAILS: Standing remains of long Viking farmhouses and outbuildings and a fortified medieval farm have been excavated. There are also the ruins of a 17th-century settlement.

23 **F1 MAES HOWE**
c. 2000 BC
LOCATION: Orkney, Scotland
TYPE OF SITE: Megalithic tomb
SITE DETAILS: A Neolithic passage grave with later runic graffiti on its walls. The nearby standing stones of Stenness may be associated with the tomb.

24 **F1 SKARA BRAE**
EARLY 3RD MILLENNIUM BC
LOCATION: Orkney, Scotland
TYPE OF SITE: Neolithic settlement
SITE DETAILS: See pp. 25, 30–1.

25 **D2 CALLANISH**
c. 1800 BC
LOCATION: Outer Hebrides, Scotland
TYPE OF SITE: Ritual monument
SITE DETAILS: A Neolithic stone circle that was never completed. The axial alignments of the rows were meant to converge at the central stone. There is a later, chambered tomb, with a short passage and two small chambers. A large bronze pot was discovered.

26 **E3 CASTLE URQUHART**
13TH–17TH CENTURIES AD
LOCATION: Scotland
TYPE OF SITE: Medieval castle
SITE DETAILS: Once one of the largest castles in Scotland, it is now in ruins. It had been a strategic site since the Pictish era. Further fortifications were added in 1509 by James IV of Scotland.

27 **F3 INCHTUTHILL**
1ST CENTURY AD
LOCATION: Scotland
TYPE OF SITE: Roman fort
SITE DETAILS: A legionary fort on the Tay River that was the most northerly of all Roman colonies. Excavations have revealed that it was never finished and,

in fact, it was deliberately dismantled. Over a million unused nails have been found in a nearby pit.

28 E4 ANTONINE WALL

AD 42
LOCATION: Scotland
TYPE OF SITE: Defensive fortification
SITE DETAILS: Running from the Firth of Forth to Clyde, the Roman wall was built of turf on a cobbled foundation. Some 19 forts were situated along its 37-mi (60-km) length.

29 D3 IONA

6TH–12TH CENTURIES AD
LOCATION: Scotland
TYPE OF SITE: Monastic settlement
SITE DETAILS: A tiny island where St. Columba founded a small settlement in the 6th century. There is a 12th-century Benedictine abbey, the Cobhan Cuilteach hut circle, and a number of cairns.

30 F4 CAIRNPAPPLE

3RD MILLENNIUM BC
LOCATION: Scotland
TYPE OF SITE: Ritual monument
SITE DETAILS: A number of remains, including a stone circle, a burial chamber, and a ritual complex that was used over a number of periods.

31 F4 NEW LANARK

LATE 18TH CENTURY AD
LOCATION: Scotland
TYPE OF SITE: Industrial village
SITE DETAILS: Built by David Dale to house his workers, the industrial complex has some impressive water-driven mills and housing tenements.

32 G4 LINDISFARNE

7TH–11TH CENTURIES AD
LOCATION: England
TYPE OF SITE: Monastic settlement
SITE DETAILS: Church and monastery site until the invasion of the Danes in AD 793, when it became the site of a fortress. The Lindisfarne priory was built in the 11th-century – its arches remain.

33 F4 YEAVERING

5TH–7TH CENTURIES AD
LOCATION: Northumbria, England
TYPE OF SITE: Saxon settlement
SITE DETAILS: Royal Northumbrian seat where an assembly grandstand and large wooden halls have been excavated.

34 F5 HADRIAN'S WALL

c. AD 122
LOCATION: England
TYPE OF SITE: Defensive fortification
SITE DETAILS: Northern boundary of Roman Empire, running from Wallsend to Bowness-on-Solway, built of stone and secured by milecastles, with forts later incorporated into the wall. (*See p. 68.*)

35 F5 VINDOLANDA

1ST–2ND CENTURIES AD
LOCATION: England
TYPE OF SITE: Roman fort
SITE DETAILS: One of the forts found along Hadrian's Wall (see left). First built in AD 90, it was also a civilian settlement with a bathhouse, an inn, and mausolea. Roman writing tablets provide details about the food, clothes, and daily life.

36 G5 HYLTON CASTLE

14TH CENTURY AD
LOCATION: Sunderland, England
TYPE OF SITE: Medieval castle
SITE DETAILS: *See pp. 90, 92–5.*

37 G5 JARROW AND MONKWEARMOUTH

7TH CENTURY AD
LOCATION: Tyne & Wear, England
TYPE OF SITE: Monastic settlement
SITE DETAILS: Two monastic churches, one of which was demolished in the 18th century. Excavations have uncovered two rectangular halls with glass windows and stone-tiled roofs.

38 F5 GREAT LANGDALE

c. 4000 BC
LOCATION: Cumbria, England
TYPE OF SITE: Neolithic mine
SITE DETAILS: Ax factory that traded for 1,000 years – unfinished axes and waste flakes have been found on the hillside.

39 G5 WHARRAM PERCY

9TH–15TH CENTURIES AD
LOCATION: Yorkshire, England
TYPE OF SITE: Medieval village
SITE DETAILS: Deserted settlement that had wattle-and-daub housing consisting of two manor houses and a 12th-century church. Every house contained a small yard called a *toft,* with an enclosed paddock called a *croft.*

40 G5 STAR CARR

8TH MILLENNIUM BC
LOCATION: Yorkshire, England
TYPE OF SITE: Mesolithic settlement
SITE DETAILS: The waterlogged nature of the site has preserved wooden tools and bone and antler artifacts. Some 187 barbed points and red deer frontlets have also been found, which may have been for ceremonial use.

41 F6 BACUP

19TH CENTURY AD
LOCATION: Lancashire, England
TYPE OF SITE: Industrial town
SITE DETAILS: The best-preserved mill town in the Rossendale Valley area, which was the center of cotton production during the Industrial Revolution. It has many 19th-century relics and archaeological finds.

42 G6 CRESWELL CRAGS

20,000–10,000 YEARS AGO
LOCATION: Derbyshire, England
TYPE OF SITE: Hunter-gatherer
SITE DETAILS: Some finds of backed blades that give their name to assemblages often found in local caves.

43 G6 ARBOR LOW

c. 2000 BC
LOCATION: Derbyshire, England
TYPE OF SITE: Ritual monument
SITE DETAILS: A stone circle sometimes referred to as the "Stonehenge of the North" – it consists of 46 stones enclosed by a ditch.

44 F7 COALBROOKDALE

18TH CENTURY AD
LOCATION: Shropshire, England
TYPE OF SITE: Industrial complex
SITE DETAILS: *See pp. 111, 118–19.*

45 G7 SOHO WORKS

18TH CENTURY AD
LOCATION: Birmingham, England
TYPE OF SITE: Industrial complex
SITE DETAILS: *See pp. 111, 112–15.*

46 H7 GRIMES GRAVES

3RD MILLENNIUM BC
LOCATION: Norfolk, England
TYPE OF SITE: Neolithic quarry
SITE DETAILS: A flint mine where red deer antlers were used to mine the galleries. The stones were shaped at the mine head and then transported over the country. Excavations have revealed chalk fertility models. (*See p. 25.*)

47 H7 THETFORD

9TH CENTURY AD
LOCATION: England
TYPE OF SITE: Medieval town
SITE DETAILS: A trading town since Anglo-Saxon times, its fortunes dipped when its priory was destroyed during the Reformation in the 16th century. In the center of the town is a big mound that was once a Norman motte-and-bailey castle.

48 H7 WEST STOW

5TH–7TH CENTURIES AD
LOCATION: Suffolk, England
TYPE OF SITE: Anglo-Saxon settlement
SITE DETAILS: Hamlet established at the end of Roman rule that has unearthed Romano-British pottery and coins. It has since been excavated and restored. (*See p. 81.*)

49 H7 SUTTON HOO

7TH CENTURY AD
LOCATION: Suffolk, England
TYPE OF SITE: Anglo-Saxon cemetery
SITE DETAILS: The burial ground of the early Anglo-Saxon kings, it is most famous for its boat burial, which contained a sword, a shield, a decorated helmet, and a gold belt buckle.

50 G8 ABBEY MILLS

19TH CENTURY AD
LOCATION: London, England
TYPE OF SITE: Industrial
SITE DETAILS: A Byzantine-style pumping station, built to house eight engines, that remains in excellent condition.

51 G7 STANTON HARCOURT

200,000 YEARS AGO
LOCATION: Oxfordshire, England
TYPE OF SITE: Hunter-gatherer
SITE DETAILS: *See pp. 13, 14–17.*

52 G7 WAYLAND'S SMITHY

c. 3500 BC
LOCATION: Oxfordshire, England
TYPE OF SITE: Megalithic monument
SITE DETAILS: A Neolithic long barrow and a megalithic passage grave, where a wooden mortuary house containing 14 burials has been excavated.

53 F8 AQUAE SULIS

1ST CENTURY AD
LOCATION: Bath, England
TYPE OF SITE: Roman settlement
SITE DETAILS: The major religious site of Roman Britain, it was situated around some hot springs and dedicated to the goddess Sulis Minerva. Excavations have revealed a temple and a gilded bronze head of the goddess.

54 F8 TOCKENHAM

1ST CENTURY AD
LOCATION: Wiltshire, England
TYPE OF SITE: Roman settlement
SITE DETAILS: *See pp. 68, 70–3.*

55 F8 AVEBURY

3RD MILLENNIUM BC
LOCATION: Wiltshire, England
TYPE OF SITE: Ritual monument
SITE DETAILS: *See pp. 35, 36–9.*

56 G8 DANEBURY

6TH–1ST CENTURIES BC
LOCATION: Hampshire, England
TYPE OF SITE: Iron Age hill fort
SITE DETAILS: A fortified Celtic settlement that had two entrances and a second line of defense. There are remains of storage pits, granary structures, and some roundhouses.

57 G8 STONEHENGE

3RD–2ND MILLENNIA BC
LOCATION: Wiltshire, England
TYPE OF SITE: Ritual monument
SITE DETAILS: The most important prehistoric site in Britain, Stonehenge consists of a circular earthwork with the famous stone settings within. A ring of 57 Aubrey Holes (used for cremation burials) – dating to c. 3000 BC – are just inside the ditch. The stones were put up from c. 2400 BC onward. The larger stones are sarsen, a local quartz sandstone, and they are arranged as a circle with lintels and a "horseshoe" setting of five trilithons – pairs with a lintel on top. There are also smaller bluestones from Wales set up as a circle and a horseshoe within the other settings. The monument was probably abandoned after 1100 BC. (*See p. 35.*)

58 F8 BUSH BARROW
3RD MILLENNIUM BC
LOCATION: Wiltshire, England
TYPE OF SITE: Burial mound
SITE DETAILS: The burial of an Early Bronze Age chieftain that was part of the Normanton Down cemetery. There have been finds of a breastplate, a gold belt fastener, bronze daggers, and an ax.

59 G8 WINTERBOURNE GUNNER
5TH–6TH CENTURIES AD
LOCATION: Wiltshire, England
TYPE OF SITE: Anglo-Saxon settlement
SITE DETAILS: *See pp. 81, 82–5.*

60 G8 BOXGROVE
500,000 YEARS AGO
LOCATION: West Sussex, England
TYPE OF SITE: Hunter-gatherer
SITE DETAILS: Acheulian site where lithic tools, debitage (waste flakes), and human remains have been excavated.

61 G8 BUTSER
3RD CENTURY BC
LOCATION: Hampshire, England
TYPE OF SITE: Iron Age settlement
SITE DETAILS: Recently reconstructed, with roundhouses, storage pits, and granaries, in an attempt to test ideas about how the people lived.

62 G8 MARY ROSE
16TH CENTURY AD
LOCATION: Portsmouth, England
TYPE OF SITE: Shipwreck
SITE DETAILS: Built in 1509, the famous ship of Henry VIII sank in 1545. The hull was raised in 1982, and thousands of objects have been salvaged.

63 F8 MAIDEN CASTLE
500 BC–AD 43
LOCATION: Dorset, England
TYPE OF SITE: Iron Age hill fort
SITE DETAILS: A large fortification where excavations have revealed large ramparts and a war cemetery. There are traces of roundhouses and streets on the interior.

64 F8 BRISTOL
17TH CENTURY AD
LOCATION: Avon, England
TYPE OF SITE: Trading port
SITE DETAILS: An important port during the colonial period with the distribution of tobacco, wine, and slaves. (*See p. 101.*)

65 F8 GLASTONBURY
3RD CENTURY BC
LOCATION: Somerset, England
TYPE OF SITE: Iron Age settlement
SITE DETAILS: A lake village with the remains of 90 circular structures that contained thick clay floors, hearths, and remains of wooden bowls and basins. There are also the remains of the 8th-century abbey.

66 F8 GOUGH'S CAVE
20,000–10,000 YEARS AGO
LOCATION: Somerset, England
TYPE OF SITE: Hunter-gatherer
SITE DETAILS: A Creswellian cave where finds of horse and red deer bone, art carvings, flint, and some human bones have been excavated. It is the site of the Cheddar Man, who was found at the beginning of the century.

67 F8 DARTMOUTH
18TH CENTURY AD
LOCATION: Devon, England
TYPE OF SITE: Industrial
SITE DETAILS: Thomas Newcomen invented the world's first atmospheric engine here, and it was used at a number of tin mines in the area. (*See p. 111.*)

68 E8 TINTAGEL
9TH CENTURY AD
LOCATION: Cornwall, England
TYPE OF SITE: Medieval castle
SITE DETAILS: Ruins of a castle built on the site of a Celtic residence that disappeared in about AD 850. Eastern Mediterranean pottery has been found.

69 E9 BOLEIGH
1ST–5TH CENTURIES AD
LOCATION: Cornwall, England
TYPE OF SITE: Iron Age settlement
SITE DETAILS: A fortified site where the inhabitants used tin. There are remains of barrows and stone circles, and a *fogou* – an underground cave for refuge or storage purposes.

70 F7 TINTERN ABBEY
11TH–15TH CENTURIES
LOCATION: Wales
TYPE OF SITE: Medieval abbey
SITE DETAILS: Founded by Cistercian monks in AD 1131, the abbey is now in ruins. Most remains date from later buildings of the 13th to 15th centuries.

71 F7 CAERLEON
c. AD 75
LOCATION: Wales
TYPE OF SITE: Roman town
SITE DETAILS: A Roman fortress settlement built for the legionary troops with 64 rows of barracks. It was also a fully fledged town with an amphitheater, a hospital, and a bathhouse complex.

72 F7 BLAENAFON
17TH–18TH CENTURIES AD
LOCATION: Wales
TYPE OF SITE: Industrial complex
SITE DETAILS: Ironworks and coal-mining site established in 1782, although there had been iron smelting in the vicinity since the 16th century. The remains revealed much about early processes of iron production and workers' lifestyles.

73 F7 HEN DOMEN
c. AD 1070
LOCATION: Wales
TYPE OF SITE: Medieval castle

SITE DETAILS: The most thoroughly excavated motte and bailey in the British Isles. Artifacts have been excavated, including potsherds.

74 F7 MARGAM
6TH–10TH CENTURIES AD
LOCATION: Wales
TYPE OF SITE: Medieval monastery
SITE DETAILS: An abbey with a Roman milestone and two 6th-century pillar stones, one of which is famous for its Ogham inscription (a Celtic script).

75 E7 PAVILAND CAVE
38,000–10,000 YEARS AGO
LOCATION: Wales
TYPE OF SITE: Hunter-gatherer
SITE DETAILS: Remains of tool industries have been excavated. The site includes the earliest British ceremonial burial; "the Red Lady of Paviland" was actually the burial of a man covered in ocher, accompanied by ivory funerary goods.

76 E7 GORS FAWR
3RD MILLENNIUM BC
LOCATION: Wales
TYPE OF SITE: Ritual monument
SITE DETAILS: A circle of 16 stones about 7 ft (2 m) apart from one another, with two entrance stones that stand apart from the main circle.

77 E6 CAERNAVON CASTLE
AD 1283
LOCATION: Wales
TYPE OF SITE: Medieval castle
SITE DETAILS: The first castle was a Norman motte and bailey. The more recent structure that stands today has an unusual polygonal tower and banded masonry.

78 E6 GRAIG LLWYD
4TH–3RD MILLENNIA BC
LOCATION: Wales
TYPE OF SITE: Neolithic quarry
SITE DETAILS: The site of a stone quarry and ax factory that have been excavated. Working floors with their debris have been found. Petrological studies can pinpoint any axe to its correct hillside.

79 E6 BRYN CELLI DDU
c. 3000 BC
LOCATION: Wales
TYPE OF SITE: Megalithic tomb
SITE DETAILS: A passage grave with a chambered cairn where a number of objects have been found, including white quartz, burned and unburned bones, post sockets, and the skeleton of an ox.

80 E6 BARCLODIAD Y GAWRES
3RD MILLENNIUM BC
LOCATION: Wales
TYPE OF SITE: Megalithic tomb
SITE DETAILS: Site of a large megalithic burial chamber that has some examples of megalithic art.

SCANDINAVIA & THE BALTIC STATES

The map for these listings is on p. 145.

1 (SEE INSET) HOFSTADIR
LATE 1ST MILLENNIUM AD
LOCATION: Iceland
TYPE OF SITE: Viking settlement
SITE DETAILS: Farmstead with a single hall and a small room, where a large cooking pot was discovered. (*See p. 81.*)

2 (SEE INSET) THINGVELLIR
10TH CENTURY AD
LOCATION: Iceland
TYPE OF SITE: Early settlement
SITE DETAILS: Site of the first Icelandic parliament, where artifacts and early building remains have been excavated.

3 (SEE INSET) SKALLAHOLT
10TH CENTURY AD
LOCATION: Iceland
TYPE OF SITE: Viking settlement
SITE DETAILS: A longhouse and an early church have been excavated here.

4 C4 VARJAREN
1ST MILLENNIUM BC
LOCATION: Norway
TYPE OF SITE: Saami cemetery
SITE DETAILS: A mound that includes the ritual burial of a bear.

5 C5 KRANKMARTENHOGEN
1ST–12TH CENTURIES AD
LOCATION: Norway
TYPE OF SITE: Saami cemetery
SITE DETAILS: Cremation graves marked by triangular stone settings, with antlers and animal bones placed on top.

6 C6 RØROS
17TH CENTURY AD
LOCATION: Norway
TYPE OF SITE: Industrial town
SITE DETAILS: Town with remains of mines, including a copper smelter and a number of other buildings.

7 B6 KAUPANG
9TH CENTURY AD
LOCATION: Norway
TYPE OF SITE: Viking center
SITE DETAILS: Houses and workshops excavated, with finds of wooden and metal objects and soapstone vessels.

8 B6 BORGUND STAVE

c. AD 1150

LOCATION: Norway

TYPE OF SITE: Medieval church

SITE DETAILS: Well-preserved stave church built only of wood.

9 B7 OSEBERG

c. AD 850

LOCATION: Norway

TYPE OF SITE: Burial mound

SITE DETAILS: A Viking ship burial filled with grave goods that may have been a royal burial. (See pp. 81, 86.)

10 B7 GOKSTAD

9TH CENTURY AD

LOCATION: Norway

TYPE OF SITE: Burial mound

SITE DETAILS: The longest Norse ship in Norway containing the ritual burial of a tall, middle-aged man.

11 B8 AGGERSBORG

LATE 1ST MILLENNIUM AD

LOCATION: Denmark

TYPE OF SITE: Viking settlement

SITE DETAILS: The largest of the Viking fortress sites that have been excavated.

12 B8 ERTEBØLLE

6TH–4TH MILLENNIA BC

LOCATION: Denmark

TYPE OF SITE: Early settlement

SITE DETAILS: Kitchen midden with finds including pottery and stone axes.

13 B8 LINDHOLM HØJE

8TH–10TH CENTURIES AD

LOCATION: Denmark

TYPE OF SITE: Burial mound

SITE DETAILS: Graves of over 700 people, including cremations and oval burial mounds, probably linked to the Vikings.

14 B8 MOLS GROUP

4TH–2ND MILLENNIA BC

LOCATION: Denmark

TYPE OF SITE: Megalithic tomb

SITE DETAILS: A site that has a number of chamber tombs and a stone circle.

15 B8 BYGHOLM NORREMARK

c. 4200 BC

LOCATION: Denmark

TYPE OF SITE: Burial mound

SITE DETAILS: A mortuary house above a pit grave, with two houses added later.

16 B8 BORUM ESHØJ

3RD MILLENNIUM BC

LOCATION: Denmark

TYPE OF SITE: Burial mound

SITE DETAILS: Three well-preserved oak coffins, one a woman's burial containing a bronze dagger and decorations.

17 B8 GRAUBALLE MAN

c. AD 310

LOCATION: Denmark

TYPE OF SITE: Bog body

SITE DETAILS: A 2,000-year-old body that was discovered in a peat bog. (See p. 58.)

18 B8 JELLING

10TH CENTURY AD

LOCATION: Denmark

TYPE OF SITE: Viking city

SITE DETAILS: The Viking capital that has some burial mounds and runestones.

19 B9 EGTVED

2ND MILLENNIUM BC

LOCATION: Denmark

TYPE OF SITE: Burial mound

SITE DETAILS: A well-preserved body of a young girl in an oak coffin wearing clothes, including a short woolen skirt and a belt with a large bronze disk.

20 B9 TYBRIND VIG

4TH MILLENNIUM BC

LOCATION: Denmark

TYPE OF SITE: Neolithic settlement

SITE DETAILS: A submerged site that has finds of a fishhook, pottery, textiles, and a boat and paddle.

21 B9 HJORTSPRING

3RD CENTURY BC

LOCATION: Denmark

TYPE OF SITE: Burial mound

SITE DETAILS: A warship buried under a mound with 150 wooden shields, 138 iron spearheads, and 20 coats of mail.

22 C9 BILDSØ

LATE 4TH MILLENNIUM BC

LOCATION: Denmark

TYPE OF SITE: Burial mound

SITE DETAILS: A set of small chambers found in a long mound.

23 C9 SKULDELEV

11TH CENTURY AD

LOCATION: Denmark

TYPE OF SITE: Shipwreck

SITE DETAILS: The excavated remains of five ships that were sunk in a fjord.

24 C8 TRELLEBORG

10TH CENTURY AD

LOCATION: Denmark

TYPE OF SITE: Viking fortress

SITE DETAILS: A settlement that housed 500 people inside its wall. Roads and a cemetery were excavated. (See p. 81.)

25 C8 LUND CATHEDRAL

12TH CENTURY AD

LOCATION: Sweden

TYPE OF SITE: Medieval church

SITE DETAILS: This Romanesque cathedral is one of the best examples of its type.

26 C8 SKATEHOLM

7TH MILLENNIUM BC

LOCATION: Sweden

TYPE OF SITE: Mesolithic cemetery

SITE DETAILS: Burials accompanied by dogs, antlers, and flint blades. There is evidence of a large ceremonial structure.

27 C9 HAGESTAD

c. 2500 BC

LOCATION: Sweden

TYPE OF SITE: Early cemetery

SITE DETAILS: Some burial mounds, one divided into nine chambers, containing the remains of about 50 people.

28 C8 KIVIK

2ND MILLENNIUM BC

LOCATION: Sweden

TYPE OF SITE: Burial mound

SITE DETAILS: A tumulus with a cist burial that has some impressive carvings.

29 D8 VALLHAGAR

6TH CENTURY AD

LOCATION: Sweden

TYPE OF SITE: Farming settlement

SITE DETAILS: Longhouses with central hearths and outbuildings were found.

30 D7 BIRKA

9TH–10TH CENTURIES AD

LOCATION: Sweden

TYPE OF SITE: Viking settlement

SITE DETAILS: Trading center with a cemetery of 2,500 graves. Goods from Russia and the Far East have been found in excavations here.

31 D7 HELGO

6TH–10TH CENTURIES AD

LOCATION: Sweden

TYPE OF SITE: Viking settlement

SITE DETAILS: Foundations of houses include smiths', bronze workers', and bead makers' houses.

32 D7 WASA SHIP

AD 1628

LOCATION: Sweden

TYPE OF SITE: Shipwreck

SITE DETAILS: The site of a sunken flagship excavated and preserved.

33 D7 GAMLA UPPSALA

5TH CENTURY AD

LOCATION: Sweden

TYPE OF SITE: Burial mound

SITE DETAILS: Early kings were buried here; it later became a Viking ritual site.

34 D7 VALSGARDE

6TH–7TH CENTURIES AD

LOCATION: Sweden

TYPE OF SITE: Norse cemetery

SITE DETAILS: A chieftain burial in which a helmet and spearheads were found.

35 D5 GRATASK

11TH–13TH CENTURIES AD

LOCATION: Sweden

TYPE OF SITE: Saami cemetery

SITE DETAILS: A Saami site of sacrifice where a metal bird has been excavated.

36 E6 KIRKONLAATTIA

c. 1500 BC

LOCATION: Finland

TYPE OF SITE: Burial mound

SITE DETAILS: A large tomb with internal decoration showing a ritual procession.

37 E7 SOUKAINEN

4TH CENTURY AD

LOCATION: Finland

TYPE OF SITE: Cairn monument

SITE DETAILS: A large round cairn with two burials – each contained a full set of weapons. A glass drinking horn and bronze bucket were also found.

38 D7 OTTERBOTE

3RD MILLENNIUM BC

LOCATION: Finland

TYPE OF SITE: Early settlement

SITE DETAILS: The floors of nine round huts found under a cliff with external fireplaces, garbage middens, and the remains of a well. About 23,000 fragments of pottery have been found.

39 E7 YLISKYLA

c. AD 600

LOCATION: Finland

TYPE OF SITE: Inhumation cemetery

SITE DETAILS: A cemetery and a mound containing a rich boat grave with finds of shield bosses and handles, and knives.

40 F7 KUNDA-LAMMASMAGI

7TH–4TH MILLENNIA BC

LOCATION: Estonia

TYPE OF SITE: Hunter-gatherer

SITE DETAILS: Finds of an elaborate antler and bone industry linked to fishing.

41 E7 TOOMPEA CASTLE

14TH CENTURY AD

LOCATION: Estonia

TYPE OF SITE: Medieval castle

SITE DETAILS: Founded by Danes, the earliest remains date to the 14th century. The Tall Hermann corner tower has 16th-century shell scars on its walls.

42 E8 SARNATE

4TH–3RD MILLENNIA BC

LOCATION: Latvia

TYPE OF SITE: Early settlement

SITE DETAILS: The remains of houses and hearths of the Narva culture were found, with a number of wooden artifacts.

43 E8 ZVEJNIEKI

7TH MILLENNIUM BC

LOCATION: Latvia

TYPE OF SITE: Mesolithic cemetery

SITE DETAILS: Over 300 graves next to a settlement with finds of perforated elks' teeth pendants, bone points, and ocher.

44 E8 IMPILTIS

11TH–13TH CENTURIES AD

LOCATION: Lithuania

TYPE OF SITE: Medieval hill fort

SITE DETAILS: Wooden fortifications that had high fences and towers. Reservoirs were discovered in its earthworks.

45 E8 SVENTOJI

3RD MILLENNIUM BC

LOCATION: Lithuania

TYPE OF SITE: Early settlement

SITE DETAILS: Narva site where finds are similar to those at Sarnate in Latvia.

CONTINENTAL EUROPE

The map for these listings is on pp. 146–7.

1 E7 ANTELAS
LATE 4TH MILLENNIUM BC
LOCATION: Portugal
TYPE OF SITE: Megalithic tomb
SITE DETAILS: A classic dolmen whose stones have some geometric designs.

2 E7 VILA NOVA DE FOZ COA
20,000 YEARS AGO
LOCATION: Portugal
TYPE OF SITE: Rock art
SITE DETAILS: A recently discovered concentration of rock carvings that may be the largest of their kind in Europe.

3 E8 VILA NOVA DO SAO PEDRO
3RD MILLENNIUM BC
LOCATION: Portugal
TYPE OF SITE: Early settlement
SITE DETAILS: A fortified Neolithic settlement occupied during three phases. Excavations revealed chisels, daggers, copper axes, and Beaker pottery. (See p. 25.)

4 E8 ZAMBUJAL
2400–1800 BC
LOCATION: Portugal
TYPE OF SITE: Ancient fortress
SITE DETAILS: A citadel that flourished for around 600 years. It was an area of copper mining, and there have been finds of Beaker pottery.

5 E8 SEIXAL
19TH CENTURY AD
LOCATION: Portugal
TYPE OF SITE: Industrial complex
SITE DETAILS: Heavy industrial center whose glass bottle and wool factories were the first industrial sites to be excavated in Portugal.

6 E8 LISBON
16TH CENTURY AD
LOCATION: Portugal
TYPE OF SITE: Medieval port
SITE DETAILS: Center of Portuguese world trade whose founding site, the Castle of St. George, still looks out over the city. There are colonial remains and treasures, despite the destruction of the city in the earthquake of 1755. (See p. 101.)

7 E8 PALMELA
EARLY 2ND MILLENNIUM BC
LOCATION: Portugal
TYPE OF SITE: Rock-cut tomb
SITE DETAILS: *Hypogea* – tombs with an antechamber – form a necropolis, whose grave goods include Beaker pottery, gold foil, and copper knives.

8 E8 ALCALA
4TH MILLENNIUM BC
LOCATION: Portugal
TYPE OF SITE: Megalithic cemetery
SITE DETAILS: A long passage leads to a corbeled chamber with a number of tombs. Tomb 3 contained Callais beads, amber pendants, and copper flat axes.

9 E7 SANTIAGO DE COMPOSTELA
11TH CENTURY AD
LOCATION: Spain
TYPE OF SITE: Medieval cathedral
SITE DETAILS: *See pp. 90, 99.*

10 E7 A GUARDA
LATE 1ST MILLENNIUM BC
LOCATION: Spain
TYPE OF SITE: Celtic settlement
SITE DETAILS: Fortified Celtic remains that contain the foundations of about 100 round stone dwellings.

11 F7 ALTAMIRA
15,000–13,000 YEARS AGO
LOCATION: Spain
TYPE OF SITE: Cave art
SITE DETAILS: A cave that is famous for its Upper Paleolithic art, especially for its painted bison figures. There have also been finds of some portable art.

12 F7 EL CASTILLO
40,000 YEARS AGO
LOCATION: Spain
TYPE OF SITE: Hunter-gatherer
SITE DETAILS: A cave with finds of a people who hunted horses, bovids, and deer. The cave art consists of 155 animal figures and 50 red hand stencils.

13 F7 AVILA
11TH CENTURY AD
LOCATION: Spain
TYPE OF SITE: Medieval town
SITE DETAILS: Impressive fortified town whose walls, some of which were built using leftover Roman material, are 39 ft (12 m) high, and have round towers every 66 ft (20 m). (See p. 90.)

14 F8 VASCOS
9TH–10TH CENTURIES AD
LOCATION: Spain
TYPE OF SITE: Islamic fortress
SITE DETAILS: A citadel that may have been a refuge site or a mining town.

15 E8 MERIDA
1ST CENTURY BC
LOCATION: Spain
TYPE OF SITE: Roman town
SITE DETAILS: The remains include a well-preserved theater, an amphitheater, a Roman house, and mosaics. There is also a 1st-century AD temple of Diana, the Los Milagros aqueduct, and the Alcazaba Moorish building.

16 E8 CANCHO RUANO
7TH CENTURY BC
LOCATION: Spain
TYPE OF SITE: Early settlement
SITE DETAILS: A center of the Tartessus civilization, where excavations have uncovered a small temple, a moat, and finds of ceramics and jewelry.

17 E8 SEVILLE
12TH CENTURY AD
LOCATION: Spain
TYPE OF SITE: Medieval town
SITE DETAILS: An important town in the Hispano-Magreb Islamic world where archaeological discoveries have been made through old engravings. Remains of a mosque are found in the cathedral.

18 E8 HUELVA
8TH CENTURY BC
LOCATION: Spain
TYPE OF SITE: Mining center
SITE DETAILS: Important mine where a large hoard of bronzes was found.

19 F8 MADINAT AZ-ZAHRA
10TH–11TH CENTURIES AD
LOCATION: Spain
TYPE OF SITE: Islamic settlement
SITE DETAILS: Remains of a royal palace, a small mosque, and wall decorations.

20 E8 ALCAZABA
8TH–11TH CENTURIES AD
LOCATION: Spain
TYPE OF SITE: Medieval castle
SITE DETAILS: A castle built into the ramparts of a Moorish city. There are remains of a Roman amphitheater and a keep. Phoenician, Roman, and Moorish artifacts have been excavated.

21 F8 GRANADA
13TH–15TH CENTURIES AD
LOCATION: Spain
TYPE OF SITE: Medieval town
SITE DETAILS: The Alhambra, a large palace and citadel, is the most impressive Moorish building. (See p. 90.)

22 F9 LOS MILLARES
c. 2400 BC
LOCATION: Spain
TYPE OF SITE: Fortified settlement
SITE DETAILS: Defensive enclosure that included rectangular houses and *tholos* (beehive) tombs. Artifacts found included imported Egyptian ware, ivory sandals, bone idols, and copper axes.

23 F8 EL ARGAR
2ND MILLENNIUM BC
LOCATION: Spain
TYPE OF SITE: Fortified settlement
SITE DETAILS: Nearly 1,000 burials have been excavated, with finds of gold and silver diadems, copper and bronze daggers and swords.

24 F8 ALMIZARAQUE
3RD MILLENNIUM BC
LOCATION: Spain
TYPE OF SITE: Early settlement
SITE DETAILS: Storage pits and oval houses where ivory sandals, alabaster idols, and copper objects were found. One tomb of 50 skeletons contained beads, buttons, tools, and carved ware.

25 F8 VILLENA
c. 1000 BC
LOCATION: Spain
TYPE OF SITE: Early settlement
SITE DETAILS: An excavated treasure that produced many gold objects, including bowls, bottles, and jewelry.

26 F7 TORRALBA
400,000–200,000 YEARS AGO
LOCATION: Spain
TYPE OF SITE: Hunter-gatherer
SITE DETAILS: Large elephant kill site that has brought finds of Acheulian tools.

27 F7 AMBRONA
400,000–200,000 YEARS AGO
LOCATION: Spain
TYPE OF SITE: Hunter-gatherer
SITE DETAILS: A site that is linked to Torralba, with similar finds of elephant kills and some Acheulian tools.

28 G8 VALTORTA
20,000 YEARS AGO
LOCATION: Spain
TYPE OF SITE: Hunter-gatherer
SITE DETAILS: A cave site that contains Stone Age art showing people holding weapons. (See p. 10.)

29 G7 TARRAGONA
3RD CENTURY BC
LOCATION: Spain
TYPE OF SITE: Roman town
SITE DETAILS: Early town that has ruins of a palace, a necropolis, and an aqueduct.

30 G7 AMPURIAS
6TH–3RD CENTURIES BC
LOCATION: Spain
TYPE OF SITE: Greco-Roman town
SITE DETAILS: Three separate settlements existed here; little remains of the Phoenician period, although the Greek town (550 BC) produced remains of several temples, an *agora*, and mosaics. There are also some Roman ruins.

31 G8 TORRE D'EN GAUMES
2ND MILLENNIUM BC
LOCATION: Spain
TYPE OF SITE: Bronze Age settlement
SITE DETAILS: A village site that has unearthed standing stones, *taulas* (T-shaped stones), and tombs.

32 F5 FRENOUVILLE

4TH–7TH CENTURIES AD

LOCATION: France

TYPE OF SITE: Large cemetery

SITE DETAILS: Previously the site of a Gallo-Roman cemetery, the orientation of the tombs changed with the arrival of the Merovingians.

33 F6 CARNAC

5TH–3RD MILLENNIA BC

LOCATION: France

TYPE OF SITE: Megalithic monument

SITE DETAILS: *See pp. 35, 42.*

34 F6 GAVRINIS

c. 3400 BC

LOCATION: France

TYPE OF SITE: Megalithic tomb

SITE DETAILS: Once a *menhir* (a single vertical standing stone), it was broken up to create a rectangular, Neolithic cairn.

35 F6 DISSIGNAC

c. 4000 BC

LOCATION: France

TYPE OF SITE: Burial mound

SITE DETAILS: Two passage graves that produced pottery and microliths.

36 F6 SAUMUR

14TH CENTURY AD

LOCATION: France

TYPE OF SITE: Medieval castle

SITE DETAILS: Impressive castle built on the base of an earlier fortification. (*See p. 90.*)

37 F6 BAGNEUX

LATE 4TH MILLENNIUM BC

LOCATION: France

TYPE OF SITE: Megalithic tomb

SITE DETAILS: The largest and most impressive of France's megalithic caverns, covered with stone slabs.

38 G6 BOUGON

MID-5TH MILLENNIUM BC

LOCATION: France

TYPE OF SITE: Megalithic cemetery

SITE DETAILS: *See pp. 25, 32.*

39 G7 LASCAUX

17,000 YEARS AGO

LOCATION: France

TYPE OF SITE: Cave art

SITE DETAILS: A cave used for ritual, with paintings of horses and bison. Bone tools, flints, and flakes were found. (*See p. 13.*)

40 F7 DURUTHY

14TH–12TH MILLENNIA BC

LOCATION: France

TYPE OF SITE: Hunter-gatherer

SITE DETAILS: A rock shelter containing Chalcolithic burials. In the Late Magdalenian period, it was a winter settlement for exploiting reindeer and bovids. There have been some finds of portable art.

41 G7 LE TUC D'AUDOUBERT

35,000–15,000 YEARS AGO

LOCATION: France

TYPE OF SITE: Paleolithic rock art

SITE DETAILS: A large cave that contains impressive clay sculptures, including pictures of bison.

42 G7 NIAUX

13,000 YEARS AGO

LOCATION: France

TYPE OF SITE: Cave art

SITE DETAILS: Prehistoric cave with some examples of Magdalenian hunting art. Paintings include black animal figures and floor engravings.

43 F5 CAEN

11TH CENTURY AD

LOCATION: France

TYPE OF SITE: Medieval town

SITE DETAILS: Two impressive abbeys of William the Conqueror's time survive, and are fine examples of Norman Romanesque architecture.

44 G5 ROUEN

12TH CENTURY AD

LOCATION: France

TYPE OF SITE: Medieval town

SITE DETAILS: Known as the "museum-town," it later became the center of French pottery. It contains a Gothic cathedral and other fine examples of medieval architecture.

45 G5 CHATEAU GUYARD

12TH CENTURY AD

LOCATION: France

TYPE OF SITE: Medieval castle

SITE DETAILS: Built by Richard (II) the Lionheart, much of it still remains.

46 G5 SAMARA

c. 5000 BC

LOCATION: France

TYPE OF SITE: Early settlement

SITE DETAILS: Prehistoric dwellings that have subsequently been reconstructed. It was occupied through the Iron Age, and there is also a Roman oppidum called "Caesar's Camp."

47 G5 CUIRY

c. 4800 BC

LOCATION: France

TYPE OF SITE: Early settlement

SITE DETAILS: A reconstructed Neolithic settlement where pits have revealed potsherds, flint tools, and waste flakes.

48 G5 ST. ACHEUL

750,000–60,000 YEARS AGO

LOCATION: France

TYPE OF SITE: Hunter-gatherer

SITE DETAILS: Several thousand artifacts were found. The site gives its name to an Early Paleolithic stone tool industry.

49 G5 ST. DENIS

12TH CENTURY AD

LOCATION: Paris, France

TYPE OF SITE: Medieval cathedral

SITE DETAILS: An impressive church that has been excavated in areas – among the finds were the sculpted heads of a queen and a prophet. (*See p. 90.*)

50 G5 PINCEVENT

35,000–12,000 YEARS AGO

LOCATION: France

TYPE OF SITE: Early settlement

SITE DETAILS: An open-air site that had at least 100 habitation structures and 20 large hearths. Artifacts include points and cooking stones.

51 G5 BIBRACTE

LATE 1ST MILLENNIUM BC

LOCATION: France

TYPE OF SITE: Iron Age settlement

SITE DETAILS: A large oppidum, currently under excavation, that was divided into residential and industrial quarters.

52 G5 POUAN

AD 450–500

LOCATION: France

TYPE OF SITE: Merovingian cemetery

SITE DETAILS: *See pp. 81, 87.*

53 H5 GRAND

1ST CENTURY BC

LOCATION: France

TYPE OF SITE: Roman settlement

SITE DETAILS: *See pp. 68, 79.*

54 G6 CITEAUX

12TH CENTURY AD

LOCATION: France

TYPE OF SITE: Medieval abbey

SITE DETAILS: The headquarters of the Cistercian order, although little remains apart from a Gothic cloister and some 18th-century buildings.

55 G6 MONT LASSOIS & VIX

6TH CENTURY BC

LOCATION: France

TYPE OF SITE: Iron Age burial

SITE DETAILS: *See pp. 58, 66.*

56 H6 AILLEVANS

4TH MILLENNIUM BC

LOCATION: France

TYPE OF SITE: Megalithic tomb

SITE DETAILS: Three Neolithic megalithic monuments containing chambers and vestibules. One grave contained 23 bodies. Another had 100 bodies with offerings including flint arrowheads, a pendant, a flint dagger blade, a copper bead, and bones of animals.

57 G6 LE CREUSOT

19TH CENTURY AD

LOCATION: France

TYPE OF SITE: Industrial complex

SITE DETAILS: Large, preserved ironworks that include a locomotive shop, an old crane, and two glass cones. There is also a rolling mill, machine assembly shops, and workers' homes. (*See p. 111.*)

58 H6 CHALAIN

LATE 4TH MILLENNIUM BC

LOCATION: France

TYPE OF SITE: Neolithic settlement

SITE DETAILS: An early lakeside village consisting of about 20 houses. Artifacts included axes and adzes with deer antler handles.

59 G6 CLUNY

10TH CENTURY AD

LOCATION: France

TYPE OF SITE: Medieval monastery

SITE DETAILS: *See pp. 90, 96.*

60 G6 AULNAT

3RD CENTURY BC

LOCATION: France

TYPE OF SITE: Early settlement

SITE DETAILS: Iron Age center that has some evidence of the working of gold, silver, bronze, coral, brass, and textiles.

61 G7 COMBE D'ARCHE

30,000 YEARS AGO

LOCATION: France

TYPE OF SITE: Cave art

SITE DETAILS: A recently discovered set of Paleolithic cave paintings containing pictures of rhinos, bison, and deer.

62 G7 NIMES

1ST CENTURY BC

LOCATION: France

TYPE OF SITE: Roman city

SITE DETAILS: Well-preserved Roman remains of the Pont du Gard aqueduct (*see p. 68*) are found nearby. Other structures include an amphitheater that seated 24,000, the *Maison Carrée* temple, and the ruined city gate. Iron Age and Roman artifacts were found.

63 G7 LA COUVERTOIRADE

11TH CENTURY AD

LOCATION: France

TYPE OF SITE: Medieval village

SITE DETAILS: A fortified settlement overlooked by a castle. The defenses remain in varying states of preservation.

64 G7 BARBEGAL WATER MILL

4TH CENTURY AD

LOCATION: France

TYPE OF SITE: Roman water mill

SITE DETAILS: A construction that provided flour for 80,000 people. It housed a loading area, a milling chamber, and an aqueduct.

65 G7 ORANGE

1ST CENTURY BC

LOCATION: France

TYPE OF SITE: Roman settlement

SITE DETAILS: A Roman town that has an impressive theater and a decorated arch.

66 G7 ARLES

1ST CENTURY BC

LOCATION: France

TYPE OF SITE: Roman town

SITE DETAILS: A settlement with an

elaborately decorated theater, and a huge amphitheater with 60 arches that held 21,000 people. Finds included antlers and boar tusks, pottery, a statue of Augustus, and a votive shield.

67 G7 CARCASSONNE
3RD CENTURY BC–17TH CENTURY AD
LOCATION: France
TYPE OF SITE: Medieval town
SITE DETAILS: *See pp. 90, 98.*

68 H7 MARSEILLES
6TH CENTURY BC–2ND CENTURY AD
LOCATION: France
TYPE OF SITE: Early settlement
SITE DETAILS: A colony where excavations revealed a Roman port, a well-preserved merchant ship, and defenses.

69 H7 ROUGIERS
12TH–15TH CENTURIES AD
LOCATION: France
TYPE OF SITE: Medieval village
SITE DETAILS: Hilltop settlement with a stone castle and terraced housing areas.

70 H7 TERRA AMATA
380,000 YEARS AGO
LOCATION: France
TYPE OF SITE: Hunter-gatherer
SITE DETAILS: A site that may have been used for 15 successive years as a camp by Acheulian hunter-gatherers.

71 H7 ENTREMONT
3RD–2ND CENTURIES BC
LOCATION: France
TYPE OF SITE: Fortified settlement
SITE DETAILS: Oppidum consisting of dry-stone houses on a grid layout. Skulls were ritually attached to some buildings.

72 H7 GROTTE DU LAZARET
150,000 YEARS AGO
LOCATION: France
TYPE OF SITE: Hunter-gatherer
SITE DETAILS: Winter camp with remains of two hearths, bed areas, seashells, and animal foot bones.

73 H7 FILITOSA
LATE 3RD MILLENNIUM BC
LOCATION: France
TYPE OF SITE: Early settlement
SITE DETAILS: Fortified center with large *torre* (circular dry-stone towers). There were also megalithic chambers and standing stones.

74 G5 TOURNAI
5TH CENTURY AD
LOCATION: Belgium
TYPE OF SITE: Burial mound
SITE DETAILS: The tomb of Childeric, a Frankish king whose grave was opened in 1653, revealing finds of jewelry.

75 G5 LE GRAND HORNU
19TH CENTURY AD
LOCATION: Belgium
TYPE OF SITE: Industrial village
SITE DETAILS: *See pp. 111, 121.*

76 G4 ZAANDAM
17TH–18TH CENTURIES AD
LOCATION: Netherlands
TYPE OF SITE: Industrial complex
SITE DETAILS: Remaining buildings include three wooden warehouses (*c.* 1875–80), cast-iron-columned warehouses, and preserved mills.

77 G4 AMSTERDAM
16TH CENTURY AD
LOCATION: Netherlands
TYPE OF SITE: Trading port
SITE DETAILS: The city was once a center of world trade and was linked to the Dutch East India Company. (*See p. 101.*)

78 H4 SCHIMMERES
EARLY 4TH MILLENNIUM BC
LOCATION: Netherlands
TYPE OF SITE: Burial mound
SITE DETAILS: Two chambers built in the shape of a large longboat.

79 G4 DORESTAD
6TH–9TH CENTURIES AD
LOCATION: Netherlands
TYPE OF SITE: Medieval town
SITE DETAILS: Early medieval port; the harbor has recently been excavated. A number of Carolingian coins were found, indicating a royal mint.

80 H4 NOVIOMAGUS
1ST CENTURY AD
LOCATION: Netherlands
TYPE OF SITE: Roman fort
SITE DETAILS: A legionary center whose *principia* (main building) acted as religious center, headquarters, and a place of assembly.

81 H4 STEIN
LATE 3RD MILLENNIUM BC
LOCATION: Netherlands
TYPE OF SITE: Burial mound
SITE DETAILS: Grave containing the cremated remains of skeletons and grave goods, including rimmed pottery and more than 100 arrowheads.

82 H3 HEDEBY
7TH–9TH CENTURIES AD
LOCATION: Germany
TYPE OF SITE: Viking town
SITE DETAILS: Trading town that became a major Viking center. It was protected by an earthen rampart and had wooden houses and workshops.

83 H4 FEDDERSEN WIERDE
1ST CENTURY BC
LOCATION: Germany
TYPE OF SITE: Iron Age settlement
SITE DETAILS: A farmstead surrounded by smaller houses and a small, raised granary near each building. Roman imports, including bronze vessels and Samian ware, were found inside.

84 I4 LEUBINGEN
17TH–16TH CENTURIES BC
LOCATION: Germany
TYPE OF SITE: Barrow burial
SITE DETAILS: Early Bronze Age tomb protected by an oak structure. One grave included two axes, three burins, and bronze daggers. The skeleton of a girl was accompanied by a number of gold burial goods.

85 H4 NEANDERTHAL
120,000–35,000 YEARS AGO
LOCATION: Germany
TYPE OF SITE: Hunter-gatherer
SITE DETAILS: The site that gives its name to the *Homo sapiens* species, with its large brain and huge brows.

86 H4 XANTEN
1ST CENTURY AD
LOCATION: Germany
TYPE OF SITE: Roman legionary fort
SITE DETAILS: A legionary base whose remains include baths, an amphitheater, a temple, *insulae* (apartment buildings), workshops, and a palace.

87 H4 ESSEN
18TH CENTURY AD
LOCATION: Germany
TYPE OF SITE: Industrial
SITE DETAILS: *See pp. 111, 120.*

88 H5 AACHEN
8TH CENTURY AD
LOCATION: Germany
TYPE OF SITE: Early city
SITE DETAILS: The Carolingian city chosen by Charlemagne as his capital. There are some impressive medieval churches, while the town hall has been built on the ruins of Charlemagne's palace.

89 H5 GONNERSDORF
12,600 YEARS AGO
LOCATION: Germany
TYPE OF SITE: Rock art
SITE DETAILS: Around 400 Paleolithic carvings of women walking in single file.

90 H5 TRIER
1ST CENTURY AD
LOCATION: Germany
TYPE OF SITE: Roman town
SITE DETAILS: Major center with remains of an amphitheater, public baths, an aqueduct, a forum, and cemeteries.

91 I5 ROTHENBURG
14TH CENTURY AD
LOCATION: Germany
TYPE OF SITE: Medieval town
SITE DETAILS: Rebuilt after an earthquake, the town includes the ruins of Staufen Castle. It has some 14th-century walls and towers, with gateways added in the 17th century, and an outlying fort. (*See p. 91.*)

92 H5 HOCHDORF
6TH CENTURY BC
LOCATION: Germany
TYPE OF SITE: Burial mound
SITE DETAILS: A chariot tomb with goods that included gold *fibulae* (pins), bracelets, a belt, a dagger, an iron drinking horn, and a bronze cauldron.

93 H5 MAGDALENBURG
c. 500 BC
LOCATION: Germany
TYPE OF SITE: Burial mound
SITE DETAILS: A huge barrow with a main chamber surrounded by a number of subsidiary chambers. It contained a funerary wagon.

94 H5 HEUNEBERG
8TH–5TH CENTURIES BC
LOCATION: Germany
TYPE OF SITE: Iron Age settlement
SITE DETAILS: A fortified hilltop with sun-dried brick walls. Artifacts point to trade with the Mediterranean area.

95 H5 WASSERBURG BUCHAU
12TH CENTURY BC
LOCATION: Germany
TYPE OF SITE: Bronze Age settlement
SITE DETAILS: Unusual settlement on an island in a lake with 50 houses protected by a wall of 15,000 wooden posts. Abandoned after waters rose, it was reoccupied with nine log cabins. Finds include pottery and bronzes. (*See p. 47.*)

96 I5 MANCHING
2ND–1ST CENTURIES BC
LOCATION: Germany
TYPE OF SITE: Celtic settlement
SITE DETAILS: Oppidum that was a major trade center and supported a large population. Excavations have revealed workshops, dwellings, and evidence of iron-working and coin minting. (*See p. 58.*)

97 H6 SISSACH
18TH–17TH CENTURIES BC
LOCATION: Switzerland
TYPE OF SITE: Bronze Age settlement
SITE DETAILS: A lakeside village with remains of Alpine log cabins. Roofs were made of shingles or wooden slates and were weighted down by heavy stones.

98 H6 CORTAILLOD
c. 4000 BC
LOCATION: Switzerland
TYPE OF SITE: Early settlement
SITE DETAILS: A lakeside village that gives its name to local Neolithic finds because of its well-preserved organic material.

99 H6 LE PETIT CHASSEUR
4TH MILLENNIUM BC
LOCATION: Switzerland
TYPE OF SITE: Early settlement
SITE DETAILS: An ancient village that included a cemetery marked by stones.

100 H8 ANGHELU RUJU
3RD MILLENNIUM BC
LOCATION: Italy
TYPE OF SITE: Early cemetery
SITE DETAILS: Passage graves decorated with horse carvings, with finds of marble and alabaster bracelets, copper tools and weapons, and figurines.

101 H7 LI LOLGHI
2ND MILLENNIUM BC
LOCATION: Italy
TYPE OF SITE: Megalithic tomb
SITE DETAILS: "Giant" graves – megalithic chambers surrounded by large Cyclopean figures with curved wings.

102 H8 BARAMUNI
c. 1800 BC
LOCATION: Italy
TYPE OF SITE: Early settlement
SITE DETAILS: A fortified *nuraghe* (large tower) with more fortifications built later to enclose the outlying village.

103 H7 ARENE CANDIDE
c. 9000 BC
LOCATION: Italy
TYPE OF SITE: Hunter-gatherer
SITE DETAILS: Mesolithic burial including a flint knife and a shell-decorated cap. There were three Neolithic burials with shell-impressed pottery, as well as later Bronze Age, Iron Age, and Roman finds.

104 H7 GENOA
12TH CENTURY AD
LOCATION: Italy
TYPE OF SITE: Medieval city
SITE DETAILS: An important trading center that expanded when its large walls were built in 1155. There are a number of medieval buildings left, including the Porta Soprana gateway.

105 I7 POPULONIA
1ST MILLENNIUM BC
LOCATION: Italy
TYPE OF SITE: Early city
SITE DETAILS: Etruscan and later Roman town that serviced the silver and iron mines of Elba.

106 I6 PADUA
13TH CENTURY AD
LOCATION: Italy
TYPE OF SITE: Medieval town
SITE DETAILS: The city has impressive numbers of medieval buildings, not least the Basilica of San Antonio and the oldest university in Italy.

107 I6 VENICE
10TH CENTURY AD
LOCATION: Italy
TYPE OF SITE: Medieval town
SITE DETAILS: A trading center that prospered through its exceptional glass trade, it has a number of impressive medieval buildings. (*See p. 91.*)

108 I6 VILLANOVA
10TH CENTURY BC
LOCATION: Italy
TYPE OF SITE: Iron Age settlement
SITE DETAILS: An Early Iron Age culture that controlled local iron and copper mines. Cemeteries show that bodies were cremated, with the ashes placed in pottery and covered with a bowl.

109 I6 RAVENNA
5TH CENTURY AD
LOCATION: Italy
TYPE OF SITE: Early town
SITE DETAILS: *See pp. 81, 88–9.*

110 I7 CLUSIUM
7TH CENTURY BC
LOCATION: Italy
TYPE OF SITE: Etruscan cemetery
SITE DETAILS: A settlement that has a number of grave chambers. It was later a Roman colony.

111 I7 ORVIETO
7TH CENTURY BC
LOCATION: Italy
TYPE OF SITE: Etruscan settlement
SITE DETAILS: Cemetery of rock-cut tombs situated above an Etruscan city.

112 I7 TARQUINIA
1ST MILLENNIUM BC
LOCATION: Italy
TYPE OF SITE: Etruscan city
SITE DETAILS: The main Etruscan settlement, with painted tombs, the base of a great temple, and the remains of a large wall. There is a large cemetery with burial mounds and frescoed rock-cut tombs.

113 I7 CAERE (CERVETERI)
9TH–4TH CENTURIES BC
LOCATION: Italy
TYPE OF SITE: Etruscan city
SITE DETAILS: A walled settlement that includes an impressive cemetery with tombs, found under tumuli or cut into tuff, and finds of Greek pottery.

114 I7 ROME
1ST MILLENNIUM BC
LOCATION: Italy
TYPE OF SITE: Early city
SITE DETAILS: First settled in the 9th century BC, Rome was later fortified by the Servian wall. It developed into the hub of the Roman Empire, and as a result there are thousands of remains, including the Forum, the Colosseum, and catacomb burials outside the city. (*See p. 68.*)

115 I7 VEII
7TH CENTURY BC
LOCATION: Italy
TYPE OF SITE: Etruscan settlement
SITE DETAILS: Currently under excavation, revealing a number of cemeteries. It was later fortified, and ruined walls and towers remain. One temple has large terra-cotta statues of gods.

116 I7 GROTTA GUATTARI
57,000–51,000 YEARS AGO
LOCATION: Italy
TYPE OF SITE: Hunter-gatherer
SITE DETAILS: Cave preserved by a landslide, where bones of deer, horses, and bovine animals, and the skull of a Neanderthal man have been excavated.

117 I7 ISERNIA LA PINETA
730,000 YEARS AGO
LOCATION: Italy
TYPE OF SITE: Hunter-gatherer
SITE DETAILS: A lake bed whose stone tools – limestone choppers and flint flakes – are the earliest-known artifacts in Europe.

118 I7 CUMAE
8TH CENTURY BC
LOCATION: Italy
TYPE OF SITE: Classical town
SITE DETAILS: An old Greek colony taken over by the Romans. Tombs revealed some early Greek pottery. Remains of the Roman town include temples to Jupiter and Apollo on the acropolis.

119 I7 POMPEII
1ST CENTURY BC
LOCATION: Italy
TYPE OF SITE: Roman town
SITE DETAILS: *See pp. 68, 78.*

120 I8 HERCULANEUM
1ST CENTURY AD
LOCATION: Italy
TYPE OF SITE: Roman town
SITE DETAILS: *See pp. 68, 79.*

121 I8 CASTELLUCCIO
3RD MILLENNIUM BC
LOCATION: Italy
TYPE OF SITE: Early cemetery
SITE DETAILS: Several hundred rock-cut tombs carved with reliefs. Funerary offerings found include pottery, copper objects, and bone plaques.

122 J8 SYRACUSE
8TH CENTURY BC
LOCATION: Italy
TYPE OF SITE: Early settlement
SITE DETAILS: The earliest Sicilian city was a Greek colony and later Roman town. The Temple of Athena was richly decorated with marble and was later incorporated into the city's cathedral.

123 I9 TARXIEN
2ND MILLENNIUM BC
LOCATION: Malta
TYPE OF SITE: Megalithic complex
SITE DETAILS: A number of temples where excavations have revealed a temple goddess statue, a decorated altar, and a flint sacrificial knife. A cremation cemetery (c. 1400 BC) has yielded pottery and metal daggers.

124 I9 GGANTIJA
3RD MILLENNIUM BC
LOCATION: Malta
TYPE OF SITE: Ceremonial center
SITE DETAILS: Cult chambers used as temples to worship a Mother Earth figure. The temples were built with massive stone blocks and pillars.

125 I9 BROCHTORFF CIRCLE
LATE 4TH MILLENNIUM BC
LOCATION: Malta
TYPE OF SITE: Ceremonial complex
SITE DETAILS: Rock-cut tomb with twin chambers containing the remains of 60 individuals. The central shrine had a large cache of idols, including a carved human head, a stone pig's head, and a large stone jar.

126 I6 HALLSTATT
c. 730 BC
LOCATION: Austria
TYPE OF SITE: Early cemetery
SITE DETAILS: The type-site of an Early Bronze Age culture that contained 2,000 graves. Artifacts included some swords with amber and ivory pommels. (*See p. 58.*)

127 J6 KLEINKLEIN
7TH CENTURY BC
LOCATION: Austria
TYPE OF SITE: Early cemetery
SITE DETAILS: A Celtic burial site where inscribed vessels and a human funerary mask have been excavated.

128 J5 CARNUNTUM
c. AD 14
LOCATION: Austria
TYPE OF SITE: Roman settlement
SITE DETAILS: A legionary fortress first built of wood, then of stone. Three of its four gates and a number of buildings still survive. An amphitheater and civilian settlements were found outside the fortress.

129 I5 PRIBAM
19TH CENTURY AD
LOCATION: Czech Republic
TYPE OF SITE: Industrial
SITE DETAILS: An old silver- and gold-mining center that was later used to mine coal. Most remains date from the 19th century, including a headstock, administrative buildings, an engineering workshop, and workers' houses.

130 I5 UNETICE
c. 2500 BC
LOCATION: Czech Republic
TYPE OF SITE: Early cemetery
SITE DETAILS: A burial site with 30 richly furnished graves, in which daggers and beads were found.

131 J5 LIBENICE
3RD CENTURY BC
LOCATION: Czech Republic
TYPE OF SITE: Celtic religious complex
SITE DETAILS: A sanctuary consisting of an oblong enclosure and sacrificial area. The remains of human beings and several animals were excavated.

132 J5 STRADONICE
2ND CENTURY BC
LOCATION: Czech Republic
TYPE OF SITE: Iron Age settlement

SITE DETAILS: A Celtic oppidum of the La Tène period, where painted ceramics and a coin mint were excavated.

133 J5 DOLNI VESTONICE
25,000 YEARS AGO
LOCATION: Czech Republic
TYPE OF SITE: Hunter-gatherer
SITE DETAILS: Early settlement famous for its "Venus" carving, sculpted using powdered bone and clay. Hearths, burins, scrapers, backed blades, and human burials have been excavated.

134 J5 STARE HRADISKO
2ND–1ST CENTURIES BC
LOCATION: Czech Republic
TYPE OF SITE: Celtic settlement
SITE DETAILS: An oppidum, subdivided within its walls, that has the remains of wooden buildings, and areas for pottery making, enameling, and metalworking.

135 K5 SPISSKY STVRTOK
2ND MILLENNIUM BC
LOCATION: Slovakia
TYPE OF SITE: Early settlement
SITE DETAILS: A center that had stone fortifications and some finds of bronze and gold. It was important for its position in interregional trade.

136 J5 ZAMECEK
c. 1800 BC
LOCATION: Slovakia
TYPE OF SITE: Early settlement
SITE DETAILS: Important hilltop center surrounded by timber-framed ramparts.

137 J4 SARNOWO
c. 3000 BC
LOCATION: Poland
TYPE OF SITE: Neolithic settlement
SITE DETAILS: Remains of dwellings and nine huge barrows have been found, linked to the Kuyavian culture.

138 J5 OLSZANICA
LATE 5TH MILLENNIUM BC
LOCATION: Poland
TYPE OF SITE: Early settlement
SITE DETAILS: Center of the Linear Pottery culture where 13 longhouses, pieces of obsidian, and potsherds were excavated.

139 K5 OBLAZOWA
23,000 YEARS AGO
LOCATION: Poland
TYPE OF SITE: Hunter-gatherer
SITE DETAILS: Cave where a mammoth ivory boomerang, probably used as a hunting weapon, has been found and is believed to be the world's oldest. Scrapers and blades were also found.

140 J6 KOSZIDER
c. 1700 BC
LOCATION: Hungary
TYPE OF SITE: Burial mound
SITE DETAILS: A Bronze Age tumulus that contained rich funerary objects, including decorated axes and swords, bracelets, anklets, and other bronzes.

141 J5 BODROKERESZTUR
EARLY–4TH MILLENNIUM BC
LOCATION: Hungary
TYPE OF SITE: Neolithic cemetery
SITE DETAILS: Remains of at least 50 people accompanied by goods that included undecorated pottery and copper axes.

142 K5 TISZAPOLGAR
LATE 4TH MILLENNIUM BC
LOCATION: Hungary
TYPE OF SITE: Early cemetery
SITE DETAILS: A Copper Age burial site with 156 graves, most excavated. The burials were socially stratified.

143 I6 MAGDALENSKA GORA
6TH–3RD CENTURIES BC
LOCATION: Slovenia
TYPE OF SITE: Fortified settlement
SITE DETAILS: A Hallstatt center where an excavated cemetery has uncovered *situlae* (bucket-shaped containers), arms and armor, and other objects.

144 J7 SPLIT
4TH CENTURY AD
LOCATION: Croatia
TYPE OF SITE: Roman town
SITE DETAILS: Site of a Diocletian palace that was built when the city was known as Spalato. The Church of St. Domnus survives.

145 K6 VINCA
5TH–MID-4TH MILLENNIA BC
LOCATION: Yugoslavia
TYPE OF SITE: Neolithic settlement
SITE DETAILS: A tell with remains of small longhouses and dark burnished pottery have been excavated.

146 K6 LEPENSKI VIR
7TH MILLENNIUM BC
LOCATION: Yugoslavia
TYPE OF SITE: Early settlement
SITE DETAILS: A Mesolithic village where trapezoidal houses, burials, and carved stone heads were found. Fish carvings were placed next to hearths, suggesting belief in a form of river god. (*See p. 25.*)

147 K6 SARMIZEGETHUSA
1ST CENTURY BC
LOCATION: Romania
TYPE OF SITE: Early settlement
SITE DETAILS: The fortified center of the Dacian state, where sanctuaries, workshops, and houses were excavated.

148 L6 HISTRIA
6TH CENTURY BC–6TH CENTURY AD
LOCATION: Romania
TYPE OF SITE: Early settlement
SITE DETAILS: A site first occupied in the 6th century BC. A Temple of Zeus and a cemetery of about 1,000 Greek tombs. It was occupied by Goths in the 3rd century AD. Most surface remains are from a 5th-century Christian occupation with a basilica and residential quarters.

149 L7 CALLATIS
LATE 6TH CENTURY BC
LOCATION: Romania
TYPE OF SITE: Early city
SITE DETAILS: A Dorian settlement that became a powerful city-state. Underwater excavation has shown that much of the port has fallen into the sea. Remains of a Roman rampart, a Christian basilica, and burial grounds have been excavated.

150 L7 BUCHAREST
15TH CENTURY AD
LOCATION: Romania
TYPE OF SITE: Medieval city
SITE DETAILS: Although there is some evidence that the site was first settled over 150,000 years ago, it was first known as a city when Vlad the Impaler built his fortress in the 15th century. Excavated artifacts are now in a local museum.

151 L7 VARNA
5TH MILLENNIUM BC
LOCATION: Bulgaria
TYPE OF SITE: Neolithic cemetery
SITE DETAILS: *See pp. 25, 33.*

152 L7 KARANOVO
6TH MILLENNIUM BC
LOCATION: Bulgaria
TYPE OF SITE: Neolithic settlement
SITE DETAILS: A huge tell where a number of villages were excavated, yielding remains of copper and gold workings.

153 K7 STOBI
1ST CENTURY BC–6TH CENTURY AD
LOCATION: Macedonia
TYPE OF SITE: Roman settlement
SITE DETAILS: A colony with church buildings, including a geometrically designed 4th-century basilica. Terra-cotta figurines were found in a pagan cemetery.

154 K7 TREBENISTE
7TH CENTURY BC
LOCATION: Macedonia
TYPE OF SITE: Early cemetery
SITE DETAILS: Celtic Hallstatt burial that has produced golden funerary masks and some early Greek imports.

155 K7 HERACLEA
EARLY 1ST MILLENNIUM AD
LOCATION: Macedonia
TYPE OF SITE: Byzantine town
SITE DETAILS: Founded as a town by Philip II of Macedonia in the 3rd century BC, its later finds included a sculpture of the goddess Nemesis. There are two Christian basilicas.

156 J8 APOLLONIA
6TH CENTURY BC
LOCATION: Albania
TYPE OF SITE: Greek town
SITE DETAILS: A coastal Corinthian colony, where remains of walls, kilns, and tombs were excavated.

157 K8 NEA NIKOMEDEIA
c. 6200 BC
LOCATION: Greece
TYPE OF SITE: Neolithic settlement
SITE DETAILS: Remains of mud houses and burials, where finds include some pottery, carved greenstone frogs, female figurines, and stone axes.

158 K8 PETRALONA
400,000–200,000 YEARS AGO
LOCATION: Greece
TYPE OF SITE: Hunter-gatherer
SITE DETAILS: Cave that yielded a human skull, an ax, and remains of giant deer, red deer, and cave bears. Scrapers and chopping tools were also excavated.

159 K8 ARGISSA
6TH MILLENNIUM BC
LOCATION: Greece
TYPE OF SITE: Early settlement
SITE DETAILS: An early farming center occupied until the Early Bronze Age. Excavations revealed huts with shallow pits and roofed with branches.

160 K8 DELPHI
6TH CENTURY BC
LOCATION: Greece
TYPE OF SITE: Ancient Greek city
SITE DETAILS: The Oracle and Sanctuary of Apollo were found here, as well as the Sanctuary of Artemis, a temple, houses with portrait sculptures, and a processional way once flanked by lions.

161 K8 OLYMPIA
6TH CENTURY BC
LOCATION: Greece
TYPE OF SITE: Ancient Greek sanctuary
SITE DETAILS: Panhellenic sanctuary with the remains of the Temple of Zeus, which contained his statue – one of the Seven Wonders of the Ancient World. Pheidias's workshop, where the statue was made, has been excavated. There is also a Temple of Hera, a treasury including bronze statue bases, and the remains of a stadium.

162 K8 CORINTH
5TH CENTURY BC
LOCATION: Greece
TYPE OF SITE: Ancient Greek city
SITE DETAILS: The business center of ancient Greece, which had two ports. Its ruins include a temple to Apollo, the Acrocorinth fortress, and the Temple of Aphrodite. It was later taken over by the Romans, whose remains include a forum, theater, and *thermae* (bath complex).

163 K8 DAPHNI
11TH CENTURY AD
LOCATION: Greece
TYPE OF SITE: Medieval monastery
SITE DETAILS: An important Byzantine site built on a sanctuary to Apollo. The church contains mosaics depicting saints and monks, and the *Christos Pantocrator* (Christ in Majesty) in the dome is a masterpiece of Byzantine art.

164 K8 MYCENAE
2ND MILLENNIUM BC
LOCATION: Greece
TYPE OF SITE: Ancient citadel
SITE DETAILS: The major center of the Mycenaean culture, with remains of the Cyclopean walls and the large Lion Gate. A granary structure contained carbonized wheat and barley. Other remains include a ramp house, mud-brick buildings, and a throne room. A large shaft-grave circle was unearthed, and nine tholos tombs yielded artifacts such as weapons, jewelry, and masks, including the Mask of Agamemnon.

165 K8 FRANCHTHI CAVE
22,000 YEARS AGO
LOCATION: Greece
TYPE OF SITE: Early settlement
SITE DETAILS: Finds include Paleolithic small-backed blades and microliths. Another layer contains Mesolithic occupations and burials. The cave was inhabited until Neolithic times, with evidence of early cattle domestication.

166 K8 MYSTRA
13TH CENTURY AD
LOCATION: Greece
TYPE OF SITE: Byzantine town
SITE DETAILS: Some good examples of the Byzantine period have been excavated, with houses, churches, and frescoes.

167 K9 EPIDAURUS
8TH CENTURY BC
LOCATION: Greece
TYPE OF SITE: Ceremonial center
SITE DETAILS: A temple-sanctuary for the god of medicine, Aesculapius, including a ruined temple, the Kategogeion (the pilgrims' sanctuary), a banqueting hall, a stadium, and a ruined theater.

168 K9 ANTIKYTHERA SHIPWRECK
c. 70–80 BC
LOCATION: Greece
TYPE OF SITE: Shipwreck
SITE DETAILS: Statues of a young man, a philosopher's head, a bronze lyre, a marble bull, and a discus thrower have been salvaged. Although many of the artifacts found were Greek, it is believed that this was a Roman ship.

169 K8 ATHENS
5TH CENTURY BC
LOCATION: Greece
TYPE OF SITE: Ancient Greek city
SITE DETAILS: The Acropolis was once a town in itself, with great theaters, the Parthenon, and the Erectheon. The Temple of Hephaestus was surrounded by foundries and metal workshops. (See p. 69.)

170 K8 LAURION
5TH–4TH CENTURIES BC
LOCATION: Greece
TYPE OF SITE: Ancient Greek town
SITE DETAILS: Major center of lead and silver mining – there are extensive remains of Classical silver working, with some long mining galleries. Later tanks, channels, and cisterns remain.

171 L8 SALIAGOS
5TH–4TH MILLENNIA BC
LOCATION: Greece
TYPE OF SITE: Neolithic settlement
SITE DETAILS: A village where excavated fish bones suggest a maritime economy. There is also evidence of farming.

172 L9 AKROTIRI
EARLY 2ND MILLENNIUM BC
LOCATION: Greece
TYPE OF SITE: Bronze Age settlement
SITE DETAILS: *See pp. 47, 57.*

173 K9 IDAEAN CAVE
2ND MILLENNIUM BC
LOCATION: Greece
TYPE OF SITE: Ceremonial center
SITE DETAILS: Celebrated as the birthplace of Zeus, there is an altar on which bronze statues stood. A smaller cave was found full of votive figures, bronze figurines, lamps, and pottery.

174 K9 KNOSSOS
2ND MILLENNIUM BC
LOCATION: Greece
TYPE OF SITE: Minoan settlement
SITE DETAILS: Inhabited since Neolithic times, it dates mostly from the Minoan period, when it was a multistoried palace based on a central court. Walls were decorated with frescoes, and shrines contained figurines and Linear A and B clay tablets. There was also an aqueduct and a chamber-tomb cemetery. (See p. 47.)

175 L9 ARKHANES
16TH CENTURY BC
LOCATION: Greece
TYPE OF SITE: Minoan settlement
SITE DETAILS: An important center where a cemetery has yielded gold, ivory, and marble funerary offerings.

176 L9 GOURNIA
3RD–2ND MILLENNIUM BC
LOCATION: Greece
TYPE OF SITE: Minoan settlement
SITE DETAILS: A well-preserved Bronze Age town where a shrine sanctuary with a terra-cotta altar was discovered. Finds included a snake goddess figurine, offertory tables, and ritual vessels.

177 L9 PALAIKASTRO
2ND MILLENNIUM BC
LOCATION: Greece
TYPE OF SITE: Minoan settlement
SITE DETAILS: *See pp. 47, 48–51.*

178 M9 PAPHOS
5TH CENTURY BC
LOCATION: Cyprus
TYPE OF SITE: Early settlement
SITE DETAILS: A town including a major shrine of Aphrodite. It has the remains of a temple and a Persian siege mound.

179 M9 SOTORI-TEPPES
LATE 5TH MILLENNIUM BC
LOCATION: Cyprus
TYPE OF SITE: Neolithic settlement
SITE DETAILS: A center, characterized by combed ware, protected by a circular stone wall. Excavations revealed a pit-grave cemetery and mud-brick houses.

180 M9 ENKOMI
17TH CENTURY BC
LOCATION: Cyprus
TYPE OF SITE: Early port
SITE DETAILS: Based on the copper trade. Ores were smelted and exported from here. It was later rebuilt on a grid plan with Cyclopean fortifications.

181 M9 SALAMIS
2ND MILLENNIUM BC
LOCATION: Cyprus
TYPE OF SITE: Bronze Age settlement
SITE DETAILS: An early city with finds of later Iron Age chamber tombs that contained the cremated remains of the elite classes.

182 L6 NOVYE RUSESHTY
4TH–3RD MILLENNIA BC
LOCATION: Moldova
TYPE OF SITE: Early village
SITE DETAILS: First occupied by the Linear Pottery culture, a Cucuteni-Tripolye village yielded an impressive hoard and a number of copper objects.

183 L5 KOLOMYSHCHINA
4TH MILLENNIUM BC
LOCATION: Ukraine
TYPE OF SITE: Early settlement
SITE DETAILS: Cucuteni-Tripolye village with traces of 39 structures, mostly arranged in a circle around a central structure. Many contained ovens, though some were probably used as shrines.

184 M5 KIEV
10TH CENTURY AD
LOCATION: Ukraine
TYPE OF SITE: Medieval city
SITE DETAILS: Founded by Viking traders, it has evidence of pottery making and gold working. It was an important trade center of medieval Europe. (See p. 81.)

185 M5 DEREIVKA
MID-5TH–MID-4TH MILLENNIA BC
LOCATION: Ukraine
TYPE OF SITE: Steppe settlement
SITE DETAILS: Village where the people lived in rectangular houses that has yielded a number of the earliest-known domesticated horse bones.

186 M5 MEZHIRICH
18,000–12,000 YEARS AGO
LOCATION: Ukraine
TYPE OF SITE: Hunter-gatherer
SITE DETAILS: *See pp. 13, 21.*

187 M6 OLBIA
7TH–6TH CENTURIES BC
LOCATION: Ukraine
TYPE OF SITE: Greek colony
SITE DETAILS: Imported pottery remains show that it flourished during the 6th century BC. It was rebuilt later with a sacred area including altars, a Temple of Apollo, a workshop, and a cemetery.

188 M6 NEAPOLIS
3RD CENTURY AD
LOCATION: Ukraine
TYPE OF SITE: Scythian settlement
SITE DETAILS: Probable Scythian capital where mounds have been excavated, one yielding 1,300 gold objects.

189 M1 NOVGOROD
9TH–10TH CENTURIES AD
LOCATION: Russian Federation
TYPE OF SITE: Medieval town
SITE DETAILS: Possibly of Viking origin, it was an important city in medieval Europe. Metalwork, leather, some frescoes, and birch-bark documents have been excavated. (See p. 91.)

190 M3 SMOLENSK FORTRESS
AD 1596–1611
LOCATION: Russian Federation
TYPE OF SITE: Fortress
SITE DETAILS: Remains include impressive 4-mi- (6-km-) long walls that are 49 m (15 m) high and 18 ft (5.5 m) thick. The Korolevsky Bastion earth rampart was built by invading Poles in AD 1611.

191 O7 MAIKOP
3RD MILLENNIUM BC
LOCATION: Russian Federation
TYPE OF SITE: Burial mound
SITE DETAILS: Circular tombs, known as *kurgans*, with finds of copper, silver, and gold objects.

192 O7 KOSTROMSKAYA BARROW
7TH–6TH CENTURIES BC
LOCATION: Russian Federation
TYPE OF SITE: Steppe burial mound
SITE DETAILS: The likely burial of a local chieftain placed in an underground chamber. Another 13 human skeletons were unearthed, as were those of horses, along with an iron shield decorated with a gold plaque.

WESTERN ASIA

The map for these listings is on pp. 148–9.

1 C1 CONSTANTINOPLE
7TH CENTURY BC–15TH CENTURY AD
LOCATION: Turkey
TYPE OF SITE: Early capital
SITE DETAILS: Istanbul was a Greek colony that grew to become the city of Byzantium. It was rebuilt in the 4th century AD by Constantine and has a number of remains, including the impressive Theodosian defensive wall. The church of Santa Sophia became a mosque and is now a museum.

2 D2 GORDION
8TH CENTURY BC
LOCATION: Turkey
TYPE OF SITE: Early city
SITE DETAILS: The capital of the Phrygian people that revealed walled structures set around a courtyard. Some goods have been unearthed from its cemetery.

3 C2 SARDIS
3RD MILLENNIUM BC
LOCATION: Turkey
TYPE OF SITE: Early settlement
SITE DETAILS: Inhabited until Classical times. There are remains of a walled settlement and an acropolis from when it was controlled by the Lydian culture.

4 C2 TROY
4TH–2ND MILLENNIA BC
LOCATION: Turkey
TYPE OF SITE: Early town
SITE DETAILS: Bronze Age citadel that was a walled settlement of seven phases. Finds include the "Treasury of Priam," consisting of precious stones and metals. There have been finds of gray Minoan and Mycenaean ware.

5 C2 PERGAMON
3RD–1ST CENTURIES BC
LOCATION: Turkey
TYPE OF SITE: Early city
SITE DETAILS: An acropolis that includes the remains of an aqueduct, a theater, and a Dionysian temple. The Altar of Zeus shows a battle of the gods in relief. There is also a Roman Temple of Trajan and a sanctuary of Aesculapius.

6 C2 IZMIR
3RD MILLENNIUM BC
LOCATION: Turkey
TYPE OF SITE: Early settlement

SITE DETAILS: A Bronze Age settlement, later settled by Greeks and Romans. A Temple of Athena has been excavated, as have a Roman *agora* and *basilica*. There are remains of Byzantine and Ottoman aqueducts, and 19th-century tobacco houses and tanneries still stand.

7 C2 EPHESUS
5TH CENTURY BC
LOCATION: Turkey
TYPE OF SITE: Early city
SITE DETAILS: *See pp. 69, 76–7.*

8 C3 PRIENE
4TH CENTURY BC
LOCATION: Turkey
TYPE OF SITE: Ancient Greek town
SITE DETAILS: A coastal town in a grid plan with remains of houses, a marketplace, a stadium, and a Temple of Athena.

9 C3 MILETUS
4TH CENTURY BC
LOCATION: Turkey
TYPE OF SITE: Greek city
SITE DETAILS: Excavations have revealed administrative buildings, a theater, and statues of lions. Greek and Fikellura ware were unearthed. It was later a Roman city and an early harbor.

10 C3 APHRODISIAS
9TH CENTURY BC
LOCATION: Turkey
TYPE OF SITE: Classical city
SITE DETAILS: A major center where a Temple of Aphrodite was built on a Bronze Age mound and was turned into a Christian church under Byzantine rule. Other ruins include an *agora*, baths, and an *odeum* (concert hall).

11 D3 KAS
14TH CENTURY BC
LOCATION: Turkey
TYPE OF SITE: Ancient shipwreck
SITE DETAILS: Bronze tools and weapons; copper, tin, and glass ingots; Egyptian and Aegean jewelry; and various wares from Cyprus, Egypt, and ancient Canaan have all been salvaged.

12 D3 HACILAR
7TH MILLENNIUM BC
LOCATION: Turkey
TYPE OF SITE: Neolithic settlement
SITE DETAILS: Farming village that had finds of rectangular courtyards and hearth ovens. Bones of cattle and sheep and crops including barley and lentils were found, as were pottery and copper from later periods.

13 D2 YAZILIKAYA
13TH CENTURY BC
LOCATION: Turkey
TYPE OF SITE: Ceremonial center
SITE DETAILS: A Hittite rock sanctuary with monumental gateways and temples. There are 60 reliefs of cone-headed gods.

14 D3 CATAL HUYUK
6TH MILLENNIUM BC
LOCATION: Turkey
TYPE OF SITE: Neolithic settlement
SITE DETAILS: *See pp. 25, 26–9.*

15 E2 BOGHAZKOY
MID-2ND MILLENNIUM BC
LOCATION: Turkey
TYPE OF SITE: Hittite city
SITE DETAILS: The Hittite capital, covering several hills, that was first settled during the Middle Bronze Age, when it had large stone and mud-brick fortifications and decorated gates. Four temples have been excavated. Over 10,000 clay tablets were found.

16 E2 KANESH
2ND MILLENNIUM BC
LOCATION: Turkey
TYPE OF SITE: Hittite city
SITE DETAILS: Cuneiform tablets excavated at this site are the oldest surviving records of Turkish history.

17 F2 CAYONU
9TH–7TH MILLENNIA BC
LOCATION: Turkey
TYPE OF SITE: Neolithic settlement
SITE DETAILS: An early agricultural center where cereals were cultivated. Other finds include stone architecture, unfired clay, and hammered, native copper.

18 E3 ALALAKH
3400–1200 BC
LOCATION: Turkey
TYPE OF SITE: Bronze Age settlement
SITE DETAILS: An occupation of 17 phases that is most important for finds of two archives that illuminate the local history.

19 E3 ANTIOCH
300 BC–AD 600
LOCATION: Turkey
TYPE OF SITE: Roman city
SITE DETAILS: One of the most important Roman cities. It remained a trading capital and later became a Byzantine town, of which the walls remain.

20 F1 ABKHASIA
3RD–2ND MILLENNIA BC
LOCATION: Georgia
TYPE OF SITE: Megalithic tomb
SITE DETAILS: Novisobodnajan culture burials have been found with an antechamber or a slab-stone door.

21 G2 METSAMOR
2ND MILLENNIUM BC
LOCATION: Armenia
TYPE OF SITE: Bronze Age settlement
SITE DETAILS: The remains include a fortified citadel, a cult area, graves, and residential and industrial areas.

22 G2 KARMIR-BLUR
8TH–7TH CENTURIES BC
LOCATION: Armenia
TYPE OF SITE: Urartian fortress

SITE DETAILS: A fortified town with a citadel containing residential rooms with storerooms beneath them.

23 G2 GARNI
1ST CENTURY AD
LOCATION: Armenia
TYPE OF SITE: Ceremonial temple
SITE DETAILS: The site of a Greek-style temple dedicated to Mithras.

24 G2 LCHASHEN
2ND MILLENNIUM BC
LOCATION: Armenia
TYPE OF SITE: Steppe settlement
SITE DETAILS: Several pit graves that included a cemetery of wagon burials with large disk wheels. (*See p. 35.*)

25 E3 UGARIT
2ND MILLENNIUM BC
LOCATION: Syria
TYPE OF SITE: Bronze Age city
SITE DETAILS: Coastal settlement where temples and an elaborate palace have been excavated. Clay cuneiform tablets were found in the palace.

26 E3 AMRIT
1ST MILLENNIUM BC
LOCATION: Syria
TYPE OF SITE: Early city
SITE DETAILS: A settlement that has ruins of a Phoenician temple, a large square, and 5th-century BC funerary towers.

27 E3 KRAK DES CHEVALIERS
12TH CENTURY AD
LOCATION: Syria
TYPE OF SITE: Crusader castle
SITE DETAILS: *See pp. 91, 96–7.*

28 E3 EBLA
2ND MILLENNIUM BC
LOCATION: Syria
TYPE OF SITE: Early city
SITE DETAILS: A once-powerful state whose remains include a palace that contained state archives covering 140 years of the city's history.

29 F3 HUREYA
7TH–6TH MILLENNIA BC
LOCATION: Syria
TYPE OF SITE: Neolithic settlement
SITE DETAILS: An important center that shows the transition from hunter-gathering to agriculture and herding. There are remains of mud-brick structures with painted walls. Stone vessels, bone tools, and clay figurines have been excavated. (*See p. 25.*)

30 F3 PALMYRA
EARLY 1ST MILLENNIUM AD
LOCATION: Syria
TYPE OF SITE: Early city
SITE DETAILS: Most remains are of the Roman colony. There is a Temple of Bel, colonnaded streets, and towers with tombs.

31 E4 YABRUD
150,000 YEARS AGO
LOCATION: Syria
TYPE OF SITE: Hunter-gatherer
SITE DETAILS: Three rock shelters that give the name to a tool industry that includes scrapers, blades, and bifaces.

32 F3 TELL HALAF
1ST MILLENNIUM BC
LOCATION: Syria
TYPE OF SITE: Early city
SITE DETAILS: Aramean palace that had finds of Halaf and Dark-Faced Burnished ware within its foundations. It was later a Neo-Assyrian capital.

33 F3 MARI
3RD–1ST MILLENNIA BC
LOCATION: Syria
TYPE OF SITE: Early city
SITE DETAILS: Mesopotamian settlement with remains, including a palace of 300 painted rooms and temples, based on two courtyards. There was an archive containing over 15,000 tablets.

34 E3 AL MINA
7TH–5TH CENTURIES BC
LOCATION: Lebanon
TYPE OF SITE: Early city
SITE DETAILS: A harbor settlement, with ten levels of excavation, that has produced Greek pottery and *amphorae*.

35 E4 BAALBEK
1ST CENTURY BC
LOCATION: Lebanon
TYPE OF SITE: Roman colony
SITE DETAILS: Religious center; the ruins include a large monumental temple.

36 E4 BYBLOS
3RD MILLENNIUM BC
LOCATION: Lebanon
TYPE OF SITE: Early city
SITE DETAILS: Fortified trade center that was later occupied by the Phoenician and Achaemenid Empires, and has the remains of a number of temples.

37 E4 SIDON
3RD–2ND MILLENNIA BC
LOCATION: Lebanon
TYPE OF SITE: Phoenician city
SITE DETAILS: A one-time prosperous settlement where temples and the necropolis have been excavated, with finds of decorated sarcophagi. It was later a Roman town, and the Crusaders built two castles that are now in ruins.

38 E4 TYRE
1ST MILLENNIUM BC
LOCATION: Lebanon
TYPE OF SITE: Early city
SITE DETAILS: The site of the chief Phoenician city, although there are now only Roman and Byzantine remains, including an aqueduct, a necropolis, and a hippodrome.

39 E4 HAZOR
18TH–8TH CENTURIES BC
LOCATION: Israel
TYPE OF SITE: Early town
SITE DETAILS: Important Canaanite town with later remains from King Solomon's time (*c.* 10th century BC), including a six-chambered gateway and a storehouse. There are later remains of a water shaft and a citadel.

40 E4 ACRE
12TH CENTURY AD
LOCATION: Israel
TYPE OF SITE: Medieval port
SITE DETAILS: A walled city and port with remains including a refectory and the Khan-el Umdan – a building that was part inn, part warehouse.

41 E4 MOUNT CARMEL
60,000–50,000 YEARS AGO
LOCATION: Israel
TYPE OF SITE: Hunter-gatherer
SITE DETAILS: *See pp. 13, 22.*

42 E4 NAZARETH
11TH–12TH CENTURIES AD
LOCATION: Israel
TYPE OF SITE: Medieval town
SITE DETAILS: Excavations have found the Church of the Annunciation, built as a synagogue and later converted into a Byzantine church.

43 E4 MEGIDDO
1ST MILLENNIUM BC
LOCATION: Israel
TYPE OF SITE: Early town
SITE DETAILS: A Bronze Age settlement that became a fortified royal city in the 9th century BC. Ruins include a palace, a lime-mortared courtyard, and a six-chambered red gateway.

44 E4 CAESAREA
1ST CENTURY BC
LOCATION: Israel
TYPE OF SITE: Roman city
SITE DETAILS: A port built during the time of King Herod, it is currently under excavation, with finds of some important texts. It has later Byzantine and medieval ruins.

45 E4 KEBARA CAVE
60,000 YEARS AGO
LOCATION: Israel
TYPE OF SITE: Hunter-gatherer
SITE DETAILS: A Neanderthal dwelling that has signs of ashes and hearths and remains of a male Neanderthal.

46 E4 KETEF HINNOM
7TH CENTURY BC
LOCATION: Israel
TYPE OF SITE: Cave burial
SITE DETAILS: Aristocratic tombs located in a set of caves. Bones and grave goods were placed together in a repository.

47 E4 BEIT GUVRIN
1ST CENTURY BC
LOCATION: Israel
TYPE OF SITE: Early settlement
SITE DETAILS: Ancient remains spanning from Roman and Byzantine times to the Crusaders and Mameluks. Houses have been excavated, as have family tombs, which were often painted with frescoes.

48 E4 LACHISH
2ND MILLENNIUM BC
LOCATION: Israel
TYPE OF SITE: Early town
SITE DETAILS: A settlement with an impressive Assyrian siege ramp, gate complex, and Canaanite temple ruins. The Lachish inscriptions, using the earliest form of alphabet, were found written on a dagger and some vessels.

49 E4 JERUSALEM
2ND MILLENNIUM BC
LOCATION: Israel
TYPE OF SITE: Early city
SITE DETAILS: First settled by a Canaanite branch, it was captured by the Israelites, who set up a town. Solomon later added a temple, which was built over by Herod. Most excavation dates to the time of Herod and Hadrian, and the later Byzantine period.

50 E4 BOKER TACHTIT
45,000 YEARS AGO
LOCATION: Israel
TYPE OF SITE: Hunter-gatherer
SITE DETAILS: A Paleolithic industry where a Levallois technique was used. Finds included points, blades, and cores.

51 E4 AVDAT
2ND CENTURY BC
LOCATION: Israel
TYPE OF SITE: Early town
SITE DETAILS: A strategic Nabataean settlement that became a Roman town. It has Byzantine remains, with several restored churches. Houses were built into cliffs, with caves used as storerooms.

52 E4 MASADA
37 BC–AD 73
LOCATION: Israel
TYPE OF SITE: Fortress-palace
SITE DETAILS: Built by Herod as a retreat, remains include large water cisterns, palaces, and bathhouses. It was the site of a stronghold against the Romans, and their remains include a containment wall, eight siege camps, and a large ramp. Finds include scrolls and *ostraka* – individually named pieces of pottery.

53 E4 ALA-SAFAT
4TH MILLENNIUM BC
LOCATION: West Bank
TYPE OF SITE: Neolithic cemetery
SITE DETAILS: Contained over 200 burials, 164 of which were in chamber tombs.

54 E4 HERODIUM
1ST CENTURY BC
LOCATION: West Bank
TYPE OF SITE: Roman settlement
SITE DETAILS: Artificial hilltop where palaces, bathhouses, and administrative buildings have been excavated.

55 E4 QUMRAN
1ST CENTURY AD
LOCATION: West Bank
TYPE OF SITE: Monastic settlement
SITE DETAILS: The site of the discovery in a cave of the 400 Dead Sea Scrolls, which contain most of the Old Testament, legal documents, and other writings.

56 E4 JERICHO
10TH MILLENNIUM BC
LOCATION: West Bank
TYPE OF SITE: Early city
SITE DETAILS: The oldest city in the world, first inhabited by the Nafutian people, who built single-roomed houses and may have planted cereals. It later had a fortified wall with a circular tower and a stone-built sanctuary. Finds include human skulls, with shells inside the eye sockets, and some plain and painted ceramics. (*See p. 24.*)

57 E4 JERASH
1ST–2ND CENTURIES AD
LOCATION: Jordan
TYPE OF SITE: Roman town
SITE DETAILS: Excavations have revealed foundations of an early Hellenistic settlement – most of the ruins are Roman, including the Street of Columns and a forum.

58 E4 GHASSUL
4TH MILLENNIUM BC
LOCATION: Jordan
TYPE OF SITE: Early settlement
SITE DETAILS: Village of four occupations that have produced mud-brick houses with painted walls. There are finds of copper exploitation and decorated ware.

59 E4 ARAQ EL-AMIR
2ND CENTURY BC
LOCATION: Jordan
TYPE OF SITE: Early settlement
SITE DETAILS: Ruins include a large tell, an aqueduct, and Qasr al-Abid (The Castle of the Slave).

60 E4 KERAK OF MOAB
12TH CENTURY AD
LOCATION: Jordan
TYPE OF SITE: Medieval castle
SITE DETAILS: Crusader castle that contains tunnels, chambers, stores, and dungeons. There are also deep wells and cisterns, and the ruined Byzantine Church of Nazareth.

61 E5 PETRA
5TH CENTURY BC–2ND CENTURY AD
LOCATION: Jordan
TYPE OF SITE: Trading settlement

SITE DETAILS: A Nabatean capital with remains including rock-cut tombs, public buildings, and temples. It also has a colonnaded Roman street.

62 G3 NINEVEH
7TH CENTURY BC

LOCATION: Iraq

TYPE OF SITE: Early city

SITE DETAILS: An Achaemenid, Assyrian, and Parthian city that has many ruins. Most of the monuments are Assyrian, with palaces, carved reliefs, cuneiform inscriptions, and winged bulls. The most important finds were libraries of clay tablets found in two palaces.

63 G3 KHORSABAD
8TH CENTURY BC

LOCATION: Iraq

TYPE OF SITE: Early city

SITE DETAILS: Assyrian capital built over an existing village, with two compounds for an arsenal and a palace.

64 G3 ARPACHIYAH
c. 3500 BC

LOCATION: Iraq

TYPE OF SITE: Early settlement

SITE DETAILS: A village of small clay huts and a cemetery of 40 graves that produced shards of 'Ubaid pottery.

65 G3 BALAWAT
9TH CENTURY BC

LOCATION: Iraq

TYPE OF SITE: Early town

SITE DETAILS: A walled settlement that has temple remains and impressive bronze gates, carved with reliefs.

66 G3 NIMRUD
9TH CENTURY BC

LOCATION: Iraq

TYPE OF SITE: Early city

SITE DETAILS: Assyrian capital of 150 years, based around a rectangular enclosure. One palace contained reliefs and headed figures. The arsenal contained a temple, a palace, and administration buildings in itself. Artifacts included carved ivories, a "book" with cuneiform text, and gold jewelry.

67 G3 ZARZI
14,000 YEARS AGO

LOCATION: Iraq

TYPE OF SITE: Hunter-gatherer

SITE DETAILS: Cave with a tool industry that included backed bladelets, microliths, and notched blades.

68 G3 ASSUR
2ND MILLENNIUM BC

LOCATION: Iraq

TYPE OF SITE: Assyrian city

SITE DETAILS: Ancient center whose remains include the Ziggurat of Enlil, the Temple of Assur, and city walls. There are also three other ziggurats, 38 temples, and two palaces.

69 G3 JARMO
c. 5800 BC

LOCATION: Iraq

TYPE OF SITE: Neolithic settlement

SITE DETAILS: A farming village with traces of houses that had clay-lined pits. Finds include pottery, consisting of stone jars and other clay vessels.

70 G4 KHEIT QASIM
3RD MILLENNIUM BC

LOCATION: Iraq

TYPE OF SITE: Early settlement

SITE DETAILS: A center that includes a cemetery, a village site, and an 'Ubaid dwelling based on a T-shaped hall.

71 G4 TELL ES-SAWWAN
6TH–5TH MILLENNIA BC

LOCATION: Iraq

TYPE OF SITE: Neolithic settlement

SITE DETAILS: Village of three occupations – the first was protected by a rectangular ditch, and the final period had elaborate fortifications. T-shaped houses were separated by streets.

72 G4 SAMARRA
8TH–10TH CENTURIES AD

LOCATION: Iraq

TYPE OF SITE: Islamic town

SITE DETAILS: An Abbasid capital noted for its examples of early Islamic architecture. Other excavations have found the distinctive prehistoric pottery that became known as Samarran ware.

73 G4 TELL KHAFAJE
3RD MILLENNIUM BC

LOCATION: Iraq

TYPE OF SITE: Early settlement

SITE DETAILS: Center with two Sumerian temples, and where ritual vessels were excavated. Graves beneath the floors of houses have yielded pottery vessels.

74 G4 TELL AGRAB
3RD MILLENNIUM BC

LOCATION: Iraq

TYPE OF SITE: Mesopotamian city

SITE DETAILS: A mound with temple remains that have revealed a sanctuary and two subsidiary chambers. Artifacts yielded include votive objects, carved amulets, thousands of beads, bronze and stone sculptures, and mace heads.

75 G4 TELL HARMAL
3RD MILLENNIUM BC

LOCATION: Iraq

TYPE OF SITE: Early town

SITE DETAILS: Center of the Eshnunna people, with administrative, ritual, and residential buildings. Texts and terra-cotta lions were found.

76 G4 BABYLON
2ND–1ST MILLENNIA BC

LOCATION: Iraq

TYPE OF SITE: Mesopotamian city

SITE DETAILS: The ancient capital was destroyed and rebuilt many times. The structures include the decorated Ishtar Gate. Other ruins include a ziggurat, temples, palaces, fortifications, and a structure that may have been the Hanging Gardens. (See pp. 47, 59.)

77 G4 UKHAIDIR
8TH CENTURY AD

LOCATION: Iraq

TYPE OF SITE: Islamic settlement

SITE DETAILS: A port that has the remains of an impressive palace containing a series of courtyards.

78 G4 KISH
3RD MILLENNIUM BC

LOCATION: Iraq

TYPE OF SITE: Early Mesopotamian city

SITE DETAILS: A city-state with remains of Early Dynastic temple platforms, a cemetery, and two ziggurats. One area had three occupation levels, with Early Dynastic pottery, graves, three chariot burials, and goods including vases, weapons, and copper artifacts.

79 G4 ABU SALABIKH
3RD MILLENNIUM BC

LOCATION: Iraq

TYPE OF SITE: Early city

SITE DETAILS: Eight mounds that contained pottery and the earliest-known works of Sumerian literature.

80 G4 AGADE
3RD MILLENNIUM BC

LOCATION: Iraq

TYPE OF SITE: Early Mesopotamian city

SITE DETAILS: Independent city-state of the Akkadian dynasty, where excavation has revealed a raised, mud-brick temple.

81 G4 NIPPUR
4TH MILLENNIUM BC

LOCATION: Iraq

TYPE OF SITE: Early Sumerian city

SITE DETAILS: Remains of a fortified religious quarter, ziggurat, and temple with a number of reliefs. Pottery and cuneiform texts have been found.

82 G4 URUK
4TH MILLENNIUM BC

LOCATION: Iraq

TYPE OF SITE: Mesopotamian city

SITE DETAILS: Occupied for 4,000 years, with a number of successive cities. Excavation was based on two Sumerian ziggurats. Objects of silver, copper, and gold; seals; and amulets were found.

83 H4 LAGASH
3RD MILLENNIUM BC

LOCATION: Iraq

TYPE OF SITE: Early Mesopotamian city

SITE DETAILS: An important city site where 50,000 cuneiform texts were found, revealing vital information about Sumerian culture.

84 H4 TELL EL 'UBAID
6TH–5TH MILLENNIA BC

LOCATION: Iraq

TYPE OF SITE: Early settlement

SITE DETAILS: The type-site that gives its name to local settlements and pottery of the period. A Sumerian temple and other mounds have been excavated.

85 H5 ERIDU
4TH MILLENNIUM BC

LOCATION: Iraq

TYPE OF SITE: Early city

SITE DETAILS: A settlement that later became a Sumerian city, with some sun-dried brick temples and a ziggurat. There is a cemetery with finds of votive pottery offerings. Eridu has the longest sequence of 'Ubaid pottery in the area.

86 H5 UR
4TH–1ST MILLENNIA BC

LOCATION: Iraq

TYPE OF SITE: Mesopotamian city

SITE DETAILS: Sumerian city with a large number of ziggurats and religious and administrative buildings. Royal tombs contained wagons, attendants, and oxen, as well as ornamented figures. About 2,500 other graves contained pottery, metal weapons, tools, stone vessels, and other goods. (See p. 35.)

87 F5 TAYMA
6TH CENTURY BC

LOCATION: Saudi Arabia

TYPE OF SITE: Neo-Babylonian city

SITE DETAILS: A settlement of several walled compounds, each enclosing a small mound. Some of the stelae found have Aramaic inscriptions.

88 H6 DOSARIYAH
5TH MILLENNIUM BC

LOCATION: Saudi Arabia

TYPE OF SITE: Early settlement

SITE DETAILS: Finds of 'Ubaid pottery and chipped stone at a site where sheep, goats, and cattle were domesticated.

89 G9 ZAFAR
11TH–12TH CENTURIES AD

LOCATION: Yemen

TYPE OF SITE: Islamic town

SITE DETAILS: Ruined mosque that has domed tombs in its courtyards and elaborately decorated ceilings.

90 H9 HAJJAR BIN HUMEID
1ST MILLENNIUM BC

LOCATION: Yemen

TYPE OF SITE: Pre-Islamic settlement

SITE DETAILS: A small pit has revealed a culture that lasted 1,500 years. Walls, parts of houses, and a pottery sequence have all been excavated.

91 H9 HUREIDAH
7TH–6TH CENTURIES BC

LOCATION: Yemen

TYPE OF SITE: Ceremonial center

SITE DETAILS: A temple built of dressed-stone blocks, modified several times. There is also a group of small tombs.

92 K7 MAYSAR
LATE 3RD MILLENNIUM BC
LOCATION: Oman
TYPE OF SITE: Early settlement
SITE DETAILS: A center that has burials, settlements, fortifications, and finds of Bronze Age copper smelting.

93 K6 QURUM
8TH MILLENNIUM BC
LOCATION: Oman
TYPE OF SITE: Hunter-gatherer
SITE DETAILS: A number of open-air sites, some of which contain cemeteries.

94 J6 UMM AN NAR
LATE 3RD MILLENNIUM BC
LOCATION: United Arab Emirates
TYPE OF SITE: Early settlement
SITE DETAILS: Bronze Age center with tombs of a type prevalent in the southeast of the Arabian Peninsula.

95 I6 BARBAR
LATE 3RD–MID-2ND MILLENNIA BC
LOCATION: Bahrain
TYPE OF SITE: Ceremonial center
SITE DETAILS: Square temples situated on an oval platform. The culture gives its name to a sequence of pottery and seals.

96 I6 BAHRAIN TUMULUS FIELD
3RD–2ND MILLENNIA BC
LOCATION: Bahrain
TYPE OF SITE: Early cemetery
SITE DETAILS: Tumuli, of both single and multiroom chambers, and cist burials, linked to the Bronze Age Barbar culture.

97 I6 BETH SHAN
4TH–2ND MILLENNIA BC
LOCATION: Bahrain
TYPE OF SITE: Early city
SITE DETAILS: Huge tell of a biblical city containing several later Egyptian forts.

98 G2 BASTAM
7TH CENTURY BC
LOCATION: Iran
TYPE OF SITE: Urartian settlement
SITE DETAILS: Remains include two gateways, flanking towers, and a tower temple. Decorated gold and red polished ware have also been excavated.

99 G3 HASANLU
6TH MILLENNIUM BC
LOCATION: Iran
TYPE OF SITE: Early settlement
SITE DETAILS: A prehistoric center that later became an Iron Age fortified city, with great walls, streets, and buildings, many of them decorated internally with glazed tiles. Artifacts include bronze lion pins and cylinder seals, gray-black ceramic ware, and other objects.

100 H3 GANJ DAREH
c. 8450 BC
LOCATION: Iran
TYPE OF SITE: Neolithic settlement
SITE DETAILS: Permanent, sun-dried-brick structures, probably used as winter quarters. Finds include flint tools and pottery, skeletons, and clay animal and modeled Venus figurines.

101 H4 ASIAB
7TH MILLENNIUM BC
LOCATION: Iran
TYPE OF SITE: Neolithic settlement
SITE DETAILS: Site of a community that lived in small semicircular structures.

102 H4 GODIN TEPE
4TH MILLENNIUM BC
LOCATION: Iran
TYPE OF SITE: Early settlement
SITE DETAILS: Early town based on an oval enclosure. Pottery and tablets have been excavated. A stained *amphora* may contain the world's earliest-known wine.

103 H4 ALI KOSH
c. 7500 BC
LOCATION: Iraq
TYPE OF SITE: Neolithic settlement
SITE DETAILS: A transitional encampment with evidence of food production, and domestication of animals that has yielded flint work and potsherds.

104 H4 SUSA
1ST MILLENNIUM BC
LOCATION: Iran
TYPE OF SITE: Achaemenid city
SITE DETAILS: Inhabited from Neolithic times, the city reached its peak as an Achaemenid capital. It consists of four mounds, which include the stela of Naram-Sin, and the palace of King Darius I. An Elamite city and poorer housing are found within the mounds.

105 I3 MARLIK
2ND–1ST MILLENNIA BC
LOCATION: Iran
TYPE OF SITE: Royal cemetery
SITE DETAILS: A necropolis of 53 tombs, some containing altars and hearths. Bronze weapons, bronze and terra-cotta figurines, and gold jewelry and ornaments have been excavated.

106 H3 ROSTAMABAD
1ST MILLENNIUM BC
LOCATION: Iran
TYPE OF SITE: Early city
SITE DETAILS: Excavations have revealed stone foundations of several wooden buildings. Tombs yielded rich funerary offerings, including gold and silver vases.

107 I3 RAYY
8TH CENTURY AD
LOCATION: Iran
TYPE OF SITE: Islamic town
SITE DETAILS: A Parthian city that later became an important Islamic settlement. Ceramics and other small objects have been excavated. The Tower of Toghril (AD 1139) and a ruined tower remain.

108 H4 CHOGA ZAMBIL
13TH CENTURY BC
LOCATION: Iran
TYPE OF SITE: Early city
SITE DETAILS: Impressive site containing a huge ziggurat, probably the largest in Mesopotamia. Remains of temples have been found surrounding this ziggurat.

109 J3 TURENG TEPE
4TH–3RD MILLENNIA BC
LOCATION: Iran
TYPE OF SITE: Early city
SITE DETAILS: Huge mound that has some evidence of structures from Sassanian, Parthian, and Islamic periods. The lowest levels have revealed sun-dried bricks of Bronze Age buildings and small shards of ancient pottery. An elaborate terra-cotta drainage system was found in a Sassanian building. Smooth gray potsherds and many lapis-lazuli beads have been excavated.

110 I5 PERSEPOLIS
LATE 6TH CENTURY BC
LOCATION: Iran
TYPE OF SITE: Achaemenid palace
SITE DETAILS: A large structure whose buildings included audience halls and decorated reception rooms. Royal residences, a walled hunting park, and an ornamental lake were excavated.

111 I5 FIRUZABAD
3RD–7TH CENTURIES AD
LOCATION: Iran
TYPE OF SITE: Early city
SITE DETAILS: Sassanian city site with extensive remains, including a palace.

112 I5 SIRAF
LATE 1ST MILLENNIUM AD
LOCATION: Iran
TYPE OF SITE: Islamic port
SITE DETAILS: A deserted Islamic city in good condition that includes a great mosque, a palatial residence, defensive and military edifices, warehouses, *hamms* (bathhouses), private houses, an *imamzedeh* (shrine of a saint), and a monumental cemetery. A Sassanian fort has been found under one building. Imported Chinese ware and large amounts of glass have also been found.

113 J5 SHAHDAD
3RD MILLENNIUM BC
LOCATION: Iran
TYPE OF SITE: Early settlement
SITE DETAILS: Three cemeteries have been excavated and have revealed a stratified society. Structures excavated include semiprecious-stone workshops, copper-smelting buildings, and a later complex of workshops and furnaces.

114 J2 DJEITUN
7TH–6TH MILLENNIA BC
LOCATION: Turkmenistan
TYPE OF SITE: Neolithic settlement
SITE DETAILS: The earliest Neolithic settlement in the area – mud-brick buildings and pottery were excavated. The inhabitants subsisted on barley, wheat, sheep, goat, and cattle.

115 K3 ALTYN TEPE
5TH–3RD MILLENNIA BC
LOCATION: Turkmenistan
TYPE OF SITE: Early city
SITE DETAILS: Remains include city walls, a ziggurat, cemeteries, and areas of pottery making and metalworking.

116 K2 YAZ DEPE
2ND–1ST MILLENNIA BC
LOCATION: Turkmenistan
TYPE OF SITE: Iron Age settlement
SITE DETAILS: A complex of many layers – a citadel has been found at the lowest level. It had an extensive irrigation system and finds of pottery.

117 K2 GONUR
LATE 3RD–MID-2ND MILLENNIA BC
LOCATION: Turkmenistan
TYPE OF SITE: Fortified settlement
SITE DETAILS: Based on a central fortress and cemetery, finds of Murghab seal amulets with possible iconographic links to the Harappan civilization.

118 M2 TESHIK TASH
44,000 YEARS AGO
LOCATION: Uzbekistan
TYPE OF SITE: Hunter-gatherer
SITE DETAILS: A Paleolithic cave where a Neanderthal child burial has been excavated. Some Mousterian tool industries have also been unearthed.

119 M2 SAMARKAND
11TH CENTURY AD
LOCATION: Uzbekistan
TYPE OF SITE: Islamic city
SITE DETAILS: Known in antiquity as Afrasiah, it was mostly built in medieval times. The basic plan of the city remains, with streets that come in from the 6-mi- (10-km-) long walls. The center has some of the most impressive Islamic monuments of Central Asia, with large, colored domes and mausoleums.

120 M2 PENDJIKENT
6TH–8TH CENTURIES AD
LOCATION: Uzbekistan
TYPE OF SITE: Sogdian settlement
SITE DETAILS: A site that has revealed much information about the Sogdian civilization – it includes a citadel and a palace with a causeway linking it to the lower town, and the city walls made of rammed earth and mud-brick. Frescoes show aspects of the town's daily life.

121 M3 SAPALLI TEPE
LATE 3RD–EARLY 2ND MILLENNIA BC
LOCATION: Uzbekistan
TYPE OF SITE: Early settlement
SITE DETAILS: A site based on a central square that covers three or four Bronze Age periods. There were burials in the residential areas, four containing textiles.

122 M1 BEZEKLIK
8TH–9TH CENTURIES AD
LOCATION: Kazakhstan
TYPE OF SITE: Buddhist settlement
SITE DETAILS: A site containing extensive remains of a Buddhist settlement.

CENTRAL ASIA
The map for these listings is on pp. 150–1.

1 I1 TILLYA-DEPE
EARLY 1ST MILLENNIUM BC
LOCATION: Afghanistan
TYPE OF SITE: Iron Age settlement
SITE DETAILS: A fortified complex where a group of tombs, with gold vessels and jewelry, were excavated. (See p. 59.)

2 I1 DASHLY
LATE 3RD MILLENNIUM BC
LOCATION: Afghanistan
TYPE OF SITE: Early city
SITE DETAILS: A number of settlements covering Bronze Age, Achaemenid, and Classical periods. Structures included square fortresses with circular towers, and a palace with a complex of rooms inside. Residential areas had a concentration of pottery kilns and an area for metalworking.

3 I1 BALKH
9TH CENTURY AD
LOCATION: Afghanistan
TYPE OF SITE: Islamic town
SITE DETAILS: Remains of a settlement that include town walls and an ancient mosque with nine domes and an oratory.

4 I1 DIL'BERGINE
3RD CENTURY BC
LOCATION: Afghanistan
TYPE OF SITE: Early town
SITE DETAILS: A small Greco-Bactrian town with a round citadel built on the site of an Achaemenid town. Remains include a temple, a Buddhist sanctuary, and two large houses.

5 I2 BAMYAN
6TH CENTURY AD
LOCATION: Afghanistan
TYPE OF SITE: Buddhist settlement
SITE DETAILS: A center of commerce and trade enclosing two impressive rock-cut Buddhas, a number of artificial caves, and some later Islamic remains.

6 J1 AI KHANUM
3RD CENTURY BC
LOCATION: Afghanistan
TYPE OF SITE: Hellenistic town
SITE DETAILS: A settlement with public buildings at its center that include a large palace, temple, mausoleum, arsenal, theater, and residential quarters.

7 J1 SHORTUGHAI
3RD–2ND MILLENNIA BC
LOCATION: Afghanistan
TYPE OF SITE: Early settlement
SITE DETAILS: Prehistoric trading colony whose lowest two levels of excavation have revealed Harappan material. Later levels belong to the Beshkent culture.

8 J2 BEGRAM
2ND CENTURY BC
LOCATION: Afghanistan
TYPE OF SITE: Greek colony
SITE DETAILS: A Hellenistic town that was similar in scale to Aï Khanum.

9 I2 GHAZNI
11TH CENTURY AD
LOCATION: Afghanistan
TYPE OF SITE: Islamic town
SITE DETAILS: Buildings include the Palace of Sultan Masoud II and the "House of Lusters" – the site of luster-painted pottery. Ceramics have revealed much about the Ghaznavid and Ghurid periods.

10 I2 MUNDIGAK
5TH MILLENNIUM BC
LOCATION: Afghanistan
TYPE OF SITE: Early settlement
SITE DETAILS: Occupied until the Iron Age, this early farming village became a walled citadel with a colonnaded palace. Carved and painted pottery have been found.

11 H3 LASHKARI BAZAR
11TH CENTURY AD
LOCATION: Afghanistan
TYPE OF SITE: Islamic town
SITE DETAILS: A settlement centered around a richly decorated palace. Potsherds found at this site provide important examples of Islamic wares.

12 J2 KATELAI CEMETERY
1ST MILLENNIUM BC
LOCATION: Pakistan
TYPE OF SITE: Early cemetery
SITE DETAILS: A burial site of the Gandhara culture that has produced a large number of funerary objects.

13 J2 CHARSADA
1ST MILLENNIUM BC
LOCATION: Pakistan
TYPE OF SITE: Achaemenid town
SITE DETAILS: Ancient capital ruins where defenses have been excavated. There is a 2nd-century BC Indo-Greek settlement.

14 J2 TAXILA
1ST MILLENNIUM BC
LOCATION: Pakistan
TYPE OF SITE: Early city
SITE DETAILS: Three major settlements – two Achaemenid cities and an Indo-Greek city – have been excavated. The first, Bhir, was a walled town; the second, Sirkap, was built on a grid layout and had monasteries, temples, stupas, and a palace. It yielded ceramics, bronze and copper ornaments, tools, and seals. The last city, Sirsukh, revealed a Zoroastrian fire temple and a number of coins.

15 K2 BURZAHOM
3RD–2ND MILLENNIA BC
LOCATION: Pakistan
TYPE OF SITE: Farming village
SITE DETAILS: Covering four periods, pit dwellings with postholes and storage pits were excavated, yielding burnished pottery and a stone industry. Mud-brick structures and a stone circle, with finds of red pottery, were built in later periods.

16 J2 RAHMAN DEHRI
4TH–3RD MILLENNIA BC
LOCATION: Pakistan
TYPE OF SITE: Harappan settlement
SITE DETAILS: A town seen as a prototype of the Mature Harappan period that was later fortified.

17 J3 HARAPPA
3RD MILLENNIUM BC
LOCATION: Pakistan
TYPE OF SITE: Early city
SITE DETAILS: The site of the ancient city with monumental architecture, a citadel mound, and a residential area with housing and granaries. Cemeteries have yielded decorated red pottery. (See p. 47.)

18 I3 MEHRGARH
7TH–3RD MILLENNIA BC
LOCATION: Pakistan
TYPE OF SITE: Early settlement
SITE DETAILS: A Neolithic and Chalcolithic settlement whose remains of mud-brick architecture and a funerary platform have been excavated. Crops cultivated included wheat and barley. (See p. 25.)

19 I3 PIRAK
2ND–1ST MILLENNIA BC
LOCATION: Pakistan
TYPE OF SITE: Early city
SITE DETAILS: A settlement of mud-brick houses that was smaller than most Harappan cities. It had grain-storage silos, and geometrically designed pottery has been found. There is also evidence of early millet and rice cultivation.

20 I4 JHUKAR
2ND MILLENNIUM BC
LOCATION: Pakistan
TYPE OF SITE: Harappan settlement
SITE DETAILS: A site that has yielded much pottery and metalworking, including copper pins and a shaft-hole ax.

21 I4 SOHR DAMB
3RD MILLENNIUM BC
LOCATION: Pakistan
TYPE OF SITE: Early settlement
SITE DETAILS: A tell where flat copper axes and chisels were discovered. There was also a rock-cut cemetery that contained skeletons and pottery.

22 I4 MOHENJO-DARO
3RD–2ND MILLENNIA BC
LOCATION: Pakistan
TYPE OF SITE: Harappan city
SITE DETAILS: See pp. 47, 54–5.

23 I4 KOT DIJI
3RD MILLENNIUM BC
LOCATION: Pakistan
TYPE OF SITE: Early Harappan settlement
SITE DETAILS: A fortified settlement with buildings made of mud-brick that has yielded a number of ceramics, many of them decorated.

24 I4 CHANHU-DARO
2ND MILLENNIUM BC
LOCATION: Pakistan
TYPE OF SITE: Harappan settlement
SITE DETAILS: A Late Harappan tell that is especially important due to excavation of its craft workshops.

25 I4 AMRI
LATE 4TH MILLENNIUM BC
LOCATION: Pakistan
TYPE OF SITE: Early village
SITE DETAILS: Remains of mud-brick buildings that have yielded a number of painted ceramics – tall beakers of fine buff ware – and chert and copper tools.

26 I4 ALLAHDINO
3RD MILLENNIUM BC
LOCATION: Pakistan
TYPE OF SITE: Harappan settlement
SITE DETAILS: A small village site where excavation has revealed much about the Mature Harappan agriculture.

27 K3 BANAWALI
3RD–2ND MILLENNIA BC
LOCATION: India
TYPE OF SITE: Harappan settlement
SITE DETAILS: The earliest occupation has yielded fragments of pottery. There are remains of typical Mature Harappan urban structures from a later period, while its last period has revealed a type of ware known as Bara.

28 K3 KALIBANGAN
3RD MILLENNIUM BC
LOCATION: India
TYPE OF SITE: Harappan settlement
SITE DETAILS: Village site that has revealed evidence of ancient farming practices. A cemetery has revealed aspects of Harappan funerary rites and contained some distinctive pottery.

29 K3 MITATHAL
3RD MILLENNIUM BC
LOCATION: India
TYPE OF SITE: Early settlement
SITE DETAILS: A double mound that was an Indus Valley site, later to become occupied by the Harappan civilization. Ceramics and other artifacts were found.

30 K3 HASTINAPURA
1ST MILLENNIUM BC
LOCATION: India
TYPE OF SITE: Early settlement
SITE DETAILS: Remains of mud-brick architecture have been excavated and show the movement toward urbanism. It is an ancient Buddhist city widely mentioned in the *Mahabharata* epic, and inhabited until medieval times. Ceramics have been found on all its levels.

31 L3 ALAMGIRPUR
3RD MILLENNIUM BC
LOCATION: India
TYPE OF SITE: Early city
SITE DETAILS: A settlement with brick and wattle-and-daub architecture where the pottery and other artifacts excavated were a regional variation of the Mature Harappan type.

32 L4 AHICHCHATRA
500 BC–AD 1100
LOCATION: India
TYPE OF SITE: Early city
SITE DETAILS: Ancient city that has the remains of ramparts 3 mi (5.5 km) in diameter. Various ceramics have been excavated, including Painted Gray and Northern Black Polished wares.

33 J4 JAISALMER
9TH CENTURY AD
LOCATION: India
TYPE OF SITE: Hindu settlement
SITE DETAILS: A desert fort with many structures, including palaces and a number of Hindu temples.

34 K4 FATEHPUR SIKRI
16TH CENTURY AD
LOCATION: India
TYPE OF SITE: Islamic palace
SITE DETAILS: Palace-city combining Hindu and Islamic architecture that was built to house rulers' courts. There is a Great Mosque, a private audience hall with several baths, and water tanks. It was deserted due to lack of water.

35 K4 RANTHAMBHOR
13TH CENTURY AD
LOCATION: India
TYPE OF SITE: Islamic settlement
SITE DETAILS: Ruins of a fort containing palaces, barracks, temples, mosques, and a Muslim tomb.

36 K5 CHITOGARH
8TH–9TH CENTURIES AD
LOCATION: India
TYPE OF SITE: Hindu fortress
SITE DETAILS: A settlement with ruined pavilions and temples, situated 500 ft (152 m) above a plain. Its walls are considered some of the finest Hindu defenses. There is an immersed palace and a nine-story tower, the Jayastambha.

37 J4 KUMBHALGARH
MID-16TH CENTURY AD
LOCATION: India
TYPE OF SITE: Hindu fort
SITE DETAILS: Fortress whose huge walls enclose a large palace and 365 temples.

38 J5 AHAR
3RD–2ND MILLENNIA BC
LOCATION: India
TYPE OF SITE: Early settlement
SITE DETAILS: Structures include large stone and mud-brick buildings, and excavation has revealed that the inhabitants practiced cattle herding, agriculture, and copper smelting.

39 J5 DEVNI MORI
3RD CENTURY AD
LOCATION: India
TYPE OF SITE: Buddhist settlement
SITE DETAILS: A stupa, decorated with terra-cotta sculptures, and a monastery that are both examples of the spread of Greco-Buddhist influence.

40 J5 LOTHAL
3RD MILLENNIUM BC
LOCATION: India
TYPE OF SITE: Harappan settlement
SITE DETAILS: A fortified port whose mud-brick structures included a citadel and a granary. It is likely to have been linked with the Persian Gulf trade.

41 K5 NAGDA
8TH–6TH CENTURIES BC
LOCATION: India
TYPE OF SITE: Iron Age settlement
SITE DETAILS: Around 60 objects have been excavated, including a dagger, an ax, arrowheads, and spearheads.

42 L4 GWALIOR
6TH CENTURY AD
LOCATION: India
TYPE OF SITE: Hindu fort
SITE DETAILS: A large fortress containing a mosque, eight water tanks, six temples, and later 11th-century Hindu buildings.

43 K5 KAYATHA
3RD–2ND MILLENNIA BC
LOCATION: India
TYPE OF SITE: Harappan settlement
SITE DETAILS: The lowest level has revealed a distinctive brown-slipped ware. Excavations revealed traces of small huts and lithic and copper tools. Later levels have yielded other wares.

44 L5 KHAJURAHO
9TH CENTURY AD
LOCATION: India
TYPE OF SITE: Hindu city
SITE DETAILS: The capital of the Chandella people, with magnificently decorated Hindu and Jain water tanks and temple ruins, including the huge 11th-century Kadariya Mahadeo Temple.

45 L4 KALINJAR
10TH CENTURY AD
LOCATION: India
TYPE OF SITE: Hindu fort
SITE DETAILS: Ancient fortress with the ruins of temples, gates, and Hindu rock carvings. There is a king's palace, which yielded a number of priceless relics.

46 L5 AJAIGARH
9TH CENTURY AD
LOCATION: India
TYPE OF SITE: Hindu fort
SITE DETAILS: A Chandella settlement where a giant cliff face is covered with Hindu rock carvings. The fortress walls have large bastions and water caverns.

47 L4 KAUSAMBI
1500 BC–MID-1ST MILLENNIUM AD
LOCATION: India
TYPE OF SITE: Early city
SITE DETAILS: Early levels have revealed ocher-colored pottery, while a mud-brick wall and palace complex is dated to the 5th century BC. It was the site of one of the earliest Buddhist monasteries. *(See p. 59.)*

48 L5 BHARHUT
2ND CENTURY BC
LOCATION: India
TYPE OF SITE: Buddhist temple
SITE DETAILS: The ruins of a Buddhist brick stupa that has fine relief carvings and stone gateways situated at the four cardinal points.

49 M4 RAJGHAT
800 BC–MID-2ND MILLENNIUM AD
LOCATION: India
TYPE OF SITE: Early city
SITE DETAILS: A variety of different, distinctive wares have been excavated, covering all periods. Traces of buildings are dated to around 200 BC. The site reveals an urban development different from that of other Ganges Basin settlements.

50 M4 SARNATH
1ST MILLENNIUM BC
LOCATION: India
TYPE OF SITE: Buddhist settlement
SITE DETAILS: The town where the Buddha preached his first sermon with a stupa and a pillar erected by the emperor Asoka in the 3rd century BC.

51 M4 CHIRAND
3RD MILLENNIUM BC
LOCATION: India
TYPE OF SITE: Early settlement
SITE DETAILS: A settlement that lasted from Chalcolithic until medieval times. The first occupation was a village of bamboo huts, with a lithic industry, and red and gray burnished pottery. There are later finds of copper objects and some mud-brick architecture.

52 M4 VAISALI
5TH CENTURY BC
LOCATION: India
TYPE OF SITE: Early city
SITE DETAILS: The capital of the Licchavi Republic, consisting of two partially excavated mounds, with finds of pre-Buddhist black and red pottery, and later finds of Buddhist Northern Black Polished ware. It is believed that it was once surrounded by three walls with gates and watchtowers.

53 M4 LAURIYA NANDANGARH
242 BC
LOCATION: India
TYPE OF SITE: Monumental column
SITE DETAILS: The best preserved of all edict columns erected by the Mauryan emperor Asoka. It is topped by a seated lion and a shaft of sandstone.

54 M4 LOMAS RISHI
3RD CENTURY BC
LOCATION: India
TYPE OF SITE: Rock-cut chamber
SITE DETAILS: Situated in a cave, the chambers were constructed to imitate wood and had a barrel-vaulted interior.

55 M4 PATALIPUTRA
5TH CENTURY BC
LOCATION: India
TYPE OF SITE: Early city
SITE DETAILS: An important trading center that has been extensively excavated. It was the capital of the Mauryan and Gupta empires. Finds include a buff sandstone capital and stone columns that may have been part of a 1,000-columned hall.

56 N4 ANTICHAK
9TH–13TH CENTURIES AD
LOCATION: India
TYPE OF SITE: Medieval settlement
SITE DETAILS: A settlement that has a central brick mound with chambers in all its sides. A monastery contained 200 monastic cells. Artifacts include Buddhist and Hindu terra-cotta objects and stone images, a few Buddhist bronzes, and some finds of pottery.

57 N4 GAUR
14TH CENTURY AD
LOCATION: India
TYPE OF SITE: Fortified settlement
SITE DETAILS: An impressive mud fort with ramparts on three sides. Remains include the Dakhil Darwaza gate and a mosque.

58 K5 SANCHI
3RD CENTURY BC–12TH CENTURY AD
LOCATION: India
TYPE OF SITE: Buddhist complex
SITE DETAILS: An impressive collection of stupas, temples, and monasteries – the most impressive monument being the Great Stupa, built by Ashoka.

Many remains date to the early 1st millennium AD.

59 L5 BHIMBETKA
40,000 YEARS AGO
LOCATION: India
TYPE OF SITE: Hunter-gatherer
SITE DETAILS: Stratified site with some Paleolithic and later rock art in seven rock shelters. It has revealed a number of Acheulian tools, including hand-axes and cleavers.

60 L5 JABALPUR
4TH MILLENNIUM BC
LOCATION: India
TYPE OF SITE: Neolithic settlement
SITE DETAILS: A district with a number of finds, including debitage flakes, blades, and other remains.

61 K5 UJJAIN
5TH CENTURY BC
LOCATION: India
TYPE OF SITE: Early city
SITE DETAILS: Sacred Indian city where black-and-red and painted gray ware were excavated. The city had a 39-ft- (12-m-) high mud rampart around it.

62 K5 MANDU
14TH–15TH CENTURIES AD
LOCATION: India
TYPE OF SITE: Fortified settlement
SITE DETAILS: First settled in the 6th century AD, most buildings date from the 15th century, after an invasion by Timur. Many structures remain, including a royal palace.

63 K5 NAVDATOLI
2ND MILLENNIUM BC
LOCATION: India
TYPE OF SITE: Early settlement
SITE DETAILS: Occupied from Paleolithic to medieval times, its Chalcolithic period has been the one most studied. It has four phases of pottery. Later periods have ruins of monumental architecture.

64 K6 AJANTA CAVES
EARLY 1ST MILLENNIUM AD
LOCATION: India
TYPE OF SITE: Rock-cut temple
SITE DETAILS: Monasteries and temples that are some of the most impressive Buddhist remains in India, many with wall reliefs. The monasteries contain pillared halls, colonnades, and verandas.

65 L6 SIRPUR
6TH–8TH CENTURIES AD
LOCATION: India
TYPE OF SITE: Early city
SITE DETAILS: Former capital containing ruins that include a Hindu temple and Buddhist monasteries and temples. Bronze and stone carvings of the Mahayana pantheon have been found.

66 K6 JORWE
LATE 2ND MILLENNIUM BC
LOCATION: India
TYPE OF SITE: Early settlement
SITE DETAILS: A cultural type-site where finds include fine red pottery and an extensive copper industry. Evidence shows that the people hunted and herded cattle, buffalo, sheep, and goats, and grew barley, wheat, and legumes.

67 K6 DAIMABAD
3RD–2ND MILLENNIA BC
LOCATION: India
TYPE OF SITE: Early settlement
SITE DETAILS: Harappan, Malwa, and Jorwe ware have all been excavated. Structures here had platforms that may have been used for rituals. A copper hoard was discovered, containing wheeled animal figures and a chariot.

68 K6 NEVASA
40,000 YEARS AGO
LOCATION: India
TYPE OF SITE: Hunter-gatherer
SITE DETAILS: An open-air river site where evidence of a Nevasan tool industry was found. A later phase revealed Jorwe ware, double-urn burials, and houses with plaster floors.

69 K6 INAMGOAN
2ND MILLENNIUM BC
LOCATION: India
TYPE OF SITE: Early settlement
SITE DETAILS: A farming site laid out in a form of social hierarchy. The wattle-and-daub houses had circular platforms, and there are finds of storage vessels.

70 K6 KARLI
2ND CENTURY AD
LOCATION: India
TYPE OF SITE: Rock-cut temple
SITE DETAILS: An excellent example of a rock-cut *chaitya* (sacred) hall that has many fine carvings and columns.

71 K6 PIKLIHAL
c. 2100 BC
LOCATION: India
TYPE OF SITE: Neolithic settlement
SITE DETAILS: Site occupied for a long period, with remains that include circular beaten-earth floors surrounded by stone rings. The pottery has some similarities with Jorwe ware.

72 J7 RATNAGIRI
8TH CENTURY AD
LOCATION: India
TYPE OF SITE: Temple complex
SITE DETAILS: Buddhist settlement whose complex includes a large star-shaped stupa, some smaller stupas, two monasteries, and rectangular shrines.

73 K7 TEKKALAKOTA
2ND MILLENNIUM BC
LOCATION: India
TYPE OF SITE: Early settlement
SITE DETAILS: The earliest two levels contained circular huts with stone and mud floors. Burials, some in urns, were found around the settlement. Copper tools and household items were found.

74 L7 AMARAVATI
1ST–2ND CENTURIES AD
LOCATION: India
TYPE OF SITE: Buddhist city
SITE DETAILS: Ancient capital whose remains include a carved Andhran stupa.

75 L7 NAGARJUNAKONDA
1ST–2ND CENTURIES AD
LOCATION: India
TYPE OF SITE: Buddhist monument
SITE DETAILS: A site of megalithic culture that has several cist graves with stone enclosures. Grave goods include red and black pottery and iron tools. Most remains are Buddhist and include two apsidal shrines. There are also the ruins of the only known stadium in India.

76 L7 UTNUR
3RD MILLENNIUM BC
LOCATION: India
TYPE OF SITE: Seasonal settlement
SITE DETAILS: "Ash mounds" that were actually piles of cattle dung, which were occasionally fired during periods of seasonal occupation.

77 K7 HALLUR
2ND–1ST MILLENNIA BC
LOCATION: India
TYPE OF SITE: Early settlement
SITE DETAILS: A stratified site, covering three periods, where finds include circular structures, copper and iron tools, and megalithic graves.

78 K7 OLD GOA
16TH CENTURY AD
LOCATION: India
TYPE OF SITE: Colonial settlement
SITE DETAILS: Ruins of the old Portuguese city, which has a number of elegant 16th- to 18th-century buildings.

79 L7 MADRAS
17TH CENTURY AD
LOCATION: India
TYPE OF SITE: Colonial settlement
SITE DETAILS: A city that was the principal English station on the east coast of the Indian subcontinent, and was subsequently fortified. (See p. 101.)

80 L8 ARIKAMEDU
1ST CENTURY BC
LOCATION: India
TYPE OF SITE: Early port
SITE DETAILS: Remains of an old town and warehouses, yielding evidence of trade with the Roman Empire. Finds have included Arrentine ware, *amphorae*, and Roman coins.

81 L8 KARAIKADU
1ST CENTURY AD
LOCATION: India
TYPE OF SITE: Early town
SITE DETAILS: A settlement linked to Rome, as shown through its ware, glass beads, fragments of *amphorae*, and jars.

82 K8 COCHIN
17TH CENTURY AD
LOCATION: India
TYPE OF SITE: Colonial settlement
SITE DETAILS: Founded by the Portuguese in the 16th century, the fort still stands. European architecture is still evident in the city. The fort contains the St. Francis Church (AD 1510). (See p. 101.)

83 L9 POMPARIPPU CEMETERY
3RD CENTURY BC
LOCATION: Sri Lanka
TYPE OF SITE: Burial complex
SITE DETAILS: A cemetery that has yielded terra-cotta urns containing smaller black and red pots, bronzes, iron blades, and a number of beads.

84 L9 ANURADHAPURA
4TH CENTURY BC
LOCATION: Sri Lanka
TYPE OF SITE: Early city
SITE DETAILS: A Buddhist center with a number of remains, including temples, sculptures, and irrigation reservoirs.

85 L9 MIHINTALE
1ST CENTURY BC
LOCATION: Sri Lanka
TYPE OF SITE: Buddhist city
SITE DETAILS: Platforms that have some of the earliest Buddhist sculptures in Sri Lanka.

86 L9 POLONNURAWA
3RD CENTURY AD
LOCATION: Sri Lanka
TYPE OF SITE: Ancient city
SITE DETAILS: Capital of ancient Ceylon, which has ruins that include temples and other Buddhist structures, many of which date from the 12th century AD.

87 L9 SIGIRYA
3RD–2ND CENTURIES BC
LOCATION: Sri Lanka
TYPE OF SITE: Fortified palace
SITE DETAILS: An almost impregnable fortified capital that has remains of a palace, a monastery, and an ornamental lake supplied by a system of channels.

88 L9 BELLAN BANDI
1ST MILLENNIUM BC
LOCATION: Sri Lanka
TYPE OF SITE: Early settlement
SITE DETAILS: Site settled by the Balangoda culture, where the bodies of inhabitants were buried in pairs.

EASTERN ASIA

The map for these listings is on pp. 152–3.

1 C1 KOKOREVO
16,000–13,000 YEARS AGO
LOCATION: Siberia, Russian Federation
TYPE OF SITE: Hunter-gatherer
SITE DETAILS: Six open-air sites linked to the Kokorevo and Afontova cultures. Wedge-shaped blades, scrapers, and faunal remains were buried in alluvium.

2 C1 ABAKAN
3RD–2ND MILLENNIA BC
LOCATION: Siberia, Russian Federation
TYPE OF SITE: Early settlement
SITE DETAILS: The site of an ancient community that contained burial mounds and stone idols.

3 D1 PAZYRYK
5TH–3RD CENTURIES BC
LOCATION: Siberia, Russian Federation
TYPE OF SITE: Barrow burial
SITE DETAILS: Well-preserved frozen tombs whose contents have revealed evidence of a horse-riding, nomadic people. Remains include charcoal-filled bronze braziers. Wooden wagons were buried with the dead bodies. (See p. 59.)

4 B2 DENISOVA CAVE
25,000–14,000 YEARS AGO
LOCATION: Siberia, Russian Federation
TYPE OF SITE: Hunter-gatherer
SITE DETAILS: Recently discovered site, with scrapers and burins excavated from its uppermost layer.

5 F2 TOLBAGA
35,000–27,000 YEARS AGO
LOCATION: Siberia, Russian Federation
TYPE OF SITE: Hunter-gatherer
SITE DETAILS: An open-air site where faunal remains and artifacts, including burins, retouched blades, and scrapers, have been excavated.

6 K4 GEOGRAPHICAL SOCIETY CAVE
32,500 YEARS AGO
LOCATION: Siberia, Russian Federation
TYPE OF SITE: Hunter-gatherer
SITE DETAILS: A Paleolithic excavation that has yielded cores, flakes, and faunal remains that include those of a mammoth, a bear, and a rhinoceros.

7 F2 NOIN ULA
2ND CENTURY BC–2ND CENTURY AD
LOCATION: Mongolia
TYPE OF SITE: Barrow burial
SITE DETAILS: Graves of the nomadic Hsiung-nu people that may have been royal tombs. Finds include bronze plaques and belt ornaments.

8 A6 THOLING
15TH–16TH CENTURIES AD
LOCATION: Tibet, China
TYPE OF SITE: Buddhist city
SITE DETAILS: The ruined former capital of the Guge kingdom that has the ruins of an important monastery. Only three temples remain, although there are a number of well-preserved murals.

9 A6 TSAPARANG
16TH CENTURY AD
LOCATION: Tibet, China
TYPE OF SITE: Tibetan city
SITE DETAILS: Ruined Guge capital that has some Buddhist ruins with many well-preserved Tantric-inspired murals.

10 C4 KONGQUE RIVER BARROW
2ND MILLENNIUM BC
LOCATION: China
TYPE OF SITE: Barrow burial
SITE DETAILS: Ancient tombs that have yielded goddess figurines – many of the dead clutched smooth stones, probably some form of talisman, in their hands. There were also concentric circles of wooden stakes.

11 C4 LOULAN
EARLY 1ST MILLENNIUM AD
LOCATION: China
TYPE OF SITE: Buddhist center
SITE DETAILS: A prosperous, fortified center on the Silk Road that once had a post office, a hospital, schools, and its own government. Artifacts have been well preserved by the desert, and include a mummy, fabrics, wooden implements, and wooden tablets and papers with letters and legal records.

12 D4 MOGAO CAVES
4TH CENTURY AD
LOCATION: China
TYPE OF SITE: Cave art
SITE DETAILS: Buddhist caves that have the best-preserved cave art in China. One cave contained a copy of the earliest-known printed book.

13 E4 JIAYUGUAN PASS
14TH CENTURY AD
LOCATION: China
TYPE OF SITE: Fortified settlement
SITE DETAILS: A fortress site that was built to protect the western end of the Great Wall of China. Restored in 1988, it includes a pavilion, a temple, and an open-air theater within its walls.

14 E5 DUNHUANG
4TH–14TH CENTURIES AD
LOCATION: China
TYPE OF SITE: Rock-carved temple
SITE DETAILS: Nearly 500 Buddhist cave temples decorated with frescoes have been cut into limestone cliffs. Paintings and manuscripts have been found.

15 F5 BINGLING SI CAVES
7TH–10TH CENTURIES AD
LOCATION: China
TYPE OF SITE: Rock-carved temple
SITE DETAILS: Buddhist cave temples containing frescoes and thousands of clay sculptures and statues.

16 F5 SHUIDONGGOU
50,000–10,000 YEARS AGO
LOCATION: China
TYPE OF SITE: Hunter-gatherer
SITE DETAILS: A Late Paleolithic site that has yielded bison remains, blades, retouched flakes, and pebble tools.

17 F6 SHIZHAO
4TH–3RD MILLENNIA BC
LOCATION: China
TYPE OF SITE: Early settlement
SITE DETAILS: Maijayao center where some elegantly painted ceramics with geometric and spiral designs were found.

18 F6 FENGHUANGSHAN
2ND CENTURY BC
LOCATION: China
TYPE OF SITE: Chou burial
SITE DETAILS: A noble lady's wooden tomb, which contained 100 lacquered objects, including large eating dishes, wine containers, and wooden servant figures. Other grave goods include silk textiles, bamboo cups, bronze mirrors, and ceramics. All the goods were listed on an inventory of 74 wooden strips.

19 F6 SANXINGDUI
3RD–2ND MILLENNIA BC
LOCATION: China
TYPE OF SITE: Early city
SITE DETAILS: See pp. 47, 56.

20 F6 XINJIN
1ST–3RD CENTURIES AD
LOCATION: China
TYPE OF SITE: Eastern Han burial
SITE DETAILS: A rock-cut tomb with four grave chambers that have revealed six coffins and 156 objects, including a model of a farmer and a rice field.

21 E8 SHIZHAISHAN
LATE 1ST MILLENNIUM BC
LOCATION: China
TYPE OF SITE: Chou cemetery
SITE DETAILS: Grave site that yielded distinctive Dian bronzes, iron weapons, animal and human motifs, bronze cowrie containers, and drums.

22 F8 HUASHAN
2ND CENTURY BC
LOCATION: China
TYPE OF SITE: Rock art
SITE DETAILS: The "mountain of flowers" has rock art that depicts human figures, bronze drums, and ring-handled swords. There are 1,819 figures in all.

23 G8 CANTON (GUANGZHOU)
16TH CENTURY AD
LOCATION: China
TYPE OF SITE: Colonial settlement
SITE DETAILS: One of the oldest cities in China, it later became a trading point – Jesuits and the British set up trading factories for tea and silk. (See p. 101.)

24 G7 MAWANGDUI
1ST CENTURY BC
LOCATION: China
TYPE OF SITE: Burial mound
SITE DETAILS: Western Han tombs that were buried 66 ft (20 m) under the ground. Wall paintings depict scenes of everyday life, while gold, bronze, and jade objects were found in the tombs.

25 F6 BEISHOULING
5TH–4TH MILLENNIA BC
LOCATION: China
TYPE OF SITE: Neolithic settlement
SITE DETAILS: Yangshao village site that has house remains and finds of some ceramics and stone implements.

26 F6 ZHANGJIAPO
10TH–9TH CENTURIES BC
LOCATION: China
TYPE OF SITE: Western Zhou cemetery
SITE DETAILS: The tombs have revealed carvings, pendants, and bronze vessels. Jade objects of the Neolithic Shijiahe culture were also found, probably handed down through the generations.

27 F6 LANGJIAGOU
2ND CENTURY BC
LOCATION: China
TYPE OF SITE: Burial mound
SITE DETAILS: Western Han site where a pottery army was found next to tombs of ministers. The 1,000 figures were positioned in battle and were shown fully clothed and carrying shields.

28 G6 BANPO
EARLY 4TH MILLENNIUM BC
LOCATION: China
TYPE OF SITE: Neolithic village
SITE DETAILS: A fortified Yangshao settlement notable for its practice of swidden agriculture. There were traces of 46 houses, 90 hearths, and 200 storage pits. Finds include clay vessels, textile imprints, utensils, ornaments, and some impressive wares. Millet was cultivated, and pigs and dogs were kept.

29 G6 LINTONG

2ND CENTURY BC

LOCATION: China

TYPE OF SITE: Barrow burial

SITE DETAILS: Burial place of the first emperor of China, Qin Shi Huangdi, which once held all kinds of treasures. It is the site of a terra-cotta army of 7,000 soldiers and horses, and is thought to contain a scale model of the palace. Many of the warriors were equipped with bronze weapons.

30 G6 LANTIAN

700,000 YEARS AGO

LOCATION: China

TYPE OF SITE: Hunter-gatherer

SITE DETAILS: An Early Paleolithic site where core tools and remains of *Homo erectus* have been found.

31 G5 FAMEN TEMPLE

3RD CENTURY AD

LOCATION: China

TYPE OF SITE: Buddhist temple

SITE DETAILS: Built to house the sacred relics of Sakyamuni, the temple was destroyed by a landslide that revealed an ancient crypt containing a huge range of artifacts – 2,400 pieces of gold and silver ware, glazed ware, porcelain and pottery, jewelry, and textiles.

32 G5 TAOSI

MID-3RD MILLENNIUM BC

LOCATION: China

TYPE OF SITE: Longshan cemetery

SITE DETAILS: Burials where artifacts found include a large shallow dish with a serpent creature painted in its bowl – a depiction of the dragon, which was revered in early China. Few graves contained goods, although one tomb revealed around 200 items, including 14 ceramic vessels.

33 G6 ERLITOU

2ND MILLENNIUM BC

LOCATION: China

TYPE OF SITE: Early city

SITE DETAILS: Two palace compounds that revealed four bronze Jue vessels – wine vessels dating from the early 2nd millennium BC.

34 G6 LONGMEN CAVES

6TH–7TH CENTURIES AD

LOCATION: China

TYPE OF SITE: Rock-carved temple

SITE DETAILS: Caves containing over 100,000 images, mostly of the Buddha and his disciples.

35 G6 JINGU PARK

2ND–3RD CENTURIES AD

LOCATION: China

TYPE OF SITE: Burial

SITE DETAILS: Big, brick-built, double-chamber grave of the Late Eastern Han period that contained many artifacts, including an inscribed mirror.

36 G6 JIANXI XILIHE

2ND CENTURY AD

LOCATION: China

TYPE OF SITE: Han burial

SITE DETAILS: A T-shaped brick tomb with a number of side rooms. Among the artifacts is a large, tree-shaped lamp with several figures.

37 G6 XICHUAN

6TH CENTURY BC

LOCATION: China

TYPE OF SITE: Early cemetery

SITE DETAILS: A large main tomb, surrounded by lesser tombs, containing bronzes. Other artifacts indicate the cemetery was for important members of the Chu state. One pit contained remains of 19 horses and six chariots. Another tomb contained over 500 bronze items, including 26 bells.

38 G6 LEIGUDUN

c. 433 BC

LOCATION: China

TYPE OF SITE: Dynastic burial

SITE DETAILS: A tomb that contained the body of Marquis Yi of the Zeng state. The wood tomb had four chambers with a large, central room. Goods include a large assemblage of bells and 70 ritual bronzes, nine *shin ding* (tripods), eight *gui* (ritual vessels), a carved bird, and a painted clothes chest.

39 H7 TONGLUSHAN

1ST MILLENNIUM BC

LOCATION: China

TYPE OF SITE: Copper mine

SITE DETAILS: Mines such as this were essential for the manufacture of bronzes and other metal objects. Copper was mined mostly during the Chou Dynasty. Bronze axes and wooden hammers and shovels were found in the shafts.

40 G5 HUIXIAN

2ND MILLENNIUM BC

LOCATION: China

TYPE OF SITE: Early city

SITE DETAILS: The site of an early city that was important during the Shang Dynasty. Excavations revealed 27 intact tombs. Some chariot remains with spoked wheels were found, predating European use by about 2,000 years.

41 G5 LIULIHE

11TH–10TH CENTURIES BC

LOCATION: China

TYPE OF SITE: Early city

SITE DETAILS: The center of the Yan state, where a large cemetery was excavated. Artifacts include inscribed Zhou bronzes and bronze humanlike masks.

42 G6 ZHENGZHOU

17TH–11TH CENTURIES BC

LOCATION: China

TYPE OF SITE: Early city

SITE DETAILS: Shang capital based on a walled palace and ritual area, surrounded by housing and workshops. Cemeteries were situated outside the city walls. (*See p. 47.*)

43 G5 CISHAN

6TH MILLENNIUM BC

LOCATION: China

TYPE OF SITE: Neolithic settlement

SITE DETAILS: A village site that has much evidence of swidden agriculture use. Storage pits, pit houses, and burials have been excavated, with finds of querns, ground-stone sickles, tripods, and bone and stone fishing implements. Pigs and dogs were domesticated here.

44 G5 ANYANG

13TH CENTURY BC

LOCATION: China

TYPE OF SITE: Early city

SITE DETAILS: A large Shang royal and ceremonial center known through Shang texts, containing the great city of Yin, surrounded by 9 sq mi (24 sq km) of settlements, workshops, and cemeteries. The royal palaces and city are south of the Huan River, with the royal cemetery to the north. Hamlets, farms, and workshops have been found, with evidence of bronze and bone working, and tombs rich in elaborate bronzes and silks. (*See pp. 35, 47.*)

45 G5 XIAOTUN

LATE 2ND MILLENNIUM BC

LOCATION: China

TYPE OF SITE: Early city

SITE DETAILS: Excavations have revealed a palace complex with several building compounds containing watchtowers, altars, and halls of ancestral worship. Platform foundations include burials that may have been human sacrifices.

46 H5 XIBEIGANG

EARLY 2ND MILLENNIUM BC

LOCATION: China

TYPE OF SITE: Burial complex

SITE DETAILS: The most impressive of the Shang Dynasty cemeteries, with the tomb of Fu Hao, 11 shaft graves, and over 2,000 small graves. Finds have shown that cowrie shells were used as currency at the time. Battleaxes with animal-face designs were also found.

47 G4 YUNGANG

5TH CENTURY AD

LOCATION: China

TYPE OF SITE: Buddhist cave

SITE DETAILS: Caves with a number of Buddhist monuments that produced a number of foreign objects, including Persian and Byzantine weapons. The walls are exquisitely carved with Buddhist images, and there are large replicas of the Buddha himself.

48 G5 SHIYU

30,000 YEARS AGO

LOCATION: China

TYPE OF SITE: Hunter-gatherers

SITE DETAILS: Excavations at this site have revealed fossilized human bone, microlithic cores and blades, and the remains of 14 animal species.

49 H5 LINGSHOU

5TH–3RD CENTURIES BC

LOCATION: China

TYPE OF SITE: Early city

SITE DETAILS: Capital of the Zhongshan state, where excavations of 30 burials have taken place, recovering over 2,000 bronze objects. The burial mound of King Cuo (323–311 BC) showed that he had been accompanied by his queen, retainers, horses, chariots, and boats. A bronze plaque shows the tomb layout, which lay under some funerary temples.

50 H5 YANXIADU

4TH CENTURY BC

LOCATION: China

TYPE OF SITE: Early city

SITE DETAILS: The Yan capital, where a late Warring States tomb has been excavated, with finds of iron armor and weapons. The excavations revealed that high-carbon steel was created for the manufacture of weapons.

51 H4 ZHOUKOUDIAN CAVE

700,000–200,000 YEARS AGO

LOCATION: China

TYPE OF SITE: Hunter-gatherer

SITE DETAILS: Cave where the remains of a *Homo erectus* known as Peking Man were excavated. A later period of occupation (c. 7500 BC) has revealed some of the first personal ornaments, worked from bones of tigers, leopards, bears, and deer. They were carved and strung together to make pendants, necklaces, and earrings. (*See p. 13.*)

52 G4 GREAT WALL

2ND CENTURY BC

LOCATION: China

TYPE OF SITE: Fortification

SITE DETAILS: Construction of the 3,125-mi- (5,000-km-) long wall started in the Qin Dynasty, to link the walls of separate kingdoms. It is at Badaling, near Beijing, that the wall has been most extensively renovated. (*See p. 59.*)

53 H4 MANCHENG

c. 100 BC

LOCATION: China

TYPE OF SITE: Dynastic burial

SITE DETAILS: Rock tomb of Liu Sheng and Du Wan, who were buried in jade burial suits. The tomb was lined with wood and tiles, and filled with bronze, silver, and gold items.

54 H4 DINGLING

c. AD 1583

LOCATION: China

TYPE OF SITE: Ming tomb

SITE DETAILS: The burial complex built for the Ming emperor Wanli. The complex consisted of underground vaults, with ritual buildings above the ground. It was equipped with marble thrones and treasure chests, and has revealed a large silver hoard. Wanli was buried with two women and burial goods, including gold and silver utensils and the "phoenix crown" of the empress.

55 H4 NIUHELIANG

MID-4TH MILLENNIUM BC

LOCATION: China

TYPE OF SITE: Ritual center

SITE DETAILS: A temple complex where there have been finds of graves and clay figurines of women. Four stone-lined graves were found under some tumuli, containing jade and other objects.

56 H4 HONGSHAN

LATE 4TH MILLENNIUM BC

LOCATION: China

TYPE OF SITE: Neolithic settlement

SITE DETAILS: A type-site with substantial remains that include stone enclosures and a temple. Jade artifacts and life-size female statues were found.

57 H4 XINGLONGWA

6TH MILLENNIUM BC

LOCATION: China

TYPE OF SITE: Neolithic settlement

SITE DETAILS: An early village site where excavations have revealed artifacts, including a stone carving of a kneeling human figure.

58 I4 XINLE

LATE 6TH MILLENNIUM BC

LOCATION: China

TYPE OF SITE: Neolithic settlement

SITE DETAILS: An early village site that is characterized by millet agriculture, shell middens, and pottery similar to that in the Korean Chulmun tradition.

59 H5 SUFUTUN

2ND MILLENNIUM BC

LOCATION: China

TYPE OF SITE: Burial complex

SITE DETAILS: A Shang burial pit with one main burial, and skeletons of sacrificial victims and dogs. Grave goods include two large bronze axes, jades, pottery, and cowrie shells.

60 H6 YINGPASHAN

EARLY 3RD MILLENNIUM BC

LOCATION: China

TYPE OF SITE: Early cemetery

SITE DETAILS: Neolithic burial complex where finds have included clay human faces and jade head and neck ornaments.

61 I6 FANSHAN

3RD MILLENNIUM BC

LOCATION: China

TYPE OF SITE: Early settlement

SITE DETAILS: A Liangzhu center where a large number of tombs have been excavated. Two ritual jades – congs – whose corners are decorated with faces wearing plumed headdresses, were found. Other artifacts include disks, ceramics, stone axes, and jade objects, one accompanying each burial.

62 I7 HEMUDU

EARLY 6TH–LATE 5TH MILLENNIA BC

LOCATION: China

TYPE OF SITE: Early settlement

SITE DETAILS: A Neolithic village site where inhabitants built dwellings on wooden stilts. There are animal remains, deposits of rice grains, and straw. Evidence reveals that dogs, pigs, and buffalo were domesticated.

63 J5 LOLANG

1ST CENTURY BC–5TH CENTURY AD

LOCATION: North Korea

TYPE OF SITE: Early settlement

SITE DETAILS: A Han Dynasty colony that has left behind a large number of administrative buildings and hundreds of tombs. Excavations have revealed goods including ceramics and a number of metal-worked objects.

64 J4 KULP'O-RI

50,000 YEARS AGO

LOCATION: North Korea

TYPE OF SITE: Hunter-gatherer

SITE DETAILS: The discovery of a rock that had supported some form of hut structure provides the first evidence of habitation during this period.

65 J5 AMSADONG

5TH–2ND MILLENNIA BC

LOCATION: South Korea

TYPE OF SITE: Early settlement

SITE DETAILS: A Chulmun village where 12 pit houses were excavated. There were finds of querns and net weights that were used for fishing.

66 J5 MONGCHOON

3RD–5TH CENTURIES AD

LOCATION: South Korea

TYPE OF SITE: Early settlement

SITE DETAILS: A site that was probably an important Paekche center. Pit houses, storage pits, iron horse trappings, and bone-plated armor were excavated.

67 J5 MURYONG

EARLY 6TH CENTURY AD

LOCATION: South Korea

TYPE OF SITE: Burial mound

SITE DETAILS: A Paekche tomb that has produced a bronze chopstick-and-spoon set, toilet goods, glass sculptures, and an inscribed plaque.

68 J5 SINAN

14TH CENTURY AD

LOCATION: South Korea

TYPE OF SITE: Shipwreck

SITE DETAILS: A trading shipwreck whose salvage has resulted in finds of over 17,000 pieces of pottery, 20 tons (18.3 tonnes) of coins, and other objects made of stone, lacquer, and metal.

69 J6 SONGGUK

5TH CENTURY BC

LOCATION: South Korea

TYPE OF SITE: Agricultural settlement

SITE DETAILS: A farming village site where two pit houses and a stone cist burial were excavated.

70 J6 WANDO

11TH CENTURY AD

LOCATION: South Korea

TYPE OF SITE: Shipwreck

SITE DETAILS: The wreck of a traditional Korean ship that contained over 30,000 Celadon vessels.

71 J5 OSAN-NI

7TH MILLENNIUM BC

LOCATION: South Korea

TYPE OF SITE: Neolithic settlement

SITE DETAILS: An early village site that produced some of the earliest finds of Chulmun pottery.

72 J5 ANAPCHI

7TH CENTURY AD

LOCATION: South Korea

TYPE OF SITE: Palace complex

SITE DETAILS: A large palace linked to the Silla Kingdom. The excavation of the garden pond revealed Buddhist images, dugout boats, pottery, wooden vessels, inscribed wooden tablets, and tiles.

73 J5 HEAVENLY HORSE TOMB

5TH CENTURY AD

LOCATION: South Korea

TYPE OF SITE: Burial mound

SITE DETAILS: A Silla tomb that consisted of an internal wooden chamber covered with earth and pebbles. It contained the coffin of a man, wearing a gold crown and girdle of decorated beads. Grave goods included horse trappings, lacquer ware, glass cups, and ox horns.

74 M4 SHIRATAKI

13,000 YEARS AGO

LOCATION: Japan

TYPE OF SITE: Hunter-gatherer

SITE DETAILS: Type-site of the Yubetsu culture and microlithic industry, whose tools have been found at sites as far west as Mongolia and as far north as Alaska.

75 L4 KAMEGAOKA

10TH–4TH CENTURIES BC

LOCATION: Japan

TYPE OF SITE: Early settlement

SITE DETAILS: A Jomon village site that

yielded lacquered wooden vessels, stone ornaments, and ceramic figurines.

76 L4 BABADAN

70,000–50,000 YEARS AGO

LOCATION: Japan

TYPE OF SITE: Hunter-gatherer

SITE DETAILS: An excavation that brought controversy, as its finds of stone tools may suggest an earlier human presence in Japan than previously thought.

77 L5 FUDODO

3RD MILLENNIUM BC

LOCATION: Japan

TYPE OF SITE: Early settlement

SITE DETAILS: A Middle Jomon village site where 16 postholes of a large, oval building were found, along with four stone-lined hearths.

78 L5 IWAJUKU

22,000–17,000 YEARS AGO

LOCATION: Japan

TYPE OF SITE: Hunter-gatherer

SITE DETAILS: A Paleolithic site that has yielded blades, scrapers, and choppers.

79 L5 UBAYAMA

3RD MILLENNIUM BC

LOCATION: Japan

TYPE OF SITE: Early settlement

SITE DETAILS: Shell mound that revealed six different types of pit, suggesting a variety of types of housing. One pit contained five skeletons. The site gives its name to a type of Jomon pottery.

80 L5 KAMIHONGO MOUNDS

LATE 4TH MILLENNIUM BC

LOCATION: Japan

TYPE OF SITE: Early settlement

SITE DETAILS: Nine mounds that have yielded some finds of Ubayama pottery.

81 L5 KASORI

5TH MILLENNIUM BC

LOCATION: Japan

TYPE OF SITE: Early settlement

SITE DETAILS: A huge midden whose alternating layers of shell and ash suggest that it was originally a seasonal camp and primarily a fishing settlement. About 50 pit houses have been excavated.

82 L5 IDOJIRI

3RD MILLENNIUM BC

LOCATION: Japan

TYPE OF SITE: Early settlement

SITE DETAILS: A scattering of 50 localities that are representative of the Middle Jomon period, yielding a large number of stone querns and food remains.

83 K5 TORIHAMA

7TH–5TH MILLENNIA BC

LOCATION: Japan

TYPE OF SITE: Early settlement

SITE DETAILS: Waterlogged site that has

preserved horticultural practices and artifacts made of bone and wood, including canoes, paddles, hunting bows, tool hafts, and bowls.

84 K6 KARAKO
LATE 1ST MILLENNIUM BC
LOCATION: Japan
TYPE OF SITE: Farming settlement
SITE DETAILS: Excavations have shown that this was a rice-farming Yayoi village, surrounded by a double moat. Some storage pits have been excavated.

85 K6 HORYUJI
7TH CENTURY AD
LOCATION: Japan
TYPE OF SITE: Monastic settlement
SITE DETAILS: An island monastery that shows the strong influence of the continental mainland. It burned down in AD 670, and was rebuilt in 711.

86 K6 HEIJO
8TH CENTURY AD
LOCATION: Japan
TYPE OF SITE: Early city
SITE DETAILS: Palace and capital of the Ritsuryo state that has many ancient Buddhist buildings, including the five-story Kofuku Temple, a large bronze statue, and the Shinto Grand Shrine of Kasuga. Many artifacts were preserved in a log storehouse for over 1,000 years.

87 K6 TAKAMATSUZUKA
LATE 7TH CENTURY AD
LOCATION: Japan
TYPE OF SITE: Burial mound
SITE DETAILS: A Kofun mounded tomb that has a stone chamber with painted murals of Chinese directional symbols.

88 K6 FUJINOKI
7TH CENTURY AD
LOCATION: Japan
TYPE OF SITE: Burial mound
SITE DETAILS: A mounded tomb in which a stone sarcophagus contained a collection of horse trappings and bronze ornaments.

89 K6 ASUKA
6TH–7TH CENTURIES AD
LOCATION: Japan
TYPE OF SITE: Palace complex
SITE DETAILS: A large Yamato site whose buildings include the earliest Japanese Buddhist temple, the Asuka-dera.

90 K6 FUJIWARA
7TH CENTURY AD
LOCATION: Japan
TYPE OF SITE: Early city
SITE DETAILS: A capital that was modeled on the Chinese city of Chang'an, but was abandoned after only 20 years.

91 K6 NINTOKU
5TH CENTURY AD
LOCATION: Japan
TYPE OF SITE: Burial mound
SITE DETAILS: A large, keyhole-shaped shrine, measuring 1,591 ft (485 m), that contains the body of Emperor Nintoku.

92 K6 KAMIKUROIWA
9TH MILLENNIUM BC
LOCATION: Japan
TYPE OF SITE: Early settlement
SITE DETAILS: Jomon rock shelter that has yielded early ceramics and some portable art depicting female figures.

93 J6 DAZAIFU
8TH CENTURY AD
LOCATION: Japan
TYPE OF SITE: Administrative center
SITE DETAILS: A bureaucratic center linked to the Risuryo state. Foundations of some of the buildings were excavated.

94 J6 FUKUI
12,700 YEARS AGO
LOCATION: Japan
TYPE OF SITE: Hunter-gatherer
SITE DETAILS: A cave that yielded the world's earliest pottery, linked to the Jomon culture. Obsidian cord-marked microliths were also excavated.

95 J6 YOSHINOGARI
1ST CENTURY BC
LOCATION: Japan
TYPE OF SITE: Yayoi settlement
SITE DETAILS: A village site that has revealed watchtowers, pit-house remains, and storage pits. Cemeteries have yielded 2,000 jars. One burial contained a bronze dagger and beads.

96 K6 SAITOBARU
5TH–6TH CENTURIES AD
LOCATION: Japan
TYPE OF SITE: Cemetery
SITE DETAILS: Together covering several square miles, many of the large burial mounds are keyhole-shaped.

97 I8 DAPENKENG
6TH–5TH MILLENNIA BC
LOCATION: Taiwan
TYPE OF SITE: Early settlement
SITE DETAILS: Village site that yielded textured pottery. Excavated evidence shows that plant cordage was used to decorate pots and to make fishnets.

98 I8 FENGBITOU
MID-3RD–EARLY 2ND MILLENNIA BC
LOCATION: Taiwan
TYPE OF SITE: Neolithic settlement
SITE DETAILS: Excavated center of the Fengbitou Neolithic culture.

99 G8 MACAO
16TH CENTURY AD
LOCATION: Macao
TYPE OF SITE: Colonial settlement
SITE DETAILS: Portuguese trading center that still contains early architecture, including the ruins of São Paulo and a hill fortress. (See p. 101.)

SOUTHEAST ASIA
The map for these listings is on pp. 154–5.

1 F1 HALIN
1ST CENTURY AD
LOCATION: Myanmar
TYPE OF SITE: Early city
SITE DETAILS: An ancient city with walled fortifications that was surrounded by a moat. It includes Buddhist and Brahmanist remains.

2 F2 PAGAN
9TH CENTURY AD
LOCATION: Myanmar
TYPE OF SITE: Early city
SITE DETAILS: Buddhist religious center, now mostly in ruins, containing around 5,000 temples and stupas. The earliest buildings standing are a couple of *nat* shrines, and the books of the monastery were stored in an 11th-century library.

3 F2 PEIKTHANO
1ST–5TH CENTURIES AD
LOCATION: Myanmar
TYPE OF SITE: Buddhist settlement
SITE DETAILS: Early Buddhist center that has remains of over 100 brick structures, surrounded by a mud-brick wall. There are brick crypt mounds that contained a number of burial urns. The remains of a monastery are found nearby.

4 F2 SRI KSETRA
7TH CENTURY AD
LOCATION: Myanmar
TYPE OF SITE: Early city
SITE DETAILS: An ancient capital and religious center that was circular in shape. It included the Shwesandaw pagoda, which was surrounded by 83 other temples.

5 F2 PADAH LIN
13,000 YEARS AGO
LOCATION: Myanmar
TYPE OF SITE: Hunter-gatherer
SITE DETAILS: Cave site that produced a number of finds, including a Hoabinhian assemblage and other remains.

6 F3 PEGU
9TH CENTURY AD
LOCATION: Myanmar
TYPE OF SITE: Early city
SITE DETAILS: The former capital of the Mon kingdom, whose ruins include an old wall and a moat.

7 F3 THATON
EARLY 1ST MILLENNIUM AD
LOCATION: Myanmar
TYPE OF SITE: Early city
SITE DETAILS: A port that was subject to Indian influence and that later became a Buddhist center.

8 G2 WAT PADEANG
15TH CENTURY AD
LOCATION: Thailand
TYPE OF SITE: Temple complex
SITE DETAILS: A religious site with 25 mounds of ruined monuments. There are also four lines of ramparts.

9 G2 SPIRIT CAVE
10TH MILLENNIUM BC
LOCATION: Thailand
TYPE OF SITE: Hunter-gatherer
SITE DETAILS: An early site of occupation, where the people lived on shellfish and fish, almonds, and butternuts. Finds include evidence of plant cultivation, stone tools, and pottery. (See p. 13.)

10 G3 SUKOTAI
13TH CENTURY AD
LOCATION: Thailand
TYPE OF SITE: Buddhist center
SITE DETAILS: Early Thai state whose ruins include several temples, Buddhist images, and glazed stoneware vessels.

11 G3 U-THONG
12TH CENTURY AD
LOCATION: Thailand
TYPE OF SITE: Buddhist center
SITE DETAILS: Large moated settlement in the middle of a trade route that has a number of Hindu and Buddhist ruins. It lends its name to a style of Buddhist art.

12 G3 BAN KAO
9TH MILLENNIUM BC
LOCATION: Thailand
TYPE OF SITE: Neolithic cemetery
SITE DETAILS: A major burial of 44 graves where polished stone adzes and pottery have been unearthed. (See p. 25.)

13 G4 NAKHON PATHOM
7TH–10TH CENTURIES AD
LOCATION: Thailand
TYPE OF SITE: Buddhist center
SITE DETAILS: Large, moated settlement that may well have been a capital of Dvaravati state, and includes substantial Buddhist monuments. Bronzes, including a chandelier, bells, and cymbals, have been found.

14 G3 PHO NOI
2ND MILLENNIUM BC
LOCATION: Thailand
TYPE OF SITE: Early cemetery
SITE DETAILS: Three areas of the site were excavated and contained 26 burials. Finds include pottery vessels, polished stone adzes, and stone and shell beads.

15 G3 LOPBURI
2ND MILLENNIUM BC
LOCATION: Thailand
TYPE OF SITE: Early settlement
SITE DETAILS: A site positioned next to areas producing copper and bronze. In the 11th–13th centuries AD, it became a major cultural center.

16 G3 THE PORTUGUESE CAMP
16TH CENTURY AD
LOCATION: Thailand
TYPE OF SITE: Colonial settlement
SITE DETAILS: A Portuguese community, resident at the capital of the time, that had three churches. Artifacts include beads, coins, pottery, and a cross.

17 G4 KHOK PHNOM DI
2ND MILLENNIUM BC
LOCATION: Thailand
TYPE OF SITE: Early settlement
SITE DETAILS: A mound with evidence of rice growing and domesticated animals. Finds include pottery, shells, beads, polished adzes, anvils, and the burials of over 150 people.

18 G4 KO SI CHANG
16TH CENTURY AD
LOCATION: Thailand
TYPE OF SITE: Shipwreck
SITE DETAILS: Salvage operations have recovered remains of the ship, earthenware, Chinese ceramics, bronze objects, and fragments of lacquerware.

19 H2 BAN CHIANG
3RD MILLENNIUM BC
LOCATION: Thailand
TYPE OF SITE: Early settlement
SITE DETAILS: Occupied over a number of periods, the site has produced bronze artifacts made using sophisticated metalworking techniques and decorated pottery, including red-on-buff and incised black ware.

20 H3 BAN NA DI
1500 BC–AD 200
LOCATION: Thailand
TYPE OF SITE: Early settlement
SITE DETAILS: Village built on a mound that has produced a number of burials. One has a male accompanied by a vessel containing fishbone fragments and clay figurines. (*See p. 35.*)

21 G3 NON PRAW
10TH–6TH CENTURY BC
LOCATION: Thailand
TYPE OF SITE: Early cemetery
SITE DETAILS: A Bronze Age inhumation cemetery of four layers, incorporating 25 burials. Goods include pottery vessels, shell bracelets, and disk beads. Later burials include bronzes, mostly bracelets and socketed axes.

22 H3 NON NOK THA
2ND MILLENNIUM BC
LOCATION: Thailand

TYPE OF SITE: Early cemetery
SITE DETAILS: The graves of 217 people, with nine revealing evidence of bronze-working. Goods include pottery, bronze-socketed axes, pig bones, stone adze heads, and grinding stones.

23 H3 PHIMAI
2ND CENTURY BC
LOCATION: Thailand
TYPE OF SITE: Early settlement
SITE DETAILS: Moated iron-working and trading center that has yielded finds of large quantities of pottery. It later became a major Khmer center in the 11th–13th centuries AD.

24 H3 BURIRAM
EARLY 1ST MILLENNIUM AD
LOCATION: Thailand
TYPE OF SITE: Early city
SITE DETAILS: A center that has a number of Buddhist and Brahmanist remains.

25 G5 TAKUAPA
EARLY 2ND MILLENNIUM AD
LOCATION: Thailand
TYPE OF SITE: Buddhist center
SITE DETAILS: Settlement and religious center with links to south India. A document was found with Tamil inscription. Its structures include the Prasat Thom temple and a reservoir.

26 H4 KOH KER
EARLY 10TH CENTURY AD
LOCATION: Cambodia
TYPE OF SITE: Early center
SITE DETAILS: The site of Jayavarman IV's capital, which once had a great pyramid at its center to identify it as a royal city.

27 H4 ANGKOR
9TH–13TH CENTURIES AD
LOCATION: Cambodia
TYPE OF SITE: Temple complex
SITE DETAILS: The Khmer capital is one of the most impressive monumental sites in Asia. Its gigantic temples incorporated both Hindu and Buddhist influences. Most of the surviving buildings date from the 12th century and are elaborately structured and carved in reliefs.

28 H4 SAMBOR PREI KUK
7TH–8TH CENTURIES AD
LOCATION: Cambodia
TYPE OF SITE: Early city
SITE DETAILS: The capital of the state of Isanapura where around 100 structures were based on three central plazas. Temples have been excavated, yielding artifacts, including a stone cat.

29 H4 LAANG SPEAN
7TH–1ST MILLENNIA BC
LOCATION: Cambodia
TYPE OF SITE: Hunter-gatherer
SITE DETAILS: A prehistoric cave site that has revealed pottery and a Hoabinhian tool industry.

30 H4 SAMRONG SEN
LATE 2ND MILLENNIUM BC
LOCATION: Cambodia
TYPE OF SITE: Early settlement
SITE DETAILS: Center with a number of burials that show the transition from stone use to bronze use. Artifacts include stone adzes, bronze sickle molds, axes, fishhooks, and bells.

31 H2 XIANG KHOUANG
300 BC–AD 300
LOCATION: Laos
TYPE OF SITE: Early settlement
SITE DETAILS: The "Plain of Jars," which has a number of ritual sites and burials in stone jars – grave goods include glass beads, bronze jewelry, arrowheads, spearheads, and cowrie shells.

32 H2 SON VI
20,000–11,000 YEARS AGO
LOCATION: Vietnam
TYPE OF SITE: Hunter-gatherer
SITE DETAILS: Site that gives its name to a tool industry that predates the Hoabinhian and that is recognized for its unifacially flaked pebbles, choppers, side scrapers, and round-edged pebbles.

33 H2 PHUNG NGUYEN
MID-3RD–MID-2ND MILLENNIA BC
LOCATION: Vietnam
TYPE OF SITE: Early settlement
SITE DETAILS: A center that shows the switch from Neolithic to Bronze Age cultures. It is characterized by its polished stone adzes and traces of bronze.

34 H2 CO LOA
1ST MILLENNIUM BC
LOCATION: Vietnam
TYPE OF SITE: Early city
SITE DETAILS: A large settlement that may have been the capital of the early state of Au Lac. There are two sets of moats and three sets of ramparts. A large bronze drum was excavated.

35 H2 DONG DAU
15TH–13TH CENTURIES BC
LOCATION: Vietnam
TYPE OF SITE: Bronze Age settlement
SITE DETAILS: Site that gives its name to an early bronze-working period with finds of a number of bronze items. A number of stone molds, used to make labor tools and weapons, were found.

36 H2 HOA BINH
10TH–3RD MILLENNIA BC
LOCATION: Vietnam
TYPE OF SITE: Hunter-gatherer
SITE DETAILS: Cave site where a stone-tool industry was discovered, and that gives its name to the Hoabinhian artifacts that are prevalent in the area.

37 I3 TRA KIEU
3RD CENTURY AD
LOCATION: Vietnam
TYPE OF SITE: Early city

SITE DETAILS: A ruined center of the Champa kingdom that had a number of religious buildings. Sculptures date to the beginning of the 10th century and are fine examples of Champa art.

38 I3 SA HUYNH
600 BC–AD 100
LOCATION: Vietnam
TYPE OF SITE: Early town
SITE DETAILS: The site of a coastal trading settlement. An urnfield cemetery has jar burials with artifacts that include earrings and double-headed, jade animal pendants.

39 I4 KAUTHARA
8TH–9TH CENTURIES AD
LOCATION: Vietnam
TYPE OF SITE: Early city
SITE DETAILS: A Champa settlement that was centered around six sanctuaries.

40 H4 OC EO
2ND–7TH CENTURIES AD
LOCATION: Vietnam
TYPE OF SITE: Trading settlement
SITE DETAILS: A port linked to the state of Fukan. It was an urban enclosure made up of five rectangular ramparts and four moats, linked to other sites by canals. Artifacts include imported Chinese bronzes, Roman medallions, Indian beads, seals, and other jewelry.

41 G5 GUNUNG JERAI
LATE 1ST MILLENNIUM AD
LOCATION: Malaysia
TYPE OF SITE: Palace complex
SITE DETAILS: The site of a palace with a sanctuary from where a flight of steps leads up to terraces and temple precincts.

42 G5 KEDAH
5TH CENTURY AD
LOCATION: Malaysia
TYPE OF SITE: Early port
SITE DETAILS: A settlement that was initially a trading post and has some shrines built along the river.

43 G6 GUA CHA
8TH MILLENNIUM BC
LOCATION: Malaysia
TYPE OF SITE: Hunter-gatherer
SITE DETAILS: Rock shelter that was used for burials, where a number of stone tools have been excavated.

44 H6 PONTIAN
EARLY 1ST MILLENNIUM AD
LOCATION: Malaysia
TYPE OF SITE: Burial mound
SITE DETAILS: A boat burial that has pottery with decorative motifs that are similar to those found at Oc Èo.

45 G6 KAMPONG SUNGEI
c. 3000 BC
LOCATION: Malaysia
TYPE OF SITE: Early settlement
SITE DETAILS: A transitional settlement

whose excavation has revealed a boat burial. The goods include two Dong Son bronze drums and iron objects.

46 G6 MALACCA
16TH CENTURY AD
LOCATION: Malaysia
TYPE OF SITE: Colonial settlement
SITE DETAILS: The former trading center of Southeast Asia, with some Dutch and Portuguese remains. The *stadthuys* (statehouse) contains a number of relics. The Porta de Santiago is all that is left of the Portuguese fort. (*See p. 101.*)

47 J6 SANTUBONG
10TH–13TH CENTURIES AD
LOCATION: Malaysia
TYPE OF SITE: Trading settlement
SITE DETAILS: Remains of a major port that had a local iron industry and exported gold and forest products. There have been finds of local pottery.

48 J6 NIAH CAVE
40,000 YEARS AGO
LOCATION: Malaysia
TYPE OF SITE: Hunter-gatherer
SITE DETAILS: Major prehistoric deposits of human remains that included one of the oldest *Homo sapiens* skulls. There are other burials from 12,000–5000 BC. There are later extended burials in wooden coffins, jar burials, and cremations (1500 BC–AD 1000).

49 J6 KOTA BATU
10TH–15TH CENTURIES AD
LOCATION: Malaysia
TYPE OF SITE: Trading settlement
SITE DETAILS: Ancient port site that became the next major port after Santubong (*see above*).

50 K6 TINGKAYU
28,000–17,000 YEARS AGO
LOCATION: Malaysia
TYPE OF SITE: Hunter-gatherer
SITE DETAILS: Site of an ancient tool industry positioned by a lake. The pebble and flake industry is characterized by bifacially flaked knives and large tabular bifaces.

51 K6 BATURONG CAVES
19,000–14,000 YEARS AGO
LOCATION: Malaysia
TYPE OF SITE: Hunter-gatherer
SITE DETAILS: Rock shelter that revealed a stone-tool industry, whose tools were characterized by long, bladelike knives.

52 K6 MADAI CAVES
9TH MILLENNIUM BC
LOCATION: Malaysia
TYPE OF SITE: Hunter-gatherer
SITE DETAILS: Prehistoric caves where flake and pebble tools were discovered. Some stone-flake tools and pottery were also found.

53 F6 LOBO TUWA
11TH–12TH CENTURIES AD
LOCATION: Indonesia
TYPE OF SITE: Trading settlement
SITE DETAILS: The remains of a Srivijaya port that has yielded Chinese stoneware and Middle Eastern glass.

54 G6 PADANG LAWAS
12TH–14TH CENTURIES AD
LOCATION: Indonesia
TYPE OF SITE: Buddhist center
SITE DETAILS: A complex of Buddhist temples that were probably connected to the Srivijaya trading empire.

55 G7 TIANGKO PANJANG
9TH MILLENNIUM BC
LOCATION: Indonesia
TYPE OF SITE: Hunter-gatherer
SITE DETAILS: Cave that revealed an obsidian microlith industry. There are remains of pottery in its upper levels.

56 H7 MAURA JAMBI
10TH–13TH CENTURIES AD
LOCATION: Indonesia
TYPE OF SITE: Trading settlement
SITE DETAILS: Largest port of the Srivijaya state that has a number of temple remains. Excavations have revealed Chinese ceramics and coins.

57 H8 PALEMBANG
7TH CENTURY AD
LOCATION: Indonesia
TYPE OF SITE: Trading settlement
SITE DETAILS: Major trading center that was the capital of the state of Srivijaya; 7th-century inscriptions were found.

58 H8 PASEMAH HIGHLANDS
1ST CENTURY AD
LOCATION: Indonesia
TYPE OF SITE: Burial complex
SITE DETAILS: Megalithic statues that depict figures, including squatting humans and animals. Nearby are stone graves, with fragments of paintings showing warriors and buffalo.

59 H8 PUGUNGRAHARJO
9TH CENTURY AD
LOCATION: Indonesia
TYPE OF SITE: Megalithic complex
SITE DETAILS: Remains of standing stones that show Buddhist and Hindu influence. Evidence shows that the area was inhabited until the 16th century.

60 I8 BATAVIA
17TH CENTURY AD
LOCATION: Indonesia
TYPE OF SITE: Colonial settlement
SITE DETAILS: A colony set up by the Dutch on the ruins of ancient Jakarta. The site was fortified and became the center of the Dutch East India Company.

61 I8 BUNI
EARLY 1ST MILLENNIUM AD
LOCATION: Indonesia
TYPE OF SITE: Trading settlement
SITE DETAILS: Early spice-trading port, where excavated burials show that there was a bronze-working industry, with finds of rouletted ware and crucibles.

62 I9 DIENG
8TH–9TH CENTURIES AD
LOCATION: Indonesia
TYPE OF SITE: Ceremonial center
SITE DETAILS: A large Hindu temple complex that contains a number of stone monuments.

63 I9 SANGIRAN
700,000 YEARS AGO
LOCATION: Indonesia
TYPE OF SITE: Hunter-gatherer
SITE DETAILS: The locality of a small flake industry that is a key site for the study of faunal and human evolution.

64 I9 BOROBUDUR
8TH–9TH CENTURIES AD
LOCATION: Indonesia
TYPE OF SITE: Temple complex
SITE DETAILS: Largest Buddhist temple in the world, it was built in the form of a stupa and incorporated 432 statues of the Buddha. It has over 3 mi (5 km) of wall carvings.

65 I9 PRAMBANAN
8TH–9TH CENTURIES AD
LOCATION: Indonesia
TYPE OF SITE: Ceremonial center
SITE DETAILS: Hindu and Buddhist temple complex that is similar to Borobudur (*see above*). Its outer court contains the ruins of 224 temples.

66 J9 TRINIL
1 MILLION–500,000 YEARS AGO
LOCATION: Indonesia
TYPE OF SITE: Hunter-gatherer
SITE DETAILS: An extremely important site where the first *Homo erectus* skull was found – the first example of this early hominid ever found.

67 J9 MOJOTOKERTO
1 MILLION YEARS AGO
LOCATION: Indonesia
TYPE OF SITE: Hunter-gatherer
SITE DETAILS: A site where the remains of a *Homo erectus* child were excavated.

68 J9 GILIMANUK
3RD MILLENNIUM BC
LOCATION: Indonesia
TYPE OF SITE: Early settlement
SITE DETAILS: First settled in Paleolithic times. Burials have been excavated and have revealed double-jar burials, with goods made of bronze, iron, gold, shell, glass, and baked clay. Pottery and beads were also found.

69 K9 SEMBIRAN
1ST–12TH CENTURIES AD
LOCATION: Indonesia
TYPE OF SITE: Early port
SITE DETAILS: A trading center that became especially important as a port on the spice routes. It has produced objects including rouletted ware and glass beads.

70 L8 ULU LEANG
8TH–7TH MILLENNIA BC
LOCATION: Indonesia
TYPE OF SITE: Early settlement
SITE DETAILS: A cave site that has produced finds of a tool and blade industry, and finds of distinctive pottery.

71 L7 KALUMPANG
c. 3000 BC
LOCATION: Indonesia
TYPE OF SITE: Neolithic settlement
SITE DETAILS: An early village site where pottery has been found.

72 M6 PASO
7TH MILLENNIUM BC
LOCATION: Indonesia
TYPE OF SITE: Shell midden
SITE DETAILS: The site of a lakeshore shell midden that has yielded obsidian flake tools and bone points.

73 K5 TABON CAVES
24,000 YEARS AGO
LOCATION: Philippines
TYPE OF SITE: Hunter-gatherer
SITE DETAILS: Cave sites that have unearthed early skeletal remains. The upper layers have revealed jar burials from the 1st millennium BC.

74 L4 KALANAY
1ST MILLENNIUM BC
LOCATION: Philippines
TYPE OF SITE: Early burial
SITE DETAILS: Cave that has unearthed jar burials. The site gives its name to a type of ware dating from 400 BC to AD 1500.

75 L4 BATUNGAN CAVES
10TH–MID-8TH CENTURIES BC
LOCATION: Philippines
TYPE OF SITE: Early settlement
SITE DETAILS: One of the earliest local appearances of decorated pottery, similar to the Lapita ware of Micronesia.

76 L3 MUSANG CAVE
10,000–3500 BC
LOCATION: Philippines
TYPE OF SITE: Hunter-gatherer
SITE DETAILS: Cave site represented by a tool-flake industry. It was followed by a later Neolithic assemblage (3500 BC.)

77 L3 DIMOLIT
c. 2500 BC
LOCATION: Philippines
TYPE OF SITE: Early settlement
SITE DETAILS: Open-air site where pottery has been found that is similar to finds at Musang Cave (*see above*).

THE PACIFIC & AUSTRALASIA

The map for these listings is on pp. 156–7.

1 A6 MANDU MANDU CREEK
34,000–19,000 YEARS AGO
LOCATION: Australia
TYPE OF SITE: Hunter-gatherer
SITE DETAILS: Rock shelter that has yielded over 5,000 artifacts, including mollusk shells and marine and terrestrial bone fragments. The site shows the earliest evidence of human exploitation of marine sources.

2 A7 VERGULDE DRAECK WRECK
AD 1656
LOCATION: Australia
TYPE OF SITE: Colonial shipwreck
SITE DETAILS: Although the ship itself was smashed to pieces, many artifacts have been salvaged – among these are 50 stoneware jugs, clay smoking pipes, a fully equipped toolbox, ballast bricks, African elephant tusks, eight chests of silver coins, 14 cannons, six anchors, and two bronze mortars.

3 A7 UPPER SWAN RIVER
38,000 YEARS AGO
LOCATION: Australia
TYPE OF SITE: Hunter-gatherer
SITE DETAILS: An extensive open-air campsite that has finds of chert flakes, some containing fossils, and small quartz scrapers.

4 A7 DEVIL'S LAIR
32,000–10,000 YEARS AGO
LOCATION: Australia
TYPE OF SITE: Hunter-gatherer
SITE DETAILS: A cave that has one of the longest occupational sequences in Australia. Excavations have revealed well-defined hearths, retouched flakes, chert and quartz adze flakes, limestone artifacts, and bone points and beads.

5 B6 WILGIE MIA
11TH CENTURY AD
LOCATION: Australia
TYPE OF SITE: Hunter-gatherer
SITE DETAILS: Aboriginal mine where ocher, highly prized by Aboriginal society, was mined by breaking the rock with stone and taking out the ocher with wooden sticks.

6 B6 PURRITJARRA
27,000–6000 YEARS AGO
LOCATION: Australia
TYPE OF SITE: Hunter-gatherer
SITE DETAILS: A red sandstone rock shelter that has a number of ocher rock-art stencils and paintings. Some stone flakes were also excavated.

7 C7 KOONDALDA
24,000–15,000 YEARS AGO
LOCATION: Australia
TYPE OF SITE: Hunter-gatherer
SITE DETAILS: A cave where flint was mined in a limestone sinkhole. Excavations have found hearths, charcoal, and residue from the quarrying process.

8 C5 FORT DUNDAS
AD 1824–9
LOCATION: Australia
TYPE OF SITE: Colonial settlement
SITE DETAILS: A short-lived fort whose remains include an embankment, traces of a wharf, a store, garbage pits, and convict barracks.

9 C5 MALANGANGERR
23,000 YEARS AGO
LOCATION: Australia
TYPE OF SITE: Hunter-gatherer
SITE DETAILS: Deep, overhanging rock shelter that contains rock paintings. Artifacts excavated include flaked-core tools, steep-edged scrapers, utilized flakes, and small ground-edge axes.

10 C5 MALAKANANJA II
52,000–18,000 YEARS AGO
LOCATION: Australia
TYPE OF SITE: Hunter-gatherer
SITE DETAILS: A shallow rock shelter that contained the oldest human occupation in Australia. Over 15,000 artifacts were uncovered from its lowest occupation level, with items including core scrapers, flakes, quartz and quartzite, and red and yellow ocher.

11 C5 NAUWALABILA
55,000–20,000 YEARS AGO
LOCATION: Australia
TYPE OF SITE: Hunter-gatherer
SITE DETAILS: Shelter that was created by the fall of a massive boulder. Excavated spear points reveal a continuous occupation during the last Ice Age.

12 C5 ANURU BAY
18TH CENTURY AD
LOCATION: Australia
TYPE OF SITE: Precolonial settlement
SITE DETAILS: Site of the Macassan culture, which was used for trepang processing (trepang is a form of sea cucumber). Several depressions mark the location of former smokehouses. There is a double grave of two Macassan men, the second buried in Islamic fashion, with the body facing west toward Mecca. Potsherds from southern Celebes, glazed wares, glass from Dutch gin bottles, and metal fishhooks were also excavated.

13 C5 LYABO
LATE 18TH CENTURY AD
LOCATION: Australia
TYPE OF SITE: Precolonial settlement
SITE DETAILS: Another Macassan site with 15 stone lines, trepang pits, and the remains of smokehouses. Artifacts include a musket ball, a lead sinker, and a Dutch East India Company coin.

14 D6 KENNIFF CAVE
19,000 YEARS AGO
LOCATION: Australia
TYPE OF SITE: Hunter-gatherer
SITE DETAILS: Large cave decorated with Aboriginal stencils in red, white, and yellow. There were finds of over 800 artifacts and 22,000 waste flakes. It is important due to its wide range of artifacts, and is representative of most Australian tool types. (See p. 13.)

15 C7 GAWLER
19TH CENTURY AD
LOCATION: Australia
TYPE OF SITE: Industrial settlement
SITE DETAILS: The remains of engineering works that produced mining equipment and locomotives. Structures that remain include a flour mill, a foundry, a gasworks, a brickworks, and lime kilns.

16 D7 ROONKA FLATS
5TH–3RD MILLENNIA BC
LOCATION: Australia
TYPE OF SITE: Early burial
SITE DETAILS: Burial at a site that was previously used for camping. The bodies were placed vertically in a shaft hole and were accompanied by ornaments. One of the skeletons had a bone dagger, while another had a headband with two rows of carved wallaby teeth.

17 D7 LAKE NITCHIE
c. 6800 BC
LOCATION: Australia
TYPE OF SITE: Early burial
SITE DETAILS: A man's burial that included 178 pierced Tasmanian devil teeth that made up a necklace.

18 D7 LAKE MUNGO
36,000–26,000 YEARS AGO
LOCATION: Australia
TYPE OF SITE: Early burial
SITE DETAILS: A site that had one of the world's oldest cremations, where there was also a male inhumation, covered in red ocher. Freshwater-shell middens and some hearths have been excavated; artifacts include blades, microliths, flakes, and backed blades. (See p. 13.)

19 D7 KOW SWAMP
13,000–9,500 YEARS AGO
LOCATION: Australia
TYPE OF SITE: Early burial
SITE DETAILS: The finds of around 40 burials that had been placed in the soft silt. Grave goods include ocher, shells, and quartz. One body lay on a bed of mussel shells, while another had a band of kangaroo teeth around his head.

20 E7 CRANEBROOK TERRACE
40,000 YEARS AGO
LOCATION: Australia
TYPE OF SITE: Hunter-gatherer
SITE DETAILS: A camp where logs were burned in a gravel area, and artifacts include pebble choppers and steep-edged scrapers.

21 E7 SYDNEY COVE
19TH CENTURY AD
LOCATION: Australia
TYPE OF SITE: Colonial settlement
SITE DETAILS: See pp. 101, 109.

22 E7 PARRAMATTA
c. AD 1790
LOCATION: Australia
TYPE OF SITE: Colonial settlement
SITE DETAILS: Little remains of one of the first colonial settlements, first inhabited by convicts and gradually replaced by settlers. There are remains of a convict's hut with postholes, pits, and shallow depressions. There are traces of another structure that may have been a barn. Artifacts include iron nails, glass fragments, and imported porcelain.

23 E7 BURRILL LAKE
20,000 YEARS AGO
LOCATION: Australia
TYPE OF SITE: Hunter-gatherer
SITE DETAILS: A huge sandstone rock shelter that has finds of a tool industry similar to that of Kenniff Cave (see left).

24 D8 CLOGGS CAVE
17,000 YEARS AGO
LOCATION: Australia
TYPE OF SITE: Hunter-gatherer
SITE DETAILS: Rock shelter where the roof has been blackened by centuries of campfires. A rock overhang was used as chipping floor, with finds of backed blades and scrapers (from the last 1,000 years). There are the remains of much older hearths (8,000 to 17,000 years old) and ground ovens, where hot stones were placed to heat food.

25 D8 CAVE BAY COVE
23,000 YEARS AGO
LOCATION: Australia
TYPE OF SITE: Hunter-gatherer
SITE DETAILS: A camp littered with shells and animal bones, and more debris of the seasonal hunting parties that frequented the site.

26 D8 WEST POINT MIDDEN
5TH CENTURY BC
LOCATION: Australia
TYPE OF SITE: Hunter-gatherer
SITE DETAILS: Site where groups of

Aborigines camped for several months each summer in dome-shaped huts to live off seals and shellfish.

27 D8 MOUNT CAMERON WEST
7TH–11TH CENTURIES AD
LOCATION: Australia
TYPE OF SITE: Rock art
SITE DETAILS: Petroglyphs found on a cliff-face shelter that depict a number of geometric forms, including circles, crosses, and a row of holes. Some pointed tools were found.

28 D8 KUTIKINI CAVE
20,000 YEARS AGO
LOCATION: Australia
TYPE OF SITE: Hunter-gatherer
SITE DETAILS: Huge cave where 250,000 animal bones and 37,000 stone flakes were found in 35 cu ft (1 cu m) of deposit alone. There were traces of hearths and finds of cutting tools that were made of glass that was created by a meteorite impact.

29 D8 PORT ARTHUR
19TH CENTURY AD
LOCATION: Australia
TYPE OF SITE: Colonial settlement
SITE DETAILS: Penal colony that includes barracks and workplaces, such as a bakery, a carpentry shop, a blacksmith, a shipyard, and a huge four-story brick mill. Excavations at the barracks have revealed many clay pipes, mostly manufactured in Glasgow, Scotland.

30 D8 RISDON CAVE
19TH CENTURY AD
LOCATION: Australia
TYPE OF SITE: Colonial settlement
SITE DETAILS: A failed attempt in AD 1803 at setting up a colony, where three structures have been excavated so far. The storehouses were abandoned after a fire, while Mountgarrett's House, made of wattle and daub, had a rammed-earth fireplace and chimney. Artifacts recovered from the site include naval buttons, a bone carving of a sailor, gunflints, lead shot, and a cannonball.

31 F7 KINGSTON
AD 1810–56
LOCATION: Australia
TYPE OF SITE: Colonial settlement
SITE DETAILS: Penal colony set up on a remote island, where stone buildings and other sites still survive. The buildings include the Governor's House, officials' accommodations, a store, underground silos, a windmill, lime kilns, and the foundations of the large jail. There is also the Murderers' Mound, which has the remains of 12 convicts buried after an uprising.

32 D4 KUK
7TH MILLENNIUM BC
LOCATION: Papua New Guinea
TYPE OF SITE: Early settlement
SITE DETAILS: A drained swamp that has revealed signs of the area's agricultural development.

33 D4 KIOWA ROCK SHELTER
9TH–3RD MILLENNIA BC
LOCATION: Papua New Guinea
TYPE OF SITE: Hunter-gatherer
SITE DETAILS: A camp that unearthed pebble and flake tools. The top level, dating to around 3000 BC, contained polished ax-adze blades and fragments of undecorated pottery.

34 D4 KAFIAVANA
c. 8700 BC
LOCATION: Papua New Guinea
TYPE OF SITE: Hunter-gatherer
SITE DETAILS: A rock shelter that was the site of a stone-flake industry, and revealed pebble tools, flakes, and fragments of ground ax-adze blades.

35 D5 MOTUPORE
16TH–18TH CENTURIES AD
LOCATION: Papua New Guinea
TYPE OF SITE: Trading settlement
SITE DETAILS: A coastal trading center that has revealed shell valuables and pottery manufacture.

36 D5 KOSIPE
24,000 YEARS AGO
LOCATION: Papua New Guinea
TYPE OF SITE: Hunter-gatherer
SITE DETAILS: The earliest camp in the country where wasted blades were found. A ground ax blade was found in its upper level (c. 8000 BC).

37 E5 MAILU ISLAND
2ND–16TH CENTURIES AD
LOCATION: Papua New Guinea
TYPE OF SITE: Trading settlement
SITE DETAILS: A midway trading point on a small island that held a monopoly on local pottery manufacture. It also made its own shell valuables.

38 E5 NUMATA
11TH CENTURY AD
LOCATION: Papua New Guinea
TYPE OF SITE: Fortified settlement
SITE DETAILS: The remains of a fort where traces of 13 round and oval stone-faced house foundations were found, surrounded by a ditch. Some impressed ceramics and a water-pot ring handle were recovered.

39 G8 GREAT MERCURY ISLAND
EARLY 2ND MILLENNIUM AD
LOCATION: New Zealand
TYPE OF SITE: Early settlement
SITE DETAILS: A number of related sites, including burials, hamlet remains, and middens. At a site called Stingray Point, storage pits, with complicated posthole arrangements, were revealed.

40 G8 OTAKANINI
15TH–17TH CENTURIES AD
LOCATION: New Zealand
TYPE OF SITE: Maori hill fort
SITE DETAILS: An important pa (fortified site) that had three phases of fortification and where some sweet-potato storage pits have been excavated.

41 G8 WAIONEKE
17TH CENTURY AD
LOCATION: New Zealand
TYPE OF SITE: Maori hill fort
SITE DETAILS: A pa that was fortified twice with a storage pit, and where the remains of several fatally wounded bodies have been found, probably the site's last inhabitants.

42 G8 SARAH'S GULLY
14TH–17TH CENTURIES AD
LOCATION: New Zealand
TYPE OF SITE: Maori hill fort
SITE DETAILS: Two adjacent sites – one was a headland used for habitation and storage, the other a stratified midden, working area, and burial site. There are remains of postholes, ovens, and pits.

43 G8 MOUNT WELLINGTON
14TH CENTURY AD
LOCATION: New Zealand
TYPE OF SITE: Maori hill fort
SITE DETAILS: A pa with ditch and bank defenses that once supported wooden pallisades, storage pits, and terraces for housing.

44 G8 ONE TREE HILL
11TH–15TH CENTURIES AD
LOCATION: New Zealand
TYPE OF SITE: Maori hill fort
SITE DETAILS: Site of a pa built on a terraced slope. Many pas were built on the slopes of extinct volcanoes.

45 G8 KAURI POINT
16TH CENTURY AD
LOCATION: New Zealand
TYPE OF SITE: Maori hill fort
SITE DETAILS: Undefended settlement with traces of storage pits, a midden, and houses, which has a nearby pa on a cliff, with three periods of occupation. Combs and other wooden artifacts were recovered from swamps below.

46 G8 AOTEA
16TH–17TH CENTURIES AD
LOCATION: New Zealand
TYPE OF SITE: Maori hill fort
SITE DETAILS: A pa that has remains of terraces, with finds of storage pits, a midden, and cooking areas.

47 G8 TE AWANGA
EARLY 2ND MILLENNIUM AD
LOCATION: New Zealand
TYPE OF SITE: Maori hill fort
SITE DETAILS: A fortified pa with the remains of houses and sweet-potato storage pits.

48 G8 PAREMATA
16TH–19TH CENTURIES AD
LOCATION: New Zealand
TYPE OF SITE: Early settlement
SITE DETAILS: A midden with prehistoric occupations that has remains of moa (a flightless bird, now extinct), Polynesian artifacts, and a 19th-century pa.

49 G8 MAKOTUKUTUKU
14TH–16TH CENTURIES AD
LOCATION: New Zealand
TYPE OF SITE: Maori settlement
SITE DETAILS: Several burials were found in a rock cleft. Stone walls and mounds were also excavated at a house site.

50 G8 PARARAKI NORTH
13TH–14TH CENTURIES AD
LOCATION: New Zealand
TYPE OF SITE: Maori hill fort
SITE DETAILS: A pa settlement that was occupied over three separate periods.

51 G8 WAIRAU BAR
12TH CENTURY AD
LOCATION: New Zealand
TYPE OF SITE: Maori camp
SITE DETAILS: Moa hunters' camp that has revealed some of the best grave goods in the country, including moa-bone reels, shark teeth, stone adzes, fishhooks, and bone harpoons.

52 G8 CLARENCE RIVER MOUTH
12TH–16TH CENTURIES AD
LOCATION: New Zealand
TYPE OF SITE: Maori settlement
SITE DETAILS: Early gardens, middens, and a pa have been excavated.

53 G8 PA BAY
EARLY 19TH CENTURY AD
LOCATION: New Zealand
TYPE OF SITE: Maori settlement
SITE DETAILS: Settlement site with remains of several houses.

54 G9 WAITAKI RIVER MOUTH
15TH CENTURY AD
LOCATION: New Zealand
TYPE OF SITE: Maori settlement
SITE DETAILS: A river site where moa remains, ovens, and artifacts, including stone adzes, have been excavated.

55 C3 BADRULCHAU
12TH CENTURY AD
LOCATION: Palau
TYPE OF SITE: Early settlement
SITE DETAILS: A large earth pyramid mound, 52 megaliths, and a bai – the men's community house on the island – have all been excavated.

56 C3 IRRAI VILLAGE
15TH CENTURY AD
LOCATION: Palau
TYPE OF SITE: Early village
SITE DETAILS: Prehistoric settlement that has the remains of 43 rectangular stone

platforms (eight are in the form of pyramids and 18 contain graves), terraced hill slopes, and three *bai*. The central *bai* contained two *chab* (fire pits).

57 D3 NOMNA BAY
14TH–17TH CENTURIES AD
LOCATION: Guam, Micronesia
TYPE OF SITE: Early settlement
SITE DETAILS: Village site consisting of 17 *latte* – large structures with megalithic foundations. Fire pits with large amounts of pottery, stone, and faunal remains; stone mortars for processing food; and burials under the *latte* were all excavated.

58 D3 TALOFOFO RIVER
3RD CENTURY BC
LOCATION: Guam, Micronesia
TYPE OF SITE: Early settlement
SITE DETAILS: Village site with a *latte*, the foundations of nine houses, and two rock shelters, one of which had a burial, containing pottery and stone tools.

59 D3 TACHOGNYA
EARLY 2ND MILLENNIUM AD
LOCATION: Guam, Micronesia
TYPE OF SITE: Early settlement
SITE DETAILS: The almost fully intact remains of a prehistoric village containing ten *latte*. Shells and shards were excavated near the structures.

60 E3 SAPWTAKAI
14TH CENTURY AD
LOCATION: Pohnpei, Micronesia
TYPE OF SITE: Early settlement
SITE DETAILS: Ancient administrative center on a mountain peak that has many stone burials, platforms, and enclosures.

61 E3 NAN MADOL
1ST–16TH CENTURIES AD
LOCATION: Pohnpei, Micronesia
TYPE OF SITE: Early settlement
SITE DETAILS: A site that was settled for a long period, although most of its megalithic architecture was built in the 12th–13th centuries AD. Walls of basalt were used to construct artificial islets, used either for residence or as burial places, the latter often consisting of a central vault with four basalt walls. The house foundations often had fire pits in the center. There are remains of large meetinghouses known as *nahs*, and enclosures that seem to have been used for ceremonial purposes.

62 E3 PANPEI
14TH CENTURY AD
LOCATION: Pohnpei, Micronesia
TYPE OF SITE: Burial complex
SITE DETAILS: Four crypts found within a large burial platform, with most of the construction using basalt blocks.

63 F3 LELA RUINS
13TH–19TH CENTURIES AD
LOCATION: Kosrae, Micronesia
TYPE OF SITE: Ceremonial settlement
SITE DETAILS: Remains of an ancient island city, which was once the administrative and ceremonial center of the island and was split into compounds. Some impressive walls were built in the 15th–17th centuries AD, and there is a sacred and royal tomb complex.

64 F5 NENUMBO
12TH CENTURY BC
LOCATION: Solomon Islands
TYPE OF SITE: Early settlement
SITE DETAILS: A Lapita site where evidence of a number of different activities were uncovered. Remains of pig, chicken, fish, and sea turtle have been found.

65 F6 ROY MATA BURIAL
13TH CENTURY AD
LOCATION: Vanuatu
TYPE OF SITE: Royal burial
SITE DETAILS: An important elitist burial, where a king, Roy Mata, was joined by sacrificial victims. There are side-by-side burials of 35 individuals, 22 of whom were men. The women were clasped around male partners. Cannibalized bones and pig bones were found. Buried goods included shell-bead necklaces, armbands, cowries, and shell adzes.

66 F6 VATCHA
1ST MILLENNIUM BC
LOCATION: New Caledonia
TYPE OF SITE: Early settlement
SITE DETAILS: Lapita site that was occupied in two periods.

67 G6 NATUNUKU
13TH CENTURY BC
LOCATION: Fiji
TYPE OF SITE: Early settlement
SITE DETAILS: Important early Lapita site that had a number of finds of complex decorated pottery.

68 H5 PULEMELEI
EARLY 2ND MILLENNIUM AD
LOCATION: Samoa
TYPE OF SITE: Ceremonial mound
SITE DETAILS: A huge mound around 39 ft (12 m) high that had a ramp leading to its summit. Postholes and stones show that a ceremonial house existed here. It is surrounded by smaller mounds. The exact date of the site is unknown, but probably dates to a period between the 11th and 15th centuries AD.

69 H5 'AOA
1ST MILLENNIUM BC
LOCATION: American Samoa
TYPE OF SITE: Ceremonial mound
SITE DETAILS: Site of 40 elevated star platforms that have raised protrusions.

70 H6 HA'AMONGA-A-MAUI
12TH CENTURY AD
LOCATION: Tonga
TYPE OF SITE: Ceremonial monument
SITE DETAILS: A trilithon (two upright stones that support a third) that was built to symbolize the two sons of the ruling Tui Tonga.

71 H6 MU'A
15TH–16TH CENTURIES AD
LOCATION: Tonga
TYPE OF SITE: Burial mound
SITE DETAILS: Chief burial mound faced with cut-coral slabs.

72 H6 PAEPAE'O TELE'A
16TH CENTURY AD
LOCATION: Tonga
TYPE OF SITE: Burial mound
SITE DETAILS: A pyramid-like structure that is exemplary of the burial style of the island's Polynesian population.

73 H6 NAMOALA
EARLY 2ND MILLENNIUM AD
LOCATION: Tonga
TYPE OF SITE: Burial mound
SITE DETAILS: A three-tiered pyramid that has a stone burial vault on top.

74 I2 NECKER ISLAND
11TH–15TH CENTURIES AD
LOCATION: Hawaiian Islands
TYPE OF SITE: Early settlement
SITE DETAILS: Small isolated island that has the remains of houses, 33 *heiau* (stone temple remains), and cultivated terraces. The *heiau* revealed a number of male carvings and stone statues.

75 I2 WAILAU
12TH CENTURY AD
LOCATION: Hawaiian Islands
TYPE OF SITE: Ceremonial center
SITE DETAILS: Hawaiian *heiau* have been found, as have a number of petroglyphs.

76 I2 BELLOWS BEACH
7TH–10TH CENTURIES AD
LOCATION: Hawaiian Islands
TYPE OF SITE: Early settlement
SITE DETAILS: One of earliest sites on the Hawaiian Islands, where excavations have revealed assemblages of fishhooks, adzes, and other artifacts.

77 J2 HALAWA VALLEY
7TH–12TH CENTURIES AD
LOCATION: Hawaiian Islands
TYPE OF SITE: Early settlement
SITE DETAILS: A village site with stone curbs that were probably the bases for round-ended houses. Artifacts unearthed include basaltic glass and adze and fishhook assemblages.

78 J2 FORT ELIZABETH
19TH CENTURY AD
LOCATION: Hawaiian Islands
TYPE OF SITE: Colonial settlement
SITE DETAILS: A Russian fort, used for only a short period of time, that has left a number of remains, including officers' quarters and interior walls with cannon placements.

79 J2 MAUNA KEA
EARLY 2ND MILLENNIUM AD
LOCATION: Hawaiian Islands
TYPE OF SITE: Early settlement
SITE DETAILS: A large adze quarry complex that has also unearthed a series of workshops and shrines.

80 J5 MAUPITI
9TH–12TH CENTURIES AD
LOCATION: French Polynesia
TYPE OF SITE: Early burial
SITE DETAILS: A large complex of graves where the finds have included fishhooks, adzes, and other ornaments.

81 J5 TAPUTAPAUTEA
17TH CENTURY AD
LOCATION: French Polynesia
TYPE OF SITE: Ceremonial center
SITE DETAILS: A religious complex that was the center of worship for the god Oro. A *marae* – Polynesian temple platform – still survives.

82 J5 VAITOOTIA
8TH–9TH CENTURIES AD
LOCATION: French Polynesia
TYPE OF SITE: Early settlement
SITE DETAILS: Habitation site with finds of *patus* – wood-and-whalebone hand clubs. There are the remains of a raised wooden storehouse and a basalt pillar that may have been part of an early temple.

83 J5 MAEVA
LATE 1ST MILLENNIUM AD
LOCATION: Society Islands
TYPE OF SITE: Ceremonial center
SITE DETAILS: Site of a number of *marae*, where upright coral stones probably marked the positions of priests.

84 N7 EASTER ISLAND
5TH–17TH CENTURIES AD
LOCATION: Easter Island
TYPE OF SITE: Early settlement
SITE DETAILS: An island with three main periods of occupation. The earliest (AD 400–1100) was characterized by the *ahu* – rectangular stone platforms – of which about 300 remain. Impressive megalithic stone statues – *moai* – often carved or with topknots, were added later. Most were carved at the Rano Raraku quarry, where a number of unfinished statues remain. The village site of Orongo has the remains of boat-shaped houses. There are a number of *tupa* (corbeled chambers). Artifacts include basalt knives, adzes, bone and stone fishhooks, and obsidian flakes.

GLOSSARY

A

ACHAEMENID
Royal dynasty of the Persian Empire from the 6th–4th centuries BC. They were defeated by Alexander the Great in 330 BC, but not before building some impressive monumental architecture.

ACHEULIAN
A widespread stone-tool industry of the Paleolithic Age that was characterized by pressure-flaked hand-axes.

ADOBE
Sun-dried (i.e., unfired) mudbricks or clay used for building construction in the Americas before colonization.

ADZE
A chopping tool, similar to an ax, with the blade set at a right angle to the shaft.

AGORA
Open space for a market or an assembly in ancient Greek towns and cities.

AHU
A Polynesian stone temple platform.

AMS (ACCELERATOR MASS SPECTOMETRY)
A new method of radiocarbon dating that allows a wider range of smaller samples to be measured in a shorter time. See *Radiocarbon dating*.

ARCHAIC
Term that describes the early stages of civilization in Egypt and Greece. In North America the term refers to a stage of development characterized by a hunting and gathering way of life.

ASSEMBLAGE
Term used to denote a specific group of objects that recur together. They are often used to identify cultures.

ASSYRIA
An empire of western Asia in the 2nd and 1st millennia BC, centered on the capitals of Nimrud and Nineveh. Its sites have left many important remains and much monumental architecture.

B

BABYLONIA
Empire of western Asia dating to the 2nd millennium BC. It was centered on the city of Babylon in the fertile valleys of Mesopotamia and was constantly at war with Assyria and Persia.

BALLCOURT
A ceremonial area that was designed for playing ritual ball games found in the cities of Mesoamerican cultures.

BARROW
A mound covering a prehistoric tomb or grave. Round and long barrows are common in many cultures, from the Neolithic and Bronze Age to Saxon and Viking times.

BASILICA
A large aisled building in the Roman world used for public functions. Later it came to be associated with the architecture of early Christian churches.

BEAKER CULTURE
A Bronze Age culture identified by its distinctive pottery, decorated with geometric patterns. Beaker burials and their artifacts have been excavated all over Europe.

BEEHIVE TOMB
See *Tholos*

BIFACIAL
A term used to describe stone tools that have been worked to create two faces, with a cutting edge between them.

BROCH
An Iron Age fortified homestead found in the north of Scotland that was built as a double-walled round tower.

BRONZE AGE
Second part of the Three-Age System that divides up prehistory, characterized by the development of bronze-working, which is linked to increasing urbanization. The discovery of bronze can be traced in different areas across the world from the 4th millennium BC onward, although in some regions, such as the Americas, it was never used. The Bronze Age itself can be split into three distinct periods: the Early Bronze Age (*c.* 3200–1900 BC), the Middle Bronze Age (*c.* 1900–1500 BC), and the Late Bronze Age (*c.* 1500–1200 BC).

BURIN
A boring tool used for punching holes and incising leather, bone, and other items. It is considered a distinctive artifact of Paleolithic times.

BURNISH
To polish an item by rubbing. For archaeologists, it is used to describe a type of pottery that was rubbed to produce a shiny finish and make it watertight.

C

CACHE
A collection of objects deliberately buried or deposited.

CAIRN
A mound of stones, usually used to cover a burial or tomb.

CARBONIZATION
When organic products, like wood and cereals are turned to carbon by heating, thus allowing radiocarbon dating.

CHOPPER
A stone tool used for cutting. Usually used to refer to early tools made from pebbles.

CIST
A burial chamber made of stone, like those found inside barrows.

CLASSIC PERIOD
A term that is used to describe the periods in Mesoamerican history that cover the eras of greatest expansion and cultural development, by cultures such as the Maya, from the 3rd century AD to its collapse at the end of the 9th century AD. The period immediately before the Classic, the Pre-Classic, relates to the period from the 2nd millennium BC during which agriculture and more settled lifestyles developed. The Post-Classic period lasted from the end of the Classic period, through the rules of the Mixtecs and the Aztecs, until the arrival of the Spanish in the 16th century.

CLEAVER
A heavy stone tool with a straight edge that was used for cutting.

COILING
A method of pot making, in which long rolls of clay are coiled and smoothed together to make a vessel.

COLOSSI
Statues depicting people or creatures far greater than life-size.

COMPOSITE TOOL
A tool, like a harpoon, which is made of several small stone or flint blades hafted together, particularly used in the Mesolithic Age.

CORBELING
A method of roof-building, in which the vault of the roof is built using layers of stones that project inward until they meet at the center.

CUNEIFORM
An early form of writing used in Mesopotamia from the 3rd to 1st millennium BC, consisting of symbols carved into clay using a wedge-shaped reed tool.

CURBSTONE
An edging stone, used to delineate archaeological features, such as houses, hearths, and burial mounds.

D

DEBITAGE
The waste flint or stone flakes that are created when stone tools are made. These can be studied to reconstruct the knapping process.

DENDROCHRONOLOGY
The study of tree-ring growth used by archaeologists to date wood from settlements and other sites.

DOLMEN
A tomb created by laying a large flat capstone on upright ones.

DYNASTY
Used to describe royal successions related by blood. It is commonly used in relation to ancient Egypt, Mesopotamia, and ancient China.

E

ECOSYSTEM
A term to describe a biological community of interacting plant and animals and their physical environment.

ELECTRON SPIN RESONANCE
A dating technique that uses the effects of changing energy levels of electrons under radiation to date teeth, bone, chert, flint, ceramics, and other objects.

F

FAIENCE
A vitrified substance, made with sand and fired like clay, which was used to make beads and decorative objects.

FIBULA
Latin word used by archaeologists for a brooch or pin used to fasten clothing.

FIELD SURVEYING
A technique used to discover any artifacts that may have been brought to the surface, such as potsherds and flint.

FLAKE
A piece of stone or flint, created as waste when tools are manufactured.

FORMATIVE
See *Pre-Classic*

FRESCO
A wall painting that is painted on plaster while it is still wet.

G

GEOMETRIC
A term used to describe regular lines and shapes that are found carved on

C

CYCLOPEAN
A term used to describe masonry using large irregular stones, especially used to describe Mycenaean fortifications.

surfaces. It is especially used to describe the lines and shapes of pottery.

GEOPHYSICAL SURVEY
Using scientific methods, such as resistance and magnetometry, to investigate features below the ground without excavation.

GLAZE
Layer of glass or minerals applied to pottery and fired to create a surface decoration or nonabsorbent finish.

GLYPH
A pictoral character or symbol used in early writing systems or decoration.

GPS
(GLOBAL POSITIONING BY SATELLITE)
An archaeological technique that is used to survey areas and make detailed maps of archaeological sites with great accuracy.

H

HELLENISTIC
The period of Greek archaeology associated with the time after the death of Alexander the Great in 323 BC until the advent of the Roman Empire in the 1st century BC.

HENGE
A circular prehistoric monument marked out with wood or stone.

HILL FORT
Fortified settlements that are usually ascribed to the Late Bronze and Early Iron Ages, which were defended by raised earthen ramparts. Many were only temporary refuges.

HITTITE
An empire of the 2nd millennium BC centered on Boghazkoy in Turkey.

HOABINHIAN
Named after the site of Hoa Binh in Vietnam, the term relates to a number of wide-ranging, prehistoric tool industries covering the period between 10,000 and 2000 BC.

HOLOCENE (POSTGLACIAL)
The modern climactic period, characterized by a warm environment, which began around 10,000 years ago

HOMINID
An abbreviation of Hominidae, the family to which modern humans and their ancestors belong.

HOMO ERECTUS
A form of early human, now extinct, who lived around half a million years ago.

HOMO HABILIS
An early human type, distinguishable from contemporary species by its bigger brain and distinct shape of the skull.

HOMO SAPIENS
The name given to modern human, who first developed around 35,000 years ago.

HOPEWELL
A North American culture that developed around the Great Lakes and was characterized by farming and trade. It dates from around 100 BC to AD 500.

HUACA
The indigenous Peruvian word for a temple, it is most often related to structures built by the Inca.

HUNTER-GATHERER
The term used to describe cultures that subsist on the hunting of animals and the collection of plants.

HYPOCAUST
A heating system developed by the Romans, using hot air, which was piped under floors and through walls.

IJK

INSULAE
A block of houses and apartments in a Roman city, similar to the grid system in modern American cities.

INUIT
The name given to the cultures of the Arctic regions of North America and Greenland. It literally means "the people."

IONIC
Style of Greek architecture characterized by columns with scrolled capitals.

IRON AGE
The last part of the Three-Age System that describes the period when iron became the priimary metal for making of weapons and tools. In many cases it replaced bronze, although in areas like Africa and the Americas, it was the first major metal to be used.

KIVA
A ceremonial room that is often found in the pueblo village remains of southwestern United States.

L

LAPITA
A culture that is defined by its distinctive stamped pottery commonly found over the Melanesian islands of the Pacific, parts of Indonesia and New Guinea, and as far away as Fiji.

LA TÈNE
An art style belonging to the Bronze Age and Early Iron Age, characterized by its delicate spirals and geometric design on metalwork.

It was the precursor to the Celtic styles of design.

LEVALLOIS
A term used to describe a method of flint flaking, whereby the face of the core is trimmed to create the shape and the size of the intended flake.

LINEAR A
A script used by the Minoan culture on Crete from 1700 to 1450 BC. It consisted of pictograms and is often found on clay tablets and other objects.

LINEAR B
The script that replaced Linear A, it was used between 1450 and 1200 BC. Once deciphered, it was found to be an earlier form of Greek.

LOST-WAX PROCESS
A method of making metal objects – a mold is made on a wax core, which is melted away to leave a casting space for the molten metal.

M

MAGDALENIAN
A culture of the Upper Paleolithic in Europe – much cave-art is assigned to this period, dating to between 18,000 and 12,000 years ago.

MAGNETOMETRY
A method of surveying for features beneath the ground, using the magnetic properties of soil and other materials.

MARAE
A Polynesian temple site.

MAUSOLEUM
A funerary monument found above ground level, usually quite elaborate in style. The word often refers to Classical and Islamic structures.

MEGALITH
Any large stone, whether freestanding or incorporated into a structure.

MEGALITHIC TOMB
A tomb built using megaliths to create passages and burial chambers.

MENHIR
Maen-hir (Celtic), meaning long stone – erected for funerary purposes.

MESOAMERICA
A term used to describe Mexico and Guatemala before the arrival of the Spanish in the 16th century AD. Also relating to the cultures that dominated the region from about 2000 BC, such as the Toltecs, the Maya, the Teothuacános, and the Aztecs, who all shared similar rather than independent traits.

MESOLITHIC
Transitional period between the Old Stone Age and the development

of farming, characterized by very small stone tools and sophisticated hunting practices.

METATE
Stone with a concave surface used in ancient America for grinding corn.

MICROCORE
A piece of stone or flint from which blades had been struck, leaving a shaped lump with several facets.

MICROLITH
A very small stone tool that was often hafted into a composite tool. It was most commonly used in the Mesolithic Age.

MIDDEN
A mound of debris, usually of domestic refuse, that includes waste from shells, food processing, and animal debris.

MIDDLE MISSISSIPPIAN CULTURE
An important culture that lived around the Mississippi valley and lasted from the 7th to 17th centuries AD. It is characterized by its impressive ceremonial centers and a dependence on corn, squash, and beans.

MINOAN
Bronze Age civilization centered on Crete, characterized by palaces and sophisticated towns and based on Mediterranean trade.

MOSAIC
A tiled pavement, particularly used in the Roman world, with designs made of small ceramic and stone blocks, known as *tessarae*.

MOTTE AND BAILEY
A Norman castle consisting of a large defended earthen mound (motte), joined to a larger area (bailey). It was within the bailey that the castle's daily activities would occur.

MYCENAEAN
A term relating to the cultures of Late Bronze Age Greece, which shared the same language, beliefs and trade. The name is taken from the ancient citadel on mainland Greece.

N

NASCA
An important culture of coastal Peru, renowned for its painted pottery and its Nasca Lines – drawings of huge geometric shapes in the desert.

NATUFIAN
An important Paleolithic culture of the Levant, which provided evidence about the development of early humans.

NEANDERTHAL
An early hominid, adapted for cold climates, found in Europe and Asia. It died out around 37,000 years ago.

NECROPOLIS
An area of burial or tombs, it is literally translated as "the city of the dead."

NEOLITHIC
Or the New Stone Age, characterized by the development of settlements and the cultivation of crops, and mostly dating from the 9th millennium BC to the 2nd millennia BC.

NEW KINGDOM
A dynastic period of ancient Egypt, relating to the period covering the 18th to 20th Dynasties.

NURAGHE
Fortified round towers found on the island of Sardinia, which date from the 2nd millennium onward.

O

OCHER
A yellow–red-colored pigment of clay and iron oxides that was used by many prehistoric groups in ritual burials or wall decoration.

OLDOWAN
Relating to the finds at Olduvai Gorge in Tanzania, the term is used to describe the stone tools and artifacts found in southern and eastern Africa.

OPPIDUM
A large fortified settlement and regional center. It is applied primarily to Roman and Celtic settlements that constituted the first towns of northern Europe.

PQ

PA
The term for a Maori hill fort.

PALEO-INDIAN
A term used to describe the hunter-gatherer cultures of North America, dating to about 5,000 years ago.

PALEOLITHIC
Also known as the Old Stone Age, it defines the earliest phase of human development, characterized by the hunter-gatherer way of life. It is divided into the Lower Paleolithic, the emergence of early humans; the Middle Paleolithic, Neanderthal humans and flaked tool industries; and the Upper Paleolithic, the emergence of modern humans, blade industries, and early art.

PAPYRI
Writing material that was used in ancient Egypt and was made from the stems of an aquatic plant.

PETROGLYPH
Any carving on rocks or stones, especially prehistoric ones.

PICTISH
Used to describe the culture of Iron Age Scottish communities.

PLATFORM MOUND
Flat-topped earthen mounds, found in ancient North America, surmounted with a ritual structure that acted as a focus for the local communities.

PLAZA
An open space in a settlement, usually used as a focal area or an assembly place.

PLEISTOCENE
Climactic era preceding the Holocene, characterized by fluctuating cold periods, dating from around 1.8 million years ago until about 8000 BC.

POSTHOLE
A hole that is dug to receive an upright timber for a building or other structure. As the structure decays, traces of the original building are left in the soil. This can be a useful tool for archaeologists when studying settlements and determining their layout.

POTSHERD
A piece of broken pottery.

PRECERAMIC
An era in South American archaeology spanning the period between the first human settlements and the advent of pottery. It is normally assumed to be between around 9000 and 2000 BC.

PREHISTORIC
Any period before the development of written records. It can vary from region to region.

PRESSURE-FLAKE
To remove pieces of stone or flint from a core by pushing with a soft antler or bone tool, rather than striking with another stone.

PUEBLO
Stone- or brick-built dwellings found in southwestern North America, often consisting of three or four rooms.

R

RADIOCARBON DATING
A method of dating organic material, which is based on the decay rate of radioactive carbon-14 atoms that are present in all living things.

REPOUSSÉ
A method of decorating metalwork produced by hammers and punches.

RUNES
Symbols first used in the 3rd century AD, and eventually taken up by the Vikings, who carved runes on their jewelry and memorial structures, known as runestones.

S

SAMIAN
Mass-produced, high-quality red ceramic tableware made in western Europe during the Roman Empire.

SARSEN
A sandstone from the west of England, which was used to build monuments in the Neolithic and Bronze Ages.

SCRAPER
A shaped hunter-gatherer tool used to scrape animal skins or to carve wood.

SHAFT TOMB
Most prevalent in ancient China, these burial structure often consisted of a very deep burial shaft with a wooden burial chamber at the bottom.

SPONDYLUS
A shell, *Spondylus gaederopus*, found in the Mediterranean, from which bangles were cut. It was widely traded in the Mediterranean in Neolithic times.

STELA
An upright stone slab, which was often carved or painted, mainly used for commemorative or ritual purposes.

STONE AGE
The term used to describe the earliest known periods of human existence, it covers the Paleolithic, Mesolithic, and Neolithic periods. It is a global archaeological term, so it is used for cultures in existence as recently as the last century.

STRATIGRAPHY
An essential archaeological theory that presupposes that the lowest layers of human and natural deposits at a site must be the oldest. The relationship between the layers uncovered in the excavation of a site can be used to reconstruct its development over time.

STUPA
Buddhist shrine in the form of a mound or a dome, often bell-shaped.

SWIDDEN AGRICULTURE
Farming system in which areas of land are cleared and burned and then farmed until the soil is exhausted. Another area is then cleared, and the process is repeated.

T

THERMOLUMINESCENCE
A method developed for dating fired archaeological material. Items buried in certain types of sediments are exposed to radiation from the decay of radioactive isotopes which causes energy from the displaced electrons to be stored within mineral crystal lattices in the structure. When heated or exposed to light the trapped electrons are freed, releasing energy in the form of light referred to as thermoluminescence.

THOLOS
A round tomb, also known as a beehive tomb, with a rectangular entrance passage that is found in ancient Greece.

THREE-AGE SYSTEM
A term used to delineate the main periods of prehistory – the Stone Age, Bronze Age, and Iron Age – through the major changes in technology.

TORC
A necklace of twisted metal, particularly associated with Celtic cultures.

TRILITHON
A monument made of a capstone resting on two uprights, such as those found at Stonehenge.

TYPE SERIES
A reference collection of typical objects, such as pottery types, that can be used for identifying unknown objects.

TYPE SITE
A site that is a typical example of a particular period, which can be used as a guide for what might be found in future excavations.

TYPOLOGY
The classification of a group of artifacts into types, and the study of their change through time, to help understand the development of civilizations.

UV

'UBAID CULTURE
An important Mesopotamian culture of the 6th to 5th millennium BC linked with the beginning of agriculture in the region, and defined by its pottery.

VENUS FIGURINE
A type of sculpture found throughout the world in prehistory that depicts a mother-goddess figure and probably had fertility connotations.

WXYZ

WILTON
An important culture of southern Africa that gives its name to a microlithic Stone Age industry and a later ceramic type.

ZIGGURAT
A stepped pyramid or temple found in western Asia.

BIBLIOGRAPHY

Aikens, C.M. *The Prehistory of Japan*, Academic Press, 1982

Bahn, P. (ed.) *Dictionary of Archaeology*, HarperCollins, 1992

Bahn, P. & Renfrew, C. *Archaeology: Theories, Methods and Practice (2nd rev. ed)*, Thames & Hudson, 1996

Barker, G. *Prehistoric Farming in Europe*, Cambridge University Press, 1985

Bass, G. *Archaeology Under Water*, Thames & Hudson, 1966

Bellwood, P. *The Polynesians*, Thames & Hudson, 1978

Bellwood, P. *Man's Conquest of the Pacific*, Collins, 1978

Bellwood, P. *Prehistory of the Indo-Malaysian Archipelago*, Academic Press, 1985

Blendin, C. & Elvin, M. *The Cultural Atlas of China*, Phaidon, 1983

Boardman, J. *Greeks Overseas: The Archaeology of Their Early Colonies and Trade*, Thames & Hudson, 1980

Boëthius, A. *Etruscan and Early Roman Architecture*, Penguin, 1978

Bordes, F. *The Old Stone Age*, Weidenfeld & Nicholson, 1968

Bowdler, S. (ed.) *Coastal Archaeology in Eastern Australia*, Department of Prehistory, 1982

Boxer, C.R. *Fort Jesus and the Portuguese in Mombasa 1593–1729*, Thames & Hudson, 1960

Branigan, K. (ed) *The Atlas of Archaeology*, Macdonald & Co Ltd, 1982

Bray, W. & Trump, D. *The Penguin Dictionary of Archaeology*, Penguin, 1982

British School at Athens, *The Annuals of the Excavations at Palaikastro*, 1979, 1982, 1983, 1984, 1986, 1987, 1988, 1989, 1991, 1992

Burton, A. *Industrial Archaeological Sites of Britain*, Weidenfeld & Nicholson, 1977

Chang, K. *Archaeology of Ancient China*, Yale University Press, 1986

Charles-Picard, G. (ed) *The Larousse Encyclopedia of Archaeology*, Hamlyn, 1972

Childe, V.G. *Skara Brae: A Pictish Village in Orkney*, Kegan Paul, 1931

Clarke, H. *Archaeology of Medieval England*, Blackwell, 1986

Coe, M. *Mexico*, Thames & Hudson, 1984

Collis, J. *The European Iron Age*, Batsford, 1984

Connah, G. *African Civilization: Precolonial Cities and States in Tropical Africa*, Cambridge University Press, 1987

Cronyn, J.M. *The Elements of Archaeological Conservation*, Routledge, 1990

Cunliffe, B. *The Oxford Illustrated Prehistory of Europe*, Oxford University Press, 1994

Daniel, G. *Encyclopedia of Archaeology*, Thomas Cromwell Company, 1977

Davidson, J. *The Prehistory of New Zealand*, Longman, 1984

Fagan, B. *People of the Earth: An Introduction to World Prehistory*, HarperCollins, 1992

Fagan, B. *Ancient Africa*, Thames & Hudson, 1994

Fagan, B. *Ancient North America*, HarperCollins, 1991

Fagan, B. *Kingdoms of Gold, Kingdoms of Jade: The Americas Before Columbus*, Thames & Hudson, 1992

Fisher, R.E. *Buddhist Art and Architecture*, Thames & Hudson, 1993

Flood, J. *Archaeology of the Dreamtime: The Story of Prehistoric Australia and Its People*, Collins, 1983

Frank, S. *Glass and Archaeology*, London Academic Press, 1982

Garlake, P. *Great Zimbabwe*, Thames & Hudson, 1973

Glob, P.V. *The Bog People: A Man Preserved*, Faber & Faber, 1969

Grant, M. *Cities of Vesuvius: Pompeii and Herculaneum*, Weidenfeld & Nicholson, 1971

Gupta, S.P. *Archaeology of Soviet Central Asia and the Indian Borderlands*, B.R. Publishing, 1979

Hawkes, J. *Atlas of Ancient Archaeology*, Michael O'Mara Books Ltd, 1994

Hayden, B. *Archaeology – The Science of Once and Future Things*, W.H. Freeman and Company, 1993

Higham, C. *The Archaeology of Mainland Southeast Asia*, Cambridge University Press, 1989

Higham, C. *The Bronze Age of Southeast Asia*, Cambridge University Press, 1996

Hood, S. *The Minoans: Crete in the Bronze Age*, Thames & Hudson, 1998

Hudson, K. *Guide to the Industrial Archaeology of Europe*, Adam & D., 1971

Hudson, K. *Industrial Archaeology: A New Introduction (3rd ed)*, Baker, 1976

Kenyon, K. *Archaeology of the Holy Land*, Benn, 1980

Kirch, P.W. *Feathered Gods and Fishhooks: An Introduction to Hawaiian Archaeology and Prehistory*, University of Hawaii Press, 1985

Lawrence, T.E. *Crusader Castles*, Immel, 1992

MacGillivray, J.A. & Sackett, L.H. (et al) *An Archaeological Survey of the Roussolakos at Palaikastro*, BSA number 79, 1984

Malone, C. *English Heritage Book of Avebury*, Batsford, 1989

Mellaart, J. *Neolithic of the Near East*, Thames & Hudson, 1988

Moseley, M.E. *The Incas and Their Ancestors: The Archaeology of Peru*, Thames & Hudson, 1992

Miller, M.E. *The Art of Mesoamerica: From Olmec to Aztec*, Thames & Hudson, 1991

Mohen, J. *The World of Megaliths*, Cassell, 1971

Morgan, W.N. *Prehistoric Architecture in Micronesia*, University of Texas Press, 1988

Mughan, T. *Chronologies in New World Archaeology*, Academic Press, 1978

Mulvaney, D. *The Prehistory of Australia*, Thames & Hudson, 1969

Mulvaney, D. & Golson, J. *Aboriginal Man and the Environment*, Australian National University Press, 1971

Neave, R. *Making Faces*, British Museum Publications, 1997

Phillipson, D.W. *African Archaeology*, Cambridge University Press, 1993

Price, T.D. & Brown, J.A. *Prehistoric Hunter-gatherers: The Emergence of Cultural Complexity*, Academic Press, 1985

Rawson, J. *Ancient China: Art and Archaeology*, British Museum Publications Ltd, 1980

Rowlands, M.J. *The Production and Distribution of Metalwork in the Middle Bronze Age in Southern Britain*, British Archaeological Reports 3 (i), 1976

Rudenko, S. *Frozen Tombs of Siberia: the Pazyryk Burials of Iron Age Horsemen*, Dent, 1970

Scarre, C. *Ancient France: Neolithic Societies and Their Landscapes 6000–2000 BC*, Edinburgh University Press, 1983

Scarre, C. *Past Worlds: the Times Atlas of Archaeology*, Times Books, 1988

Sear, F. *Roman Architecture*, Batsford, 1982

Srejovic, D. *Europe's First Monumental Sculpture: New Discoveries At Lepenski Vir*, Thames & Hudson, 1972

Stuart G.E. & G.S. *Lost Kingdoms of the Maya*, National Geographic Society, 1992

Talbot Rice, T. *The Scythians*, Thames & Hudson, 1957

Throckmorton, P. *History from the Sea: Shipwrecks and Archaeology*, Mitchell Beazley, 1987

Tylecote, R.F. *A History of Metallurgy*, London Institute of Materials, 1992

Ucko, P. (et al) *Avebury Reconsidered: From the 1660s to the 1990s*, University College London Press, 1991

Warren, P. *The Aegean Civilizations*, Phaidon, 1989

West, S. *West Stow, The Anglo-Saxon Village*, East Anglia Archaeology Report 24, Suffolk Archaeology Unit, 1985

Wheeler, M. *The Indus Civilization (3rd ed.)*, Cambridge University Press, 1968

Willey G. & Sabloff J. *A History of American Archaeology*, W.H. Freeman, 1980

INDEX

ACKNOWLEDGMENTS

Dorling Kindersley would like to thank:
The whole team, including Joanna Wa[...]
David Williams, and Paul Greenl[...]
especially Joanne Mitchell, wh[...]
to bring the design side of the [...]
conclusion single-handedly a[...]
and good-humor. Jenni Bu[...]
without whom this book [...]
have been possible, Victori[...]
Lavery, Steve Speake St[...]
Piercy, Hugh Sack[...]
MacGillivray, [...]
Nich[...]

[...]ynn,
[...]Gale, [...]s, Malcolm Fry,
[...]y Sutherland, Dr [...]ristopher Scarre,
G[...]physical Surveys of [...]adford, Dr. Mark
Horton, Dr. John Gater, Stewart Ainsworth
and Bernard Thomason of RCHME, Dr Kate
Scott, Helena Cave-Penny, Tony Corey,
James Finnigan, Nick Pearson at BSS,
Gail Boyle, Sue Giles, Sarah Ashun,
and Carinne Allinson.

Additional design assistance:
Heather McCarry, Dooty Williams,
Ngaio Ballard, Richard Sinclair,
Tracy Timson, Jo Long.

Additional editorial assistance:
Jo Marceau, Susannah Steel, Irene Pavitt.

Index:
Richard Raper and Margaret Binns.

**Dorling Kindersley would like to thank
the following for their permission to
reproduce the photographs:**

Key: l=left, r=right, t=top, c=center,
a=above, b=below.

AKG: 24bl, 76c, 86bl, 120bl; Alexis
Rosenfeld: 5cr, 127bl, 127br; Ancient Art
and Architecture Collection: 12bl, 33c,
50bl, 99br; Architectural Association:
Paul Simpon 68bc; Ashmolean Museum,
Oxford: 109tr; Birmingham City Museum:
46br; Bridgeman Art Library, London:
Hereford Cathedral 91tr British Library,
London 100bl; Museo Conde, Chantilly,
France/Lauros-Giraudon 90bl Museo e
Gallerie Nazionali di Capodimonte
Naples/Giraudon 69tr; National Musuem
of India, New Delhi 35br, British Museum:
front cla, bc, br, spine, 10–11, 35bc, 40bl,
40bc, 43br, 46tr, 47tr, 52bc, 53br, 53bl,
53bc; 59tl, 63c, 69cr, 74bc, 75bl, 90tr,
91cl,104tr, 116–17, 128tl, 128tr, 129tl,
129tr, 131bl; Chelmsford Museums Service:
Paul Strarr 75bc; Christie's Images: 119cr;
CM Dixon: 21bc, 29tr, 57cr, 59tr, 68bl;
Conseil General des Vosges: Laurenson
79bl; 79bc; Department of Arts, Culture
and the Gaeltacht, Ireland: 67tl; English
Heritage Photographic Library: 7b, 9bl,
36bl; E.T. Archive: front tr, 58bl, 81tr,
109bl; Giraudon: 33tc, 66bc; Grand Hornu
Images: 121bl, 121br, 121tr ; Hulton
Getty: 118br, Hutchison Library: 31tr;
Images Colour Library: 43tr, 43cr, 118tl;
Institute of Archaeology Cultural Relics
Bureau, Sichuan Province, China:1c, 56bl,
56tl, 56tr, 56cr; Instituto Nacional de
Antropologia e Historia, Mexico: Michael
Zabe 34cr; Institute of Archaeology,
University College London: 27bc, 131cr;
Joe Cornish Photographer: 42tr, 42bl;
La Belle Aurore: Juliette Coombe 45tr;
MacQuitty International Photographic
Collection: 4br, 54cl; Mary Evans Picture
Library: 105tr, 106bc, 110tr, 111bl;
Maryland Historical Society, Museum and
Library of Maryland History: 102bl (24.7.1
Landing of Leonard Calvert and the First
Maryland Colonists, artist: DA Woodward,
medium: oil on canvas); Michael Holford:
46bl; Musee des Tumulus de Bougon,
Conseil General des Deux-Sevres, France:
32cr, 32bl, 32tl; Musees d'art et d'histoire
de Troyes: Jean-Marie Protte 87cr; Musee
du Chatillonais - Chatillon-sur-Seine: 66tr,
66bl; Museo del Duomo di Monza: 89tl;
Museum of London: Dave King 25br, 25tr;
73tc, 100cr; Museum of the City of New
York: 106cl, 107; National Museum of
Denmark: 80tr, front tl; National Museum
of Ireland, Dublin: 58cl, 63tr; National
Museums of Scotland: 31bc, 63tc; National
Trust: 39tr, 73bc, 73tr; Natural History
Museum, London: 13c, 20tl, 21cr, 40tl;
Novosti (London): 21tr; Oseberg Ship
Museum, Oseberg: 86tr; Parks Canada:
87tr; Photo Editions Gaud: 42bc, 96tr;
Pictor International: 119t; RCHME ©
Crown Copyright: 35c; Robert Estall: 34bl;
Robert Harding Picture Library: 20bl, 22bl,
30bl, 30cr, 45br, 54tr, 55tr, 55bl, 55br, 64tl,
65tc, Teresa Black 67bc, 96bl, 97br, 99cl;
Robert O'Dea: 78br, 79tl, Scala: Museum
of Archaeology, Florence front tc,
Museum of Archaeology, Belgrade 3c; Iraq
Museum Baghdad 47bc; 69bc; Museum
of Archaeology, Florence 81br; Bargello
Florence 87br; 88; S Apoollinaire Nuovo
Ravenna 89br; Bibliotheca Estense, Modena
132-197; Science Museum, London: Dave
King 101tc, 114bl; Science Photo Library:
John Reader 6tc, James King Holmes
126br; Smithsonian Institution: Mark
Gulezian 12c; Sonia Halliday: 76bl; 80bl;
Jane Taylor 97cr; 97tr; Studio Kontos: front
bl, 57tr; The Image Bank: Obremski back bl,
8bl, 77; By permission of Tyne and Wear
Museums Archaeology Department: 93bl,
95tc; The Metropolitan Museum of Art, New
York: 35tr; Tony Stone Images: 65bl,
76tr, 106tr, 109br; Trip: T Bognar 44; W
Jacobs 65tr; Photographs courtesy of the
Trustees of the Imperial War Museum:
120tr, 120bc; Reproduced by courtesy of the
Trustees, The National Gallery, London:
95cr; University College Dublin: 67cr;
Utah Museum of Natural History:
23bc, 23tr; Vladimir Vitanov: 33tr,
33br; Werner Forman Archive: 23cl,
64cl, 87tl; Zev Radovan: 22tr, 22br.

With additional thanks to:
Alexander Keiller Museum, Avebury: 37cr,
37br, 37tr, 37tc, 38 tr, 40bc; Mick Aston:
60bc, 60 cl, 82cl, 83 tl, 83c, 92c, bl, 93 tl,
93c, 93br, 102c, 103tl, 103 tc, 112cl, 113tl,
113tc; Reproduced by Birmingham City
Archives/Birmingham Library Services:
112–13cb, cr; Bristol City Museum and Art
Gallery: 40bl, 40br, 41bl, 41bc, 41br, 83br;
The British School at Athens: 49r; The
design of the contour bottle is reproduced
by kind permission of The Coca-Cola
Company. "Coca-Cola," Coke, and the
design of the contour bottle are registered
trademarks of the Coca-Cola Company;
Creative TV Facilities: 14c, 70cb, 70cl;
Environment and Heritage Service, Belfast:
60c; Rowena Gale: 16tr; Geophysical Surveys
of Bradford: 14cl, 38bl, 82–3, 92–3; Phil
Harding: 4tc, 18-19; INA: 108tr, 108cl,
108bc; INAH: 34cr; Arlette Mellaart: 26cl,
26bl, 27tl, 27cl, 27tc, 27tr, 28cl; © Mark
Moak: 51tr; Museo Archaeological di Napoli:
78tl; The Navan Centre: 60br, 61c, 61tl,
61r, 61bc; Oslo Ship Museum: 86bc,
Hugh Sackett: 48cl, 48bl, 49tl, 49cl, 49bl,
49tc, 49bc, 49bl, 51bc; Salisbury and
South Wiltshire Museum: 4cl, 9tr, 13tl,
18-19, 37c, 40bc, 41bc, 41br, 52br, 62bl,
62bl, 71tc, 72bl, 81bl, 83tr, 83br, 84bl,
84tr, 85tr, 85tl, 93tc, 93cr, 94tr, 99tl;
Soho House Museum: 111tl, 113cr,
113bl, 113bc,113cr; Time Team: 5bl,
6bl, 14cl, 15c, 17tr, 102br, 103br,
103cr, 103c, 103tr, 103bl.

With thanks to the illustrators:
Nick Hewetson: 5tl, 12-13, 16-17, 16cr,
24–5, 28–9, 34–5, 46–7, 62–3, 68–9, 80–1,
84–5, 90-91, 100–1, 104–5, 110–11,
Richard Draper: 72–3, 94–5, 114-5,
Stephen Conlin: 38–9, 50–1,
Royston Knipe: 19tr, 41tr,
53 tr, 75 tr, 117tr.

Additional Special Photography:
DK© sp James Finnigan, 60br, 61c,
61tl, 61r, 61bc.

Dorling Kindersley Photography:
13cr, 25cr, 41bl, 101cr, 101bl, 16bl,
21br, 24tc, 24bc; Peter Anderson: 86bc;
M. Dunning: 111tr, Andreas Einsedel:
14bl; Christi Graham/Nick Nicholls:
69br;Peter Hayman: 35bl; Dave King:
12tr, 59br, 104c; 14bl; James Stevenson:
2c, 78tl; Michel Zabé: 34cr.

Every effort has been made to trace
the copyright holders. Dorling Kindersley
apologizes for any unintentional omissions,
and would be pleased, if any such case
should arise, to add an appropriate
acknowledgment in future editions.